Thermophiles and Thermozymes

Thermophiles and Thermozymes

Special Issue Editor

María-Isabel González-Siso

MDPI • Basel • Beijing • Wuhan • Barcelona • Belgrade

MDPI

Special Issue Editor
María-Isabel González-Siso
Universidade da Coruña
Spain

Editorial Office
MDPI
St. Alban-Anlage 66
4052 Basel, Switzerland

This is a reprint of articles from the Special Issue published online in the open access journal *Microorganisms* (ISSN 2076-2607) from 2018 to 2019 (available at: https://www.mdpi.com/journal/microorganisms/special_issues/thermophiles)

For citation purposes, cite each article independently as indicated on the article page online and as indicated below:

LastName, A.A.; LastName, B.B.; LastName, C.C. Article Title. *Journal Name* **Year**, *Article Number*, Page Range.

ISBN 978-3-03897-816-9 (Pbk)
ISBN 978-3-03897-817-6 (PDF)

Contents

About the Special Issue Editor vii

María-Isabel González-Siso
Editorial for the Special Issue: Thermophiles and Thermozymes
Reprinted from: *Microorganisms* **2019**, *7*, 62, doi:10.3390/microorganisms7030062 1

Alba Blesa, Mercedes Sánchez, Eva Sacristán-Horcajada, Sandra González-de la Fuente,
Ramón Peiró and José Berenguer
Into the *Thermus* Mobilome: Presence, Diversity and Recent Activities of Insertion Sequences
Across *Thermus* spp.
Reprinted from: *Microorganisms* **2019**, *7*, 25, doi:10.3390/microorganisms7010025 3

Anders Schouw, Francesca Vulcano, Irene Roalkvam, William Peter Hocking,
Eoghan Reeves, Runar Stokke, Gunhild Bødtker and Ida Helene Steen
Genome Analysis of *Vallitalea guaymasensis* Strain L81 Isolated from a Deep-Sea Hydrothermal
Vent System
Reprinted from: *Microorganisms* **2018**, *6*, 63, doi:10.3390/microorganisms6030063 18

Juan Miguel Gonzalez
Molecular Tunnels in Enzymes and Thermophily: A Case Study on the Relationship to Growth
Temperature
Reprinted from: *Microorganisms* **2018**, *6*, 109, doi:10.3390/microorganisms6040109 39

María-Efigenia Álvarez-Cao, Roberto González, María A. Pernas and María Luisa Rúa
Contribution of the Oligomeric State to the Thermostability of Isoenzyme 3 from *Candida rugosa*
Reprinted from: *Microorganisms* **2018**, *6*, 108, doi:10.3390/microorganisms6040108 49

Mohit Bibra, Venkat Reddy Kunreddy and Rajesh K. Sani
Thermostable Xylanase Production by *Geobacillus* sp. Strain DUSELR13, and Its Application in
Ethanol Production with Lignocellulosic Biomass
Reprinted from: *Microorganisms* **2018**, *6*, 93, doi:10.3390/microorganisms6030093 62

Mohamed Amine Gomri, Agustín Rico-Díaz, Juan-José Escuder-Rodríguez,
Tedj El Moulouk Khaldi, María-Isabel González-Siso and Karima Kharroub
Production and Characterization of an Extracellular Acid Protease from Thermophilic
Brevibacillus sp. OA30 Isolated from an Algerian Hot Spring
Reprinted from: *Microorganisms* **2018**, *6*, 31, doi:10.3390/microorganisms6020031 87

Hiroyuki Hori, Takuya Kawamura, Takako Awai, Anna Ochi, Ryota Yamagami,
Chie Tomikawa and Akira Hirata
Transfer RNA Modification Enzymes from Thermophiles and Their Modified Nucleosides
in tRNA
Reprinted from: *Microorganisms* **2018**, *6*, 110, doi:10.3390/microorganisms6040110 103

Anthony J. Finch and Jin Ryoun Kim
Thermophilic Proteins as Versatile Scaffolds for Protein Engineering
Reprinted from: *Microorganisms* **2018**, *6*, 97, doi:10.3390/microorganisms6040097 149

Juan-José Escuder-Rodríguez, María-Eugenia DeCastro, María-Esperanza Cerdán,
Esther Rodríguez-Belmonte, Manuel Becerra and María-Isabel González-Siso
Cellulases from Thermophiles Found by Metagenomics
Reprinted from: *Microorganisms* **2018**, *6*, 66, doi:10.3390/microorganisms6030066 162

About the Special Issue Editor

María-Isabel González-Siso, Ph.D., is a Full Professor of Biochemistry and Molecular Biology at the University of A Coruña (Spain). Dr. González-Siso gained her PhD in Biology in 1989 from the University of Santiago de Compostela, Spain. Following a postdoctoral period in the CSIC (Vigo, Spain), since 1990, Prof. González-Siso has served as an academician and has held various academic and administrative positions. Moreover, Prof. González-Siso has over 100 research articles and reviews published in JCR journals, 20 books and book chapters, and eight patents to her name, and has supervised the writing of PhD and Masters' degree theses as well as reviewing works by invited researchers. Her recent investigations are focused on enzyme bioengineering, metagenomics of thermophiles, bioprospection of new thermozymes, and biotechnological valorisation of food industry byproducts.

microorganisms

MDPI

Editorial

Editorial for the Special Issue: Thermophiles and Thermozymes

María-Isabel González-Siso ⓘ

Grupo EXPRELA, Centro de Investigacións Científicas Avanzadas (CICA), Facultade de Ciencias, Universidade da Coruña, 15071 A Coruña, Spain; isabel.gsiso@udc.es

Received: 22 February 2019; Accepted: 22 February 2019; Published: 27 February 2019

Heat-loving microorganisms or thermophiles arouse noticeable scientific interest nowadays, not only with the aim to elucidate the mystery of life at high temperatures, but also due to the huge field of biotechnological applications of the enzymes they produce or thermozymes, able to function under industrial harsh conditions.

This Special Issue contains nine papers that study diverse aspects of thermophiles biology and their enzymes.

Two research articles deal with the genomics of thermophilic microorganisms. Blesa et al. [1] describe the characterization of active and inactive insertion sequences spanning the genus *Thermus*. This work represents an interesting contribution both to the construction of new genetic engineering tools and to the knowledge of the genomic plasticity and capacity of adaptation of thermophilic microorganisms. Schouw et al. [2] analyze the genome of a fermentative new strain isolated from a deep-sea hydrothermal vent, which was reassigned to the genus *Vallitalea* and designated *V. guaymasensis* strain L81, showing interesting features for industrial application. Furthermore, the potential of these marine ecosystems for the bioprospection of new enzymes and antimicrobials is revealed in this work.

Four other research articles deal with different aspects of thermozymes. J.M. González [3] presents a structural analysis of substrate tunnels in two types of enzymes (with low and high tunneling) from microorganisms living optimally at 15 °C to 100 °C. Molecular tunnel dimensions are reduced with increasing optimum growth temperatures, minimizing unnecessary spaces within the molecule. From this work, molecular channeling appears as a mechanism that helps to understand how thermophiles are adapted to live under high temperatures. Álvarez-Cao et al. [4], using the lipase LipE from *Candida rugosa*, show that oligomerization-dimerization is another structural feature that causes protein stabilization against temperature and pH, also expanding substrate specificity on soluble substrates. Domain swapping is the mechanism proposed to explain LipE homodimerization. Bibra et al. [5] report the statistical optimization of the production of a thermostable xylanase from a *Geobacillus* sp. strain isolated from a gold mine, its comparison (favorable) with commercial counterparts for lignocellulosic biomass hydrolysis, and the use of the strain in co-culture for ethanol fermentation of the biomass. Gomri et al. [6] depict the characterization of a new acid protease produced extracellularly by a thermophilic bacterium that was isolated from an Algerian hot spring and affiliated with *Brevibacillus thermoruber* species. The purified 32-F38 protease resulted to be thermostable and highly stable in the presence of different detergents and solvents, suggesting potential biotechnological applications.

Three reviews complete this monograph. The one by Hori et al. [7] is a comprehensive study about modified nucleosides in tRNA and tRNA modification enzymes from thermophiles, in view of strategies to stabilize tRNA structures including RNA-binding proteins and polyamines. The one by Finch and Kim [8] focuses on the application of thermophilic proteins as scaffolds for protein engineering, proposed due to their robustness and evolvability, without forgetting the trade-off between protein activity and stability. Finally, the one by Escuder-Rodríguez et al. [9] introduces the metagenomics approach, in this case for the search of thermostable cellulases, a complex group of enzymes which is here unraveled.

Altogether these papers allow us to go one step forward to explain how the thermophilic microorganisms and their enzymes are stable and functional at high temperatures, and to envisage new biotechnological applications and fields for bioprospection.

Acknowledgments: Thank you to all the authors and reviewers for their excellent contributions to this Special Issue. Also thank you to the Microorganisms Editorial Office for their professional assistance and continuous support.

Conflicts of Interest: The author declares no conflict of interest.

References

1. Blesa, A.; Sánchez, M.; Sacristán-Horcajada, E.; González-de la Fuente, S.; Peiró, R.; Berenguer, J. Into the *Thermus* Mobilome: Presence, Diversity and Recent Activities of Insertion Sequences Across *Thermus* spp. *Microorganisms* **2019**, *7*, 25. [CrossRef] [PubMed]
2. Schouw, A.; Vulcano, F.; Roalkvam, I.; Hocking, W.; Reeves, E.; Stokke, R.; Bødtker, G.; Steen, I. Genome Analysis of *Vallitalea guaymasensis* Strain L81 Isolated from a Deep-Sea Hydrothermal Vent System. *Microorganisms* **2018**, *6*, 63. [CrossRef] [PubMed]
3. Gonzalez, J. Molecular Tunnels in Enzymes and Thermophily: A Case Study on the Relationship to Growth Temperature. *Microorganisms* **2018**, *6*, 109. [CrossRef] [PubMed]
4. Álvarez-Cao, M.; González, R.; Pernas, M.; Rúa, M. Contribution of the Oligomeric State to the Thermostability of Isoenzyme 3 from *Candida rugosa*. *Microorganisms* **2018**, *6*, 108. [CrossRef] [PubMed]
5. Bibra, M.; Kunreddy, V.; Sani, R. Thermostable Xylanase Production by *Geobacillus* sp. Strain DUSELR13, and Its Application in Ethanol Production with Lignocellulosic Biomass. *Microorganisms* **2018**, *6*, 93. [CrossRef] [PubMed]
6. Gomri, M.; Rico-Díaz, A.; Escuder-Rodríguez, J.; El Moulouk Khaldi, T.; González-Siso, M.; Kharroub, K. Production and Characterization of an Extracellular Acid Protease from Thermophilic *Brevibacillus* sp. OA30 Isolated from an Algerian Hot Spring. *Microorganisms* **2018**, *6*, 31. [CrossRef] [PubMed]
7. Hori, H.; Kawamura, T.; Awai, T.; Ochi, A.; Yamagami, R.; Tomikawa, C.; Hirata, A. Transfer RNA Modification Enzymes from Thermophiles and Their Modified Nucleosides in tRNA. *Microorganisms* **2018**, *6*, 110. [CrossRef] [PubMed]
8. Finch, A.; Kim, J. Thermophilic Proteins as Versatile Scaffolds for Protein Engineering. *Microorganisms* **2018**, *6*, 97. [CrossRef] [PubMed]
9. Escuder-Rodríguez, J.; DeCastro, M.; Cerdán, M.; Rodríguez-Belmonte, E.; Becerra, M.; González-Siso, M. Cellulases from Thermophiles Found by Metagenomics. *Microorganisms* **2018**, *6*, 66. [CrossRef] [PubMed]

microorganisms

MDPI

Article

Into the *Thermus* Mobilome: Presence, Diversity and Recent Activities of Insertion Sequences Across *Thermus* spp.

Alba Blesa [1], Mercedes Sánchez [2], Eva Sacristán-Horcajada [2], Sandra González-de la Fuente [2], Ramón Peiró [2] and José Berenguer [2,*]

[1] Department of Biotechnology, Faculty of Experimental Sciences, Universidad Francisco de Vitoria, Madrid 28223, Spain; alba.blesa@ufv.es

[2] Centro de Biología Molecular Severo Ochoa (CBMSO), Universidad Autónoma de Madrid-Consejo Superior de Investigaciones Científicas, Madrid 28049, Spain; mercedes.sanchez@cbm.csic.es (M.S.); esacristan@cbm.csic.es (E.S.-H.); sandra.g@cbm.csic.es (S.G.-d.l.F.); rpeiro@cbm.csic.es (R.P.)

* Correspondence: jberenguer@cbm.csic.es; Tel.: +34-911-964-498

Received: 27 November 2018; Accepted: 17 January 2019; Published: 21 January 2019

Abstract: A high level of transposon-mediated genome rearrangement is a common trait among microorganisms isolated from thermal environments, probably contributing to the extraordinary genomic plasticity and horizontal gene transfer (HGT) observed in these habitats. In this work, active and inactive insertion sequences (ISs) spanning the sequenced members of the genus *Thermus* were characterized, with special emphasis on three *T. thermophilus* strains: HB27, HB8, and NAR1. A large number of full ISs and fragments derived from different IS families were found, concentrating within megaplasmids present in most isolates. Potentially active ISs were identified through analysis of transposase integrity, and domestication-related transposition events of ISTth7 were identified in laboratory-adapted HB27 derivatives. Many partial copies of ISs appeared throughout the genome, which may serve as specific targets for homologous recombination contributing to genome rearrangement. Moreover, recruitment of IS1000 32 bp segments as spacers for CRISPR sequence was identified, pointing to the adaptability of these elements in the biology of these thermophiles. Further knowledge about the activity and functional diversity of ISs in this genus may contribute to the generation of engineered transposons as new genetic tools, and enrich our understanding of the outstanding plasticity shown by these thermophiles.

Keywords: insertion sequence; transposons; transposases; HGT; *Thermus*; thermophiles; mobilome

1. Introduction

The amazing diversity of the prokaryotic world is a result of the countless strategies with which they deal with an unpredictable, ever-changing environment. Indeed, microbes display a great capacity for rapid adaptation, accelerating their evolution via the acquisition of novel DNA through horizontal gene transfer (HGT) mechanisms [1–3]. Phylogenetic studies have revealed that molecular machines that confer such adaptive advantages are frequently encoded in mobile genetic elements (MGEs), such as plasmids, integrons, transposons, and phages, among others [4,5]. An MGE can be defined as a DNA segment encoding proteins that mediate the movement of DNA either within a cell or between different cellular genomes [4,6,7]. Intracellular movement of DNA is mainly driven by promiscuous recombining DNA, primarily insertion sequences (ISs). ISs consist of up to 2–5 kb-long DNA segments which only encode the minimum information required to enable their mobility [8], including a transposase, an enzyme that recognizes specific short inverted repeats (IRs) located at the extremes of the ISs, and catalyzes its mobilization into a new site. These insertions may inactivate genes or activate transcription of downstream genes, leading to DNA rearrangements when acting as targets for homologous recombination [8]. As ISs can also "land" within conjugative plasmids, integrative and conjugative elements (ICE), or (pro)phages, they can "ride along" these transferable elements, increasing their dissemination and impact on diverse microbial genomes [9–11]. Therefore, there is great interest in the role of these ISs, not only in microbial evolution, but also in the spread and expression of virulence factors, antimicrobial resistance, or biotransformation of xenobiotics in many bacteria [12–15].

The growing availability of complete genome sequences has increased the number of ISs identified in all domains of life [16], showing a common structure where an IR flanks one or two genes encoding the transposase. The molecular mechanisms that catalyze and control transposition are notably heterogeneous, leading to significant diversity among ISs. In order to set a framework for the systematic classification of ISs, Mahillon and Chandler (1998) established a scheme based on the main features of transposase structure and function, dividing them into families and subfamilies (ISFinder database; www-is.biotoul.fr) [17].

Thermophilic environments seem to favor genetic plasticity, with frequent genome rearrangements and DNA movement between phylogenetically unrelated microorganisms, even those from different domains [18–20]. This intense gene shuffling is partly orchestrated by MGEs which mediate either intracellular or intercellular DNA mobility, the latter headed by frequent HGT events [4,20–22]. In fact, HGT is common among bacteria but especially relevant in thermophilic organisms, where replication burden limits the size of the genomes. For instance, the highly efficient conjugative plasmids of *Sulfolobus* spp. and the extraordinary competence capacity of *Thermus thermophilus* have been proposed as the major drivers of these organisms' adaptability to unexpected environmental shifts. Indeed, IS propagation is thought to be frequent among prokaryotes and maximal among thermophiles [23]. Complete genome sequences of thermophilic organisms have revealed an outstanding abundance and diversity of ISs, sometimes disproportional to their genome size, which would suggest positive selection [24–26].

The abundance of truncated ISs scattered throughout the genomes of thermophilic bacteria suggests a high frequency of genomic rearrangement, likely related to the evolutionary history and speciation of these strains, as reported for the alkalo-thermophilic bacteria *Bacillus halodurans* [26] or the cyanobacterial *Synechococcus* spp. isolated from hot springs [25]. Thermophilic archaea are also characterized by intense transpositional activity, with the order Sulfolobales being one of the major representatives of such diversity; *S. neocofandices* has a genome with ISs making up more than 10% of its genome [16,27].

In addition to large genomic rearrangements, homologous recombination between IS copies spread across mobile elements may facilitate acquisition of new DNA sequences. In this way, ISs may be considered to be active contributors to pan-genome systems.

The genus *Thermus* belongs to the ancient clade *Deinococcus–Thermus* [28,29], having several representatives isolated worldwide from geothermal origins, in self-heating material, such as compost, and industrial heating systems. The metabolic capabilities of these strains are very diverse, despite their small genome size, with some strictly aerobic strains and others with anaerobic respiration capacity based on nitrogen oxides or metals, and different abilities to utilize polysaccharides. This assortment of phenotypic traits is frequently encoded by megaplasmids and, thus, may be laterally transferred by different mechanisms, contributing to the fast ecological adaptation of these bacteria to sudden changes in a sort of shared pan-genome [30].

Sequencing studies have revealed the presence of MGEs in *Thermus* species, in particular, in strains that are regarded as excellent laboratory models [31], allowing the identification of recent transpositional events [1,32]. In this way, a few cases of active ISs have previously been reported in *T. thermophilus*, including IS1000B in strain HB8 [33,34] and ISTth7 in strain HB27 [35]. However, more recent genomic data suggest that a much greater frequency of transpositional events than initially though takes place in the genus [30].

In this article, we aim to gain insights into the MGEs present in *Thermus* spp., reporting the diversity and abundance of ISs identified in the published *Thermus* genomes. To further study their activity, we focused on three *T. thermophilus* isolates, including a newly sequenced strain, and analyze recent domestication-related jumps of one of the most abundant ISs of *T. thermophilus* HB27.

2. Materials and Methods

Strains. Subcultures of *T. thermophilus* HB27 (DSM7039), used for long periods in different laboratories in Germany (hereafter HB27A) and Spain (HB27E), were analyzed for domestication-related genomic changes. The nitrate respiring strain *T. thermophilus* NAR1 [36] was sequenced de novo. All strains were grown at 65 °C under aerobic conditions (150 rpm rotational shaking) in *Thermus* broth (TB) as described previously [37].

DNA isolation and sequencing. Erlenmeyer flasks filled with 250 mL pre-warmed TB (1/5 volume) and a 1/1000 dilution of overnight culture were grown until stationary phase (1.2 OD_{550nm}). Cells were harvested by centrifugation (10 min, 5000 g) for genomic DNA extraction and purification with the Qiagen DNeasy Blood and Tissue Extraction Kit (cat.no. 69504, Hilden, Germany) following manufacturers' indications. Purity and concentration of the DNA was checked with a NanoDrop ND-1000 (Thermo Scientific, Wilmington, DE, USA) spectrophotometer, and integrity was checked by electrophoresis in 1% agarose gels.

Genomic DNA from *T. thermophilus* HB27A and HB27E were adjusted to 30 ng/μL in Tris-HCl pH 8.8 and sent to Microbes NG (UK) for whole genome sequencing employing Illumina DNAseq technique. Genomic DNA (20 μg/μL) from *T. thermophilus* NAR1 was sequenced de novo at the Norwegian Sequencing Centre (www.sequencing.uio.no) using Pacific Biosciences technology (PacBio RS II).

Bioinformatic analysis of T. thermophilus NARI. A total of 60,929 PacBio SMRT reads were provided by the sequencing center. De novo genome assembly was performed following a hierarchical genome-assembly process [38], using the software HGAP v3 (Pacific Biosciences, SMRT Analysis Software v2.3.0, Menlo Park, CA, USA) and circularization with Minimus2 (using the workflow recommended by Pacific Biosciences at https://github.com/PacificBiosciences/Bioinformatics-Training/wiki/Circularizing-and-trimming College Park, MD, USA) [39]. As a result, the genome was assembled into four contigs, corresponding to the bacterial chromosome, two megaplasmids, and one small plasmid.

Due to the detection of systematic 1 bp deletions in homopolymeric regions in PacBio reads, the assembly was corrected with Illumina reads combining PacBio-Utilities [40] and Pilon [41]. Sequence deletions were corrected when they were supported by an Illumina coverage higher than 10 and the indel was present in at least 50% of the reads. Through this workflow, a final assembly was obtained, consisting of four contigs of 2,021,843; 370,865; 77,135; and 9799 bp. Lastly, annotation of the final assembly was obtained using PROKKA with default parameters [42].

Nucleotide sequence accession number. The sequence from *T. thermophilus* NARI genome has been submitted at the European Nucleotide Archive (ENA; http://www.ebi.ac.uk/ena/) with the accession PRJEB29203.

Bioinformatic analysis of T. thermophilus HB27A and HB27E. The assembly and annotation files from the Illumina reads of HB27A and HB27E samples were provided by the sequencing center. FASTQC (http://www.bioinformatics.babraham.ac.uk/projects/fastqc/) was applied for quality checking of reads, which were subsequently aligned to the reference genome of *T. thermophilus* HB27 (AE017221.1 and AE017222.1 for the chromosome and megaplasmid, respectively) using BWA aligner [43].

Identification of ISs. Identification of ISs in the sequences of *T. thermophilus* strains HB27 (AE017221.1 and AE017222), HB8 (GCA_000091545.1 and AB677526), and NAR1 (PRJEB29203) was carried out at the NGS computing facility of the CBMSO. An in-house script was designed to download the ISFinder database (www-is.biotoul.fr) and locally emulate the ISFinder BLAST service, with additional parameters to personalize the alignment. Next, IS detection was performed across the genomes of

T. thermophilus HB8, HB27, and NARI, using 80% nucleotide identity by BLASTn and 30% amino acid identity by BLASTX as thresholds to identify IS-associated ORFs. Genes identified and annotated as transposases within other *Thermus* genomes were downloaded from the JGI database (https://img.jgi.doe.gov/) and classified according to their IS family, as above. When required, specialized websites such as ACLAME, MITE-Hunter, RepSeek, Repeat Finder, or Repeat Masker were used. Consensus sequences of *T. thermophilus* IS typologies (*ISTth1* to *8* and *IS1000A/B*) were searched using BLAST against the genomes. Wherever possible, this information was checked with the IS index in the ISFinder database for the different organisms. Multiple alignments of copies belonging to the same *ISTth* family, found within the same strain or in different *Thermus* strains, were performed using MUSCLE [44]. *ISTth* copies found in other bacterial genera were also aligned using this software to check for conserved sequences.

Putatively active IS copies were identified by comparison of recognized ORFs with the sequence of the corresponding transposase in GenBank.

Phylogenetic analysis. For each full-length consensus sequence of *ISTth* reported by the ISfinder, BLASTn analysis was performed either within or excluding the *Thermus* genomes (taxi:270). Nucleotide sequences from the best hits (E-value < 10^{-5}) were clustered and a neighbor-joining tree was performed using PHYLIP [45].

Transposase mobilization in T. thermophilus HB27. In order to detect recent movement of the ISTth7 transposase, we extracted FASTQ files from the soft-clipped regions (a sequence that may not be aligned from the first residue to the last) of the HB27A and HB27E alignments using the extractSoftclipped8 tool, comparing them against the ISTth7 sequence. With these soft-clipped reads, we performed a SoftClip alignment to extract ISTth7 coordinate positions in HB27A and HB27E.

3. Results

3.1. Incidence of ISs among Members of the Genus Thermus

The availability of the complete genomes of several *Thermus* spp. enabled an exhaustive analysis of IS presence within these thermophiles. Preliminary exploration of genes annotated as transposases across 18 *Thermus* genomes revealed a higher frequency compared to controls, with an average number of 16.21 ± 1.99 IS copies per Mbp, compared with 8.84 in *E. coli* K12 (Figure 1). Notably, more than twice as many hits were found in *T. scotoductus* K1 and KI2 strains, representing more than 4% of its coding sequence. However, the numbers should be cautiously analyzed as data for these strains was extracted from permanent draft genomes available at JGI database, and it is possible that contig boundaries could have been disregarded by the annotation programs. Implementation of an in silico multi-approach looking for IS scars (defined as IS fragments left in the genome as evidence of previous transposition events), IRs, and transposase motifs in the model organisms *T. thermophilus* HB8 and HB27, and in the partial denitrifying strain NAR1, revealed that, in many cases, partial IS copies were more abundant than full copies (Table 1). However, these results include the identification of several tandem copies of a 32 bp-long DNA segment identical to the 5' extreme of IS1000A/B (positions 1165–1196) in *T. thermophilus* HB8 and HB27 which, after manual analysis, were finally identified as CRISPR repeats separated by 30–40 bp spacers. Actually, 52 copies of this repeated sequence were found in 6 clusters in *T. thermophilus* HB8 and 38 copies divided among 5 clusters in *T. thermophilus* HB27. This same 32 bp DNA sequence also appeared isolated (and, thus, independent of any CRISPR system) in the genome of other *Thermus* spp., suggesting their recruitment for CRISPR-Cas defense in the aforementioned strains.

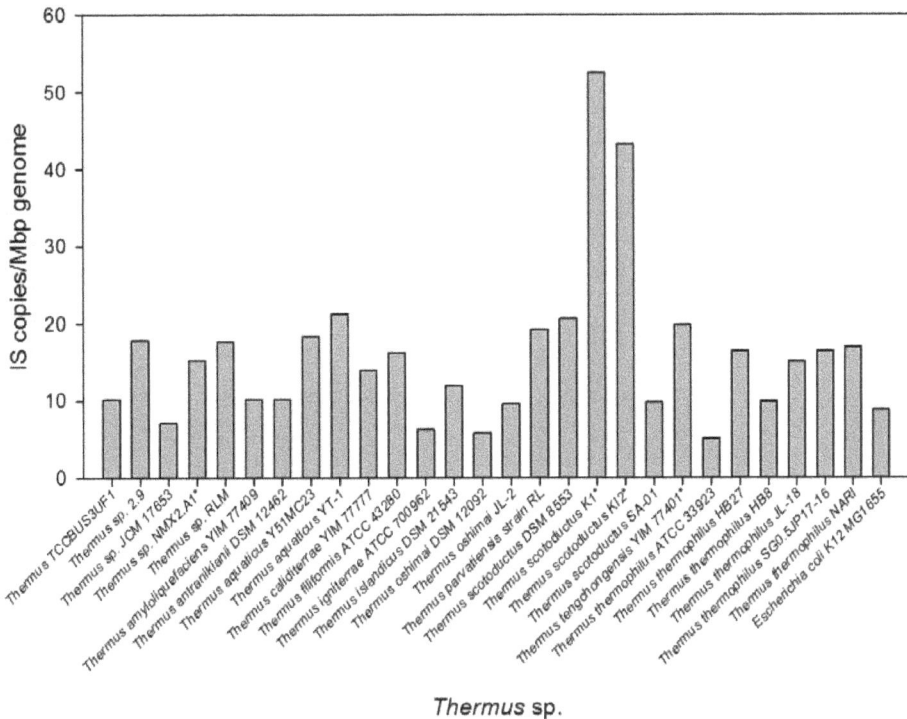

Thermus sp.

Figure 1. Distribution of insertion sequences (ISs) in *Thermus* spp. Number of genes annotated as transposases were determined in 18 available genomes at NCBI, RefSeq, and JGI. The genome of *T. thermophilus* NAR1 is described in this article. The number of IS copies was normalized to the genome size in Mbp; a strain of *E. coli* was included for comparison.

Table 1. Presence of complete (C) and partial (P) IS sequences in the genomes of *T. thermophilus* strains HB27 (AE017221.1 and AE017222), HB8 (GCA_000091545.1 and AB677526), and NAR1 (PRJEB29203). Distribution among the chromosome (Ch) or the megaplasmids M1 and M2 from NAR1, pTT27 from HB27 and HB8, and pVV8 from HB8 is shown. Note that most partial copies of IS1000A/B actually correspond to CRISPR repeats (see the text for details).*

| | NAR1 | | | | | | HB27 | | | | HB8 | | | | | |
| | Ch | | M1 | | M2 | | Ch | | pTT27 | | Ch | | pTT27 | | pVV8 | |
	C	P	C	P	C	P	C	P	C	P	C	P	C	P	C	P
ISTth1			1						1						1	
ISTth2							2		1							
ISTth3		1	1				2	1	1							1
ISTth4	5		11						4				4		2	
ISTth5							1		1							
ISTth6	1				1		3		1		1		1	1	1	
ISTth7	13		5			2	2	2	4	1	6	2	1	2		
ISTth8							3		1	1			2			
IS1000A/B*					2		1	8	1	34	1	14	3	39		

3.2. Diversity of ISs among Thermus spp. ISTth Families

Not only were ISs abundant within *Thermus* isolates, but they also showed an outstanding diversity (Table 2). As expected, closely related species harbored ISs that were almost identical in sequence, whereas phylogenetically divergent isolates carried IS sequence variants. BLASTn analysis revealed strain-specific ISs in *T. thermophilus* and *T. scotoductus* genomes, whereas a more generalized description was provided for IS4-type (found in *T. thermophilus* SG0.5 and *T. tengchongensis*) and IS30-type sequences (found in *Thermus* TCCBUS3UF1 and *T. oshimai*; data not shown). An exception to this was the conspicuous presence of an IS3-like transposase, frequently found as a full copy termed OrfB in ISFinder. Copies of this IS belonging to the IS605 subfamily were found in *T. aquaticus*, a set of *T. thermophilus* strains such as HB27, JL-18, and SG0.5, and *T. oshimai* strains, showing reasonable conservation (data not shown).

Table 2. Diversity of insertion sequences (ISs) across *Thermus* spp. Most relevant ISs found were characterized by the family and subgroup to which they belong, their main features (size, direct repeats (DRs), and number of ORFs), and their relative frequency among members of the genus *Thermus*. Frequency was graded as high (+++; present in more than 6 different isolates), moderate (++; present in 3–6 isolates) and low (+; present in 2 or fewer isolates). Presence across the genus was determined by BLASTn, with a threshold of E-value <10^{-10}, query coverage >40%, and identity >70%.

Name	Family	Sub-Group	Size Range (bp)	DRs (bp)	N° ORF	Frequency in *Thermus* spp.
ISTth1	IS3	IS150	1200–1600	3–4	2	+
ISTth2	IS4	IS10	1200–1350	9	1	+
ISTth3	IS1634	IS4	1500–2000	5–6	1	+++
ISTth4	IS256	-	1200–1500	8–9	1	+++
ISTth5	IS256		1200–1500	0	1	+
ISTth6	IS630	-	1000–1400	2	1–2	++
ISTth7	IS5	ISH1	900–1150	8	1	+++
ISTth8	IS701	-	1400–1550	4	1	++
IS1000A	IS110	-	1136–1558	2	1	+++
IS1000B	IS110		954–1558	0	1	++
IS421	IS4	IS231	1450–5400	10–12	1	+

Despite some strain-specific traits, common aspects could be established regarding the copies found in the model organism *T. thermophilus* (reviewed in [8]) and available in the ISfinder database. The ISs were classified into typologies by features, including similarity of their terminal IR sequences, length of the direct repeats, marked identity, or similarity in the transposase sequence, as well as the chemistry of transposition [17]. As shown in Table 1, there was great IS diversity within *T. thermophilus* HB27 and HB8, including 11 different types of IS, 8 of which were specific to *Thermus* (ISTth1-8), and 2 types (IS1000A and its derivative IS1000B) belonging to the IS110 family. Additionally, *T. thermophilus* SG0.5 harbored different IS copies, the majority of them belonging to the heterogeneous IS4 family. Contrary to the normally low intra-species IS diversity [46], many *Thermus* isolates contained a great diversity of ISs belonging to different families (Table 2), suggesting coexistence in the same cell of different mechanisms of transposition and induction signaling.

All ISTth types were of similar size (1214.10 ± 60.14 bp) and encoded transposases of approx. 360 amino acids in length. On average, conservation of the DDE motif (a common transposase active site) suggested a similar biochemical pathway of transposition for six of them (ISTth1, 2, 4, 5, 6, 7, and 8). By contrast, copies of IS1000A/B harbored DEDD motifs, thus supporting a different transposition mechanism. Interestingly, all typologies except ISTth4 were found in the laboratory model strain *T. thermophilus* HB27, whereas its related strain HB8 lacked ISTth2 and ISTth5 types (Table 1), and the NAR1 strain lacked ISTth2, ISTth5, and ISTth8.

Types ISTth4 and ISTth5 showed up to 65% identity, indicating their close relationship. An even greater degree of identity was found for IS1000A/IS1000B, which differed by only 14 bp. On the other hand, the greatest IS variability was observed in the 3′ ends, along the IS boundary (i.e., terminal IR regions) regardless of the genomic localization or bacterial strain. High variability was also observed in the target site duplication (TSD), which represented a unique hallmark for each IS.

It is also interesting to note that, on average, the GC content of these complete and partial IS copies was similar to that of the host genome (mean 68.1%) regardless of the ISTth diversity, supporting a long-term co-evolution in these strains.

3.3. Evidence of ISTth Propagation to Other Microorganisms

We performed serial BLASTn analysis to examine if any of ISTth copies have contributed to horizontal gene transfer to other microorganisms. Within the *Thermus* genus, identical copies of ISTth3 were detected in plasmids of *T. aquaticus*, *T. oshimai*, and *T. parvatiensis*, clearly supporting the lateral exchange between these strains. Likewise, nine almost identical ISTth7 copies (1–3 bp mismatch) were found in isolates of *T. parvatiensis*, and several highly conserved ISTth4 copies (98% identical) were found in plasmids from a wide variety of *Thermus* members, suggesting its chief role in helping the spread of the *Thermus* pan-genome.

Beyond this putative transfer among *Thermus* strains, phylogenetically distant organisms, such as *Meiothermus* spp. may have acquired some ISTth copies (or vice versa). A total of 16 ISTth1 homologues (>86% sequence identity) were found in the genome of *M. silvanus* DSM9946, 12 of which were localized to the pMESIL plasmid, where they were renamed as ISMesil (ISfinder database). Furthermore, ISTth2, ISTth3, or ISTth8 were more than 50% identical to ISs present in other genera, including ISXaca1(*Xanthomonas campestris*), ISPlu4 (*Photorhabdus luminescens*), and ISDha5 (*Desulfitobacterium hafniense*). This great similarity may be a consequence of recent transfer events between these genera. Finally, copies of IS421 found in *T. thermophilus* SG0.5 showed 99% DNA sequence identity to *E. coli* IS186, supporting a transfer between these distant phylogenetic groups, despite their functionality at very different temperatures.

3.4. Distribution of Putatively Active vs. Inactive IS Copies across Selected T. thermophilus Genomes

The ISs present in aerobic (HB27 and HB8) and nitrate respiring (NAR1) strains of *T. thermophilus* are detailed in Figure 2. Excluding the 32 bp CRISPR repeats identical to IS1000A/B fragments described above, the *T. thermophilus* HB27 reference genome contained 25 complete copies of different ISs, of which 20 encode putatively active transposases, and 8 IS fragments. Figures were similar in *T. thermophilus* HB8, with 27 full copies, 21 of which were putatively active, and 9 IS fragments. It is worth mention that several copies of ISTth7 encoding two ORFs corresponding to the N- and C-terminal domains of the corresponding transposases were classified as active, according to the proposal of Gregory and Dhalberg [35]. At least one copy of every IS identified in each analyzed *T. thermophilus* strain was potentially active (Figure 2). ISTth7, which belongs to the IS5 family, was the most abundant in terms of number of copies in both strains (9 and 11 copies in HB27 and HB8, respectively), and also in the other sequenced *Thermus* spp. (Table 2). Also, numerous copies of ISTth4 were found in *T. thermophilus* HB8 [10] and NAR1 [19] strains, in contrast to its absence from HB27. On the contrary, ISTth1, ISTth2, and ISTth5 were the least abundant in these and all other published *Thermus* spp. genome sequences. In all the cases studied, the highest IS density and diversity was associated with the pTT27-related megaplasmids, with one member of each IS in both strains except from 4 copies of ISTth7 in HB27 pTT27, and 4 and 11 copies of ISTth4 in HB8 pTT27 and NAR1 M1 megaplasmid (Table 1).

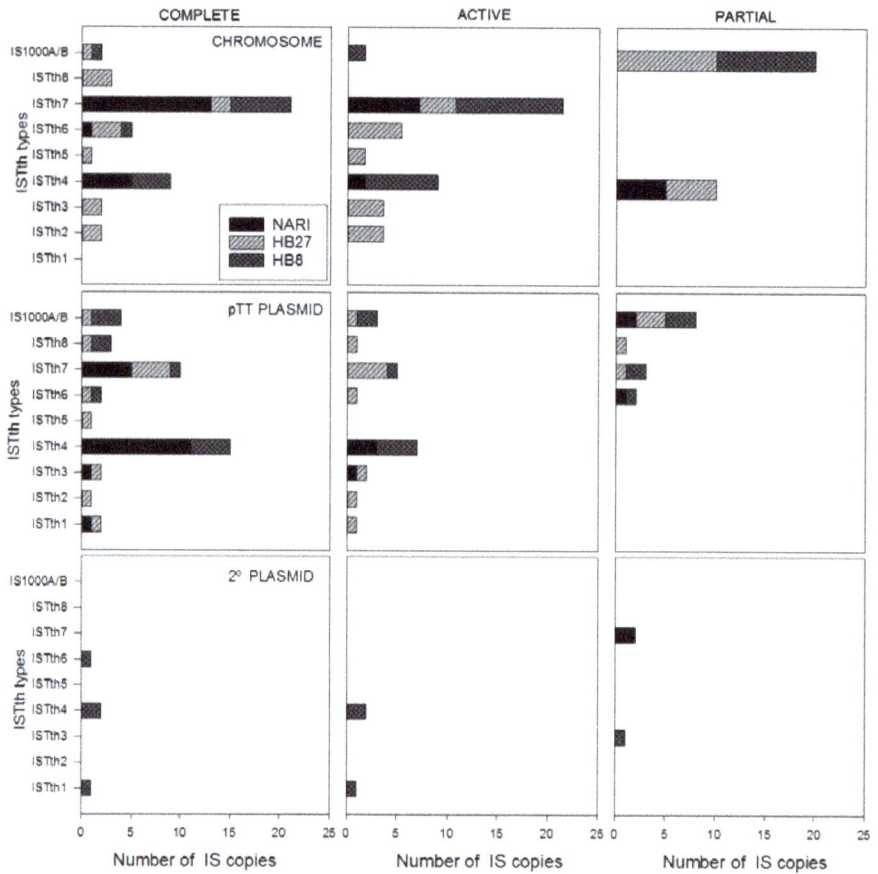

Figure 2. Abundance of insertion sequences across *Thermus thermophilus*. Number of full copies (complete), potentially active (middle column), and incomplete (partial) IS sequences found in the genomes of *T. thermophilus* HB27, HB8, and NAR1 strains. Sequences were localized to the chromosome (first row), pTT27 (HB27 and HB8), and M1 (pTT PLASMID; NAR1) megaplasmids (middle row), or in pVV8 (HB8) and M2 megaplasmid (NAR1) (2° PLASMID, third row).

Distribution of complete and partial copies of the ISs within the genome of these three strains revealed a higher concentration in megaplasmids (6.5%–7% of total sequence) than in the chromosome (0.8% of total sequence). This suggests a major role for the chromosome in coding for conserved core functions, which would act as a major counterselection factor if genes were interrupted by IS transposition. Also, our data are in agreement with the more flexible character of the megaplasmids, which encode adaptive and HGT-prone faculties that are only needed under specific conditions [30]. Our data are consistent with other observations of higher incidence of transposable elements within plasmids [47,48], and also that pTT27 harbors the highest plasticity with low synteny among strains [49].

Even within the megaplasmids, there were regions which concentrated many complete and partial IS sequences (Figure 3), serving as hotspots for the integration of new genomic traits. One such example is the nitrate respiration island, whose insertion into pTT27 of strain HB27 after conjugative transfer from *T. thermophilus* NAR1 appears associated to a copy of ISTth7 [50].

HB8

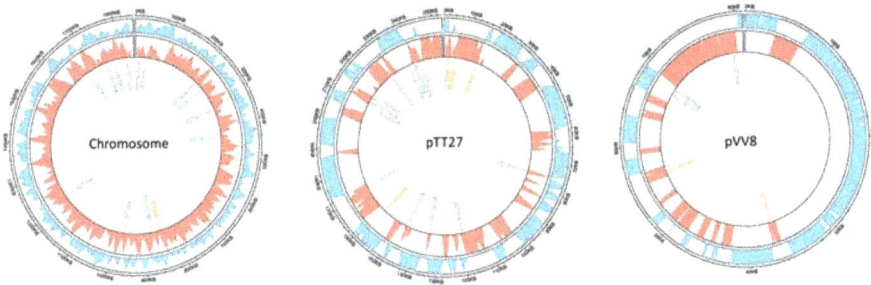

HB27

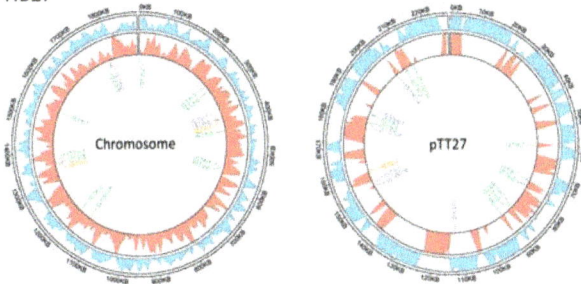

NAR1

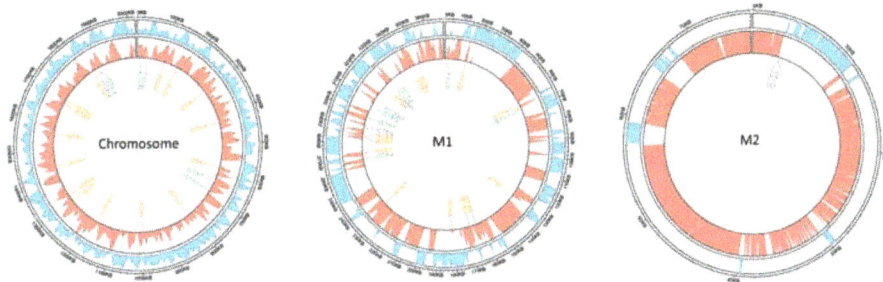

Figure 3. Location of ISs in the genomes of selected strains. The chromosome and megaplasmids (pTT, pTT27, PVV8, M1, and M2) of the *T. thermophilus* strains HB8, HB27, and NAR1 are represented as circles, with blue and orange shading indicating the coding densities of both DNA strands. Positions of the indicated IS are labeled with color codes indicating putatively active (green), inactive but complete (orange), or incomplete (purple). The replicons are drawn at different scales.

3.5. Potentially Active Transposases

The above data support the existence of a significant number of putatively functional ISs in the genomes of the three *T. thermophilus* strains analyzed. In order to further test this hypothesis in silico, we searched for recent integration events of ISTth7 in the genome of strain HB27. For this, two laboratory-adapted strains derived from *T. thermophilus* HB27 (strains HB27A and HB27E) were subjected to Illumina-based full genome sequencing (see Materials and Methods). In order to find recent movements of ISTth7, a procedure was followed in which we recruited those sequences of the HB27A and HB27E genomes, that overlapped the 3′ and 5′ extremes of ISTth7, searching for matches to the ISTth7 boundary sequences found in the reference HB27 genome. This comparison confirmed maintenance of 2 expected chromosomal copies (reference GenBank accession AE017221.1) and 4 copies expected in the megaplasmid (reference GenBank accession AE01722.2), and detected additional copies in both domesticated strains. Strain HB27E contained 7 additional copies in the chromosome and 2 additional copies in the megaplasmid (Tables 3 and 4), of which 6 chromosomal copies and 1 megaplasmid copy were also found in HB27A (Figure 4).

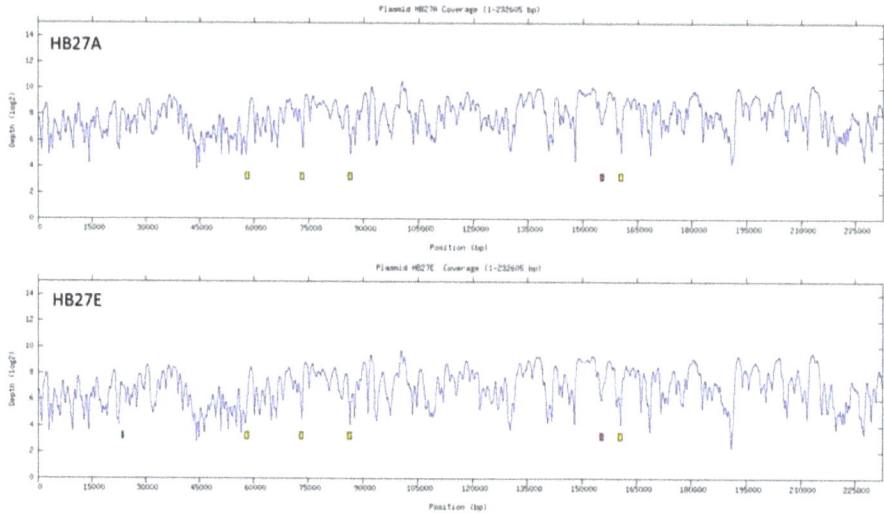

Figure 4. Identification of recent events of transposition by ISTth7 in the megaplasmid of *T. thermophilus* HB27. Coverage plot of HB27A (upper), HB27E (lower) reads plotted against the *T. thermophilus* HB27 reference pTT27 plasmid. Yellow boxes show the location of the ISTth7 transposase common between the reference HB27 genome and to the laboratory-adapted strains HB27A and HB27E. The pink box labels the position of an ISTth7 copy common to both HB27A and HB27E which is absent from the reference HB27, while the green box indicates the location of an ISTth7 copy specific to the HB27E strain.

Of the two additional ISTth7 copies found in HB27E compared to HB27A, one had already been identified in a previous study where it was shown to confer higher transformability. This was a consequence of the inactivation of the TTP0026 gene, which encodes a homologue to the eukaryotic Argonaute protein implicated in defense against invading DNA [51]. Its identification in the current study validated the procedure that was followed. In conclusion, our data demonstrate that ISTth7 is a very active insertion sequence in *T. thermophilus*, possibly playing a role in adaptation of this strain to laboratory growth conditions.

Table 3. Coordinate locations of ISTth7 soft-clipped reads in the chromosome of the reference strain *T. thermophilus* HB27 (GenBank accession AE017221.1).

Start	End	Found in Reference	Found in HB27A	Found in HB27E
259,293	259,789	NO	YES	YES
386,772	389,127	NO	YES	YES
562,354	562,716	NO	YES	YES
1,082,351	1,082,693	NO	NO	YES
1,133,885	1,134,345	NO	YES	YES
1,370,579	1,371,607	YES	YES	YES
1,533,343	1,534,271	YES	YES	YES
1,550,847	1,551,323	NO	YES	YES
1,655,564	1,655,957	NO	YES	YES

Table 4. Coordinates location of ISTth7 soft-clipped reads in the megaplasmid of the reference strain *T. thermophilus* HB27 (GenBank accession (AE017222).

Start	End	Found in Reference	Found in HB27A	Found in HB27E
23,402	23,834	NO	NO	YES
57,597	58,625	YES	YES	YES
72,589	73,617	YES	YES	YES
85,677	86,705	YES	YES	YES
154,975	155,865	NO	YES	YES
159,958	160,986	YES	YES	YES

4. Discussion and Conclusions

Despite the extreme conditions and relatively isolated habitat of *Thermus* spp., several attributes leading to extraordinary genome plasticity enhance their evolutionary success. The first aspect of this plasticity is related to the presence (in most strains) of a constitutively expressed natural competence apparatus that allows cells to import environmental DNA (eDNA) at very high efficiencies, and also to incorporate genomic DNA from specific *Thermus* spp. donor strains (tDNA) through a novel conjugation-like mechanism (transjugation) [52]. In most cases, eDNA is degraded by different pathways, including restriction enzymes and nucleic acid interference, mediated either by diverse CRISPR/Cas systems or by Argonaute. However, tDNA acquired by transjugation has a greater chance to overcome these surveillance systems, which allows the recombination apparatus to incorporate new genes if they are bordered by sequences that are homologous to the host genome or to self-catalytic elements, such as ISs.

Recently, the dynamics of ISs within bacterial populations has come to the forefront, as the spread of pathogen-related genes and antibiotic resistances has been associated with transposable elements [13,24,53]. Although no actual transposons encoding selectable genes were detected in our analysis, horizontal gene transfer via ISs seems increased in several extreme environments, evidenced by the large numbers identified in the genomes of extremophilic microorganisms [23,25]. In particular, multiple reports on thermophiles have highlighted a surprisingly high diversity and number of ISs, suggesting an association with transfer of genes involved in thermophilic adaptation. For instance, *Deinococcus geothermalis* harbors more than 113 IS copies, classified into 6 types entailing up to 12% of its genome, whereas the closely related mesophilic *Deinococcus radiodurans* only presents 4 IS copies. Additionally, Blount and Grogan [54] reported an extensive diversity of ISs among isolated *Sulfolobus* populations (thermoacidophile) from a common sampling site, confirming that despite relative isolation, these thermophilic niches are notoriously rich in these elements. This surprising IS enrichment has been observed in other extremophiles, with similar figures described for the halophile *Haloarcula marismortui*, in which 20% of its pNG500 megaplasmid corresponded to IS DNA [55], and more than 169 IS sequences have been detected across the *Bacillus halodurans* genome [56]. Indeed, various metagenomic studies focused on extreme environments have revealed the preferred location of ISs to be within autonomous extrachromosomal elements [57]. This implies their role in the spread of adaptive genes through flexible genetic platforms like the megaplasmids present in many *Thermus* spp., which provides the cells a way to assess their selective value without risking essential chromosomal genes. Chromosomal IS copies may further facilitate the intra-genomic transfer of important traits, possibly explaining the diverse transposition frequencies, patterns, and abundance of IS scars and partial copies spread across thermophilic genomes [54].

Here, we have analyzed the abundance, diversity, and recent mobility events of IS elements in *Thermus* spp., with a special emphasis on the laboratory strains HB27, HB8, and NAR1. Our data support many of the aforementioned features common to many extremophiles: a high number of ISs with broad diversity and concentration of ISs within extrachromosomal elements, with densities of 69 IS copies/Mbp compared to 9 copies/Mbp found in the chromosome. In this regard, the significant increase in chromosomal ISTth7 copies found in the laboratory-adapted HB27 strains, compared to their parent, was very surprising. This could be explained by the dispensability of the targeted genes when grown in laboratory growth media, or the positive selection of these mutations under such conditions. A clear example of this would be selection of high transformation efficiency mutants (HB27E) by knockout of the *ago* gene in the megaplasmid (*TTP026::ISTth7*). Alternatively, it is also possible that the apparent increase in number of ISTth7 sequences found and the shared localization in both strains, despite their evolution in separate laboratories, is a consequence of contig collapse in the reference genome during assembly of the published genome.

Another consequence of the greater abundance of ISs in the megaplasmids is that it provides recombination sites for the insertion of DNA acquired via HGT. In this sense, it is relevant to note that genes associated with megaplasmids are tenfold more likely to be transferred by transjugation and integrated into other *T. thermophilus* recipient strains than chromosomal genes [57]. Thus, the presence of homologous ISs in donor and recipient strains may contribute to the integration step and, consequently, to the spread of adaptive traits such as nitrate respiration [50]. Also, ISs flanking genomic regions may mobilize groups of genes and generate phenotypic diversity, expanding genetic variability within the population and ultimately speeding its evolution.

Author Contributions: Conceptualization, A.B. and J.B.; Methodology, A.B. and J.B.; Software, A.B., R.P., E.S.-H. and S.G.-d.l.F.; Validation, A.B. and J.B.; Formal Analysis, A.B. and J.B.; Sample preparation, A.B. and M.S.; Resources, A.B. and J.B.; Data Curation, A.B., M.S. and J.B.; Writing—Original Draft Preparation, A.B. and J.B.; Writing—Review & Editing, A.B. and J.B.; Visualization, A.B. and J.B.; Supervision, J.B.; Project Administration, J.B.; Funding Acquisition, J.B.

Funding: This work was supported by grants BIO2016-77031-R from the Spanish Ministry of Economy and Competitiveness and grant n° 685474 from Horizon 2020 research and innovation program of the European Union. An institutional grant from Fundación Ramón Areces to the CBMSO is also acknowledged.

Acknowledgments: We acknowledge the supervision and advice of Didier Mazel and the technical help of Esther Sánchez. The NGS experimental design of NAR1, HB27A, and HB27E genome sequencing, follow up and data analysis service were provided by the Genomics and NGS Core Facility at the Centro de Biología Molecular Severo Ochoa (CBMSO, CSIC-UAM) which is part of the CEI UAM+CSIC, Madrid, Spain. Genome sequencing of HB27A and HB27E were provided by MicrobesNG (http://www.microbesng.uk), which is supported by the BBSRC (grant number BB/L024209/1). The PacBio based sequencing service was provided by the Norwegian Sequencing Centre (www.sequencing.uio.no), a national technology platform hosted by the University of Oslo and supported by the "Functional Genomics" and "Infrastructure" programs of the Research Council of Norway and the Southeastern Regional Health Authorities.

Conflicts of Interest: The authors declare no conflict of interest.

References

1. Gogarten, J.P.; Townsend, J.P. Horizontal gene transfer, genome innovation and evolution. *Nat. Rev. Microbiol.* **2005**, *3*, 679–687. [CrossRef] [PubMed]
2. Treangen, T.J.; Rocha, E.P.C. Horizontal Transfer, Not Duplication, Drives the Expansion of Protein Families in Prokaryotes. *PLoS Genet.* **2011**, *7*, e1001284. [CrossRef] [PubMed]
3. Gogarten, J.P.; Doolittle, W.F.; Lawrence, J.G. Prokaryotic evolution in light of gene transfer. *Mol. Biol. Evol.* **2002**, *19*, 2226–2238. [CrossRef] [PubMed]
4. Frost, L.S.; Leplae, R.; Summers, A.O.; Toussaint, A. Mobile genetic elements: The agents of open source evolution. *Nat. Rev. Microbiol.* **2005**, *3*, 722. [CrossRef] [PubMed]
5. Lang, A.; Beatty, J.T.; Rice, P.A. *Guest Editorial: Mobile Genetic Elements and Horizontal Gene Transfer in Prokaryotes*; Elsevier: New York, NY, USA, 2017.
6. Wozniak, R.A.; Waldor, M.K. Integrative and conjugative elements: Mosaic mobile genetic elements enabling dynamic lateral gene flow. *Nat. Rev. Microbiol.* **2010**, *8*, 552. [CrossRef]
7. Shapiro, J. (Ed.) *Mobile Genetic Elements*; Elsevier: New York, NY, USA, 2012.
8. Mahillon, J.; Chandler, M. Insertion sequences. *Microbiol. Mol. Biol. Rev.* **1998**, *62*, 725–774. [PubMed]
9. Shintani, M. The behavior of mobile genetic elements (MGEs) in different environments. *Biosci. Biotechnol. Biochem.* **2017**, *81*, 854–862. [CrossRef] [PubMed]
10. Van Elsas, J.D.; Bailey, M.J. The ecology of transfer of mobile genetic elements. *FEMS Microbiol. Ecol.* **2002**, *42*, 187–197. [CrossRef] [PubMed]

11. Syvanen, M. The evolutionary implications of mobile genetic elements. *Annu. Rev. Genet.* **1984**, *18*, 271–293. [CrossRef] [PubMed]
12. Partridge, S.R.; Kwong, S.M.; Firth, N.; Jensen, S.O. Mobile genetic elements associated with antimicrobial resistance. *Clin. Microbiol. Rev.* **2018**, *31*, e00088-17. [CrossRef]
13. Carraro, N.; Rivard, N.; Burrus, V.; Ceccarelli, D. Mobilizable genomic islands, different strategies for the dissemination of multidrug resistance and other adaptive traits. *Mob. Genet. Elem.* **2017**, *7*, 1–6. [CrossRef] [PubMed]
14. Stokes, H.W.; Gillings, M.R. Gene flow, mobile genetic elements and the recruitment of antibiotic resistance genes into Gram-negative pathogens. *FEMS Microbiol. Rev.* **2011**, *35*, 790–819. [CrossRef] [PubMed]
15. Top, E.M.; Springael, D. The role of mobile genetic elements in bacterial adaptation to xenobiotic organic compounds. *Curr. Opin. Biotechnol.* **2003**, *14*, 262–269. [CrossRef]
16. Siguier, P.; Filée, J.; Chandler, M. Insertion sequences in prokaryotic genomes. *Curr. Opin. Microbiol.* **2006**, *9*, 526–531. [CrossRef]
17. Siguier, P.; Pérochon, J.; Lestrade, L.; Mahillon, J.; Chandler, M. ISfinder: The reference centre for bacterial insertion sequences. *Nucleic Acids Res.* **2006**, *34* (Suppl. S1), D32–D36. [CrossRef] [PubMed]
18. Brochier-Armanet, C.; Forterre, P. Widespread distribution of archaeal reverse gyrase in thermophilic bacteria suggests a complex history of vertical inheritance and lateral gene transfers. *Archaea* **2006**, *2*, 83–93. [CrossRef]
19. Nelson, K.E.; Clayton, R.A.; Gill, S.R.; Gwinn, M.L.; Dodson, R.J.; Haft, D.H.; McDonald, L.; Utterback, T.R.; Malek, J.A.; Linher, K.D.; et al. Evidence for lateral gene transfer between Archaea and bacteria from genome sequence of Thermotoga maritima. *Nature* **1999**, *399*, 323–329. [CrossRef] [PubMed]
20. Akanuma, S.; Yokobori, S.-I.; Yamagishi, A. Comparative genomics of thermophilic bacteria and archaea. In *Thermophilic Microbes in Environmental and Industrial Biotechnology*; Springer: Dordrecht, The Netherlands, 2013; pp. 331–349.
21. Aminov, R.I. Horizontal gene exchange in environmental microbiota. *Front. Microbiol.* **2011**, *2*, 158. [CrossRef] [PubMed]
22. Lossouarn, J.; Dupont, S.; Gorlas, A.; Mercier, C.; Bienvenu, N.; Marguet, E.; Forterre, P.; Geslin, C. An abyssal mobilome: Viruses, plasmids and vesicles from deep-sea hydrothermal vents. *Res. Microbiol.* **2015**, *166*, 742–752. [CrossRef] [PubMed]
23. Krupovic, M.; Gonnet, M.; Hania, W.; Forterre, P.; Erauso, G. Insights into Dynamics of Mobile Genetic Elements in Hyperthermophilic Environments. *PLoS ONE* **2013**, *8*, e49044. [CrossRef] [PubMed]
24. Siguier, P.; Gourbeyre, E.; Chandler, M. Bacterial insertion sequences: Their genomic impact and diversity. *FEMS Microbiol. Rev.* **2014**, *38*, 865–891. [CrossRef] [PubMed]
25. Nelson, W.C.; Wollerman, L.; Bhaya, D.; Heidelberg, J.F. Analysis of insertion sequences in thermophilic cyanobacteria: Exploring the mechanisms of establishing, maintaining and withstanding high insertion sequence abundance. *Appl. Environ. Microbiol.* **2011**, *77*, 5458–5466. [CrossRef] [PubMed]
26. Takami, H.; Takaki, Y.; Chee, G.J.; Nishi, S.; Shimamura, S.; Suzuki, H.; Matsui, S.; Uchiyama, I. Thermoadaptation trait revealed by the genome sequence of thermophilic Geobacillus kaustophilus. *Nucleic Acids Res.* **2004**, *32*, 6292–6303. [CrossRef] [PubMed]
27. Filee, J.; Siguier, P.; Chandler, M. Insertion sequence diversity in archaea. *Microbiol. Mol. Biol. Rev.* **2007**, *71*, 121–157. [CrossRef] [PubMed]
28. Omelchenko, M.V.; Wolf, Y.I.; Gaidamakova, E.K.; Matrosova, V.Y.; Vasilenko, A.; Zhai, M.; Daly, M.J.; Koonin, E.V.; Makarova, K.S. Comparative genomics of Thermus thermophilus and Deinococcus radiodurans: Divergent routes of adaptation to thermophily and radiation resistance. *BMC Evol. Biol.* **2005**, *5*, 57. [CrossRef] [PubMed]
29. Madigan, M.T.; Martinko, J.M.; Parker, J. *Brock Biology of Microorganisms*; Prentice Hall: Upper Saddle River, NJ, USA, 1997; Volume 11.
30. Blesa, A.; Averhoff, B.; Berenguer, J. Horizontal Gene Transfer in Thermus spp. *Curr. Issues Mol. Biol.* **2018**, *29*, 23–36. [CrossRef]
31. Cava, F.; Hidalgo, A.; Berenguer, J. Thermus thermophilus as biological model. *Extremophiles* **2009**, *13*, 213. [CrossRef] [PubMed]
32. Touchon, M.; Bobay, L.-M.; Rocha, E.P. The chromosomal accommodation and domestication of mobile genetic elements. *Curr. Opin. Microbiol.* **2014**, *22*, 22–29. [CrossRef] [PubMed]

33. Ashby, M.; Bergquist, P. Cloning and sequence of IS1000, a putative insertion sequence from *Thermus thermophilus* HB8. *Plasmid* **1990**, *24*, 1–11. [CrossRef]

34. Tabata, K.; Hoshino, T. Mapping of 61 genes on the refined physical map of the chromosome of *Thermus thermophilus* HB27 and comparison of genome organization with that of *T. thermophilus* HB8. *Microbiology* **1996**, *142*, 401–410. [CrossRef]

35. Gregory, S.T.; Dahlberg, A.E. Transposition of an insertion sequence *ISTth7* in the genome of the extreme thermophile *Thermus thermophilus* HB8. *FEMS Microbiol. Lett.* **2008**, *289*, 187–192. [CrossRef] [PubMed]

36. Cava, F.; Laptenko, O.; Borukhov, S.; Chahlafi, Z.; Blas-Galindo, E.; Gómez-Puertas, P.; Berenguer, J. Control of the respiratory metabolism of *Thermus thermophilus* by the nitrate respiration conjugative element NCE. *Mol. Microbiol.* **2007**, *64*, 630–646. [CrossRef] [PubMed]

37. Ramírez-Arcos, S.; Fernández-Herrero, L.A.; Marín, I.; Berenguer, J. Anaerobic growth, a property horizontally transferred by an Hfr-like mechanism among extreme thermophiles. *J. Bacteriol.* **1998**, *180*, 3137–3143. [PubMed]

38. Chin, C.S.; Alexander, D.H.; Marks, P.; Klammer, A.A.; Drake, J.; Heiner, C.; Clum, A.; Copeland, A.; Huddleston, J.; Eichler, E.E.; et al. Nonhybrid, finished microbial genome assemblies from long-read SMRT sequencing data. *Nat. Methods* **2013**, *10*, 563–569. [CrossRef] [PubMed]

39. Sommer, D.D.; Delcher, A.L.; Salzberg, S.L.; Pop, M. Minimus: A fast, lightweight genome assembler. *BMC Bioinform.* **2007**, *8*, 64. [CrossRef]

40. PacBio Utilities. Available online: https://github.com/douglasgscofield/PacBio-utilities (accessed on 11 November 2018).

41. Walker, B.J.; Abeel, T.; Shea, T.; Priest, M.; Abouelliel, A.; Sakthikumar, S.; Cuomo, C.A.; Zeng, Q.; Wortman, J.; Young, S.K.; et al. An Integrated Tool for Comprehensive Microbial Variant Detection and Genome Assembly Improvement. *PLoS ONE* **2014**, *9*, e112963. [CrossRef] [PubMed]

42. Seemann, T. Prokka: Rapid prokaryotic genome annotation. *Bionformatics* **2014**, *30*, 2068–2069. [CrossRef] [PubMed]

43. Li, H.; Durbin, R. Fast and accurate short read alignment with Burrows-Wheeler Transform. *Bioinformatics* **2009**, *25*, 1754–1760. [CrossRef]

44. Edgar, R.C. MUSCLE: Multiple sequence alignment with high accuracy and high throughput. *Nucleic Acids Res.* **2004**, *32*, 1792–1797. [CrossRef]

45. Felsenstein, J. *PHYLIP (Phylogeny Inference Package) Distributed by the Author*; Version 3.c; Department of Genome Sciences, University of Washington: Seattle, WA, USA, 2005.

46. Touchon, M.; Rocha, E.P.C. Causes of Insertion Sequences Abundance in Prokaryotic Genomes. *Mol. Biol. Evol.* **2007**, *24*, 969–981. [CrossRef]

47. Da Cunha, V.; Guerillot, R.; Brochet, M.; Glaser, P. Integrative and conjugative elements encoding DDE transposases. In *Bacterial Integrative Mobile Genetic Elements*; Roberts, A.P., Mullany, P., Eds.; Landes Bioscience: Austin, TX, USA, 2013; pp. 250–260.

48. De la Cruz, F.; Davies, J. Horizontal gene transfer and the origin of species: Lessons from bacteria. *Trends Microbiol.* **2000**, *8*, 128–133. [CrossRef]

49. Brüggemann, H.; Chen, C. Comparative genomics of *Thermus thermophilus*: Plasticity of the megaplasmid and its contribution to a thermophilic lifestyle. *J. Biotechnol.* **2006**, *124*, 654–661. [CrossRef] [PubMed]

50. Alvarez, L.; Bricio, C.; Gómez, M.J.; Berenguer, J. Lateral transfer of the denitrification pathway genes among *Thermus thermophilus* strains. *Appl. Environ. Microbiol.* **2011**, *77*, 1352–1358. [CrossRef] [PubMed]

51. Swarts, D.C.; Jore, M.M.; Westra, E.R.; Zhu, Y.; Janssen, J.H.; Snijders, A.P.; Wang, Y.; Patel, D.J.; Berenguer, J.; Brouns, S.J.J.; et al. DNA-guided DNA interference by a prokaryotic Argonaute. *Nature* **2014**, *507*, 258. [CrossRef]

52. Blesa, A.; Baquedano, I.; Quintáns, N.G.; Mata, C.P.; Castón, J.R.; Berenguer, J. The transjugation machinery of *Thermus thermophilus*: Identification of TdtA, an ATPase involved in DNA donation. *PLoS Genet.* **2017**, *13*, e1006669. [CrossRef]

53. Rankin, D.J.; Rocha, E.P.; Brown, S.P. What traits are carried on mobile genetic elements, and why? *Heredity* **2011**, *106*, 1. [CrossRef] [PubMed]

54. Blount, Z.D.; Grogan, D.W. New insertion sequences of *Sulfolobus*: Functional properties and implications for genome evolution in hyperthermophilic archaea. *Mol. Microbiol.* **2005**, *55*, 312–325. [CrossRef] [PubMed]

55. Baliga, N.S.; Bonneau, R.; Facciotti, M.T.; Pan, M.; Glusman, G.; Deutsch, E.W.; Shannon, P.; Chiu, Y.; Weng, R.S.; Gan, R.R.; et al. Genome sequence of *Haloarcula marismortui*: A halophilic archaeon from the Dead Sea. *Genome Res.* **2004**, *14*, 2221–2234. [CrossRef]
56. Takami, H.; Han, C.G.; Takaki, Y.; Ohtsubo, E. Identification and distribution of new insertion sequences in the genome of alkaliphilic *Bacillus halodurans* C-125. *J. Bacteriol.* **2001**, *183*, 4345–4356. [CrossRef] [PubMed]
57. Brazelton, W.J.; Baross, J.A. Abundant transposases encoded by the metagenome of a hydrothermal chimney biofilm. *ISME J.* **2009**, *3*, 1420. [CrossRef]

microorganisms

MDPI

Article

Genome Analysis of *Vallitalea guaymasensis* Strain L81 Isolated from a Deep-Sea Hydrothermal Vent System

Anders Schouw [1], Francesca Vulcano [1], Irene Roalkvam [1], William Peter Hocking [1], Eoghan Reeves [2], Runar Stokke [1], Gunhild Bødtker [3] and Ida Helene Steen [1,*]

[1] Department of Biological Sciences and KG Jebsen Centre for Deep Sea Research, University of Bergen, N-5020 Bergen, Norway; anders.schouw@uib.no (A.S.); F.vulcano@uib.no (F.V.); irene.roalkvam@uib.no (I.R.); William.hocking@metis.no (W.P.H.); Runar.stokke@uib.no (R.S.)
[2] Department of Earth Science and KG Jebsen Centre for Deep Sea Research, University of Bergen, N-5020 Bergen, Norway; Eoghan.reeves@uib.no
[3] Centre for Integrated Petroleum Research (CIPR), Uni Research AS, Nygårdsgaten 112, N-5008 Bergen, Norway; Gunhild.bodtker@uni.no
* Correspondence: ida.steen@uib.no; Tel.: +47-55588375

Received: 13 June 2018; Accepted: 29 June 2018; Published: 4 July 2018

Abstract: *Abyssivirga alkaniphila* strain L81^T, recently isolated from a black smoker biofilm at the Loki's Castle hydrothermal vent field, was previously described as a mesophilic, obligately anaerobic heterotroph able to ferment carbohydrates, peptides, and aliphatic hydrocarbons. The strain was classified as a new genus within the family *Lachnospiraceae*. Herein, its genome is analyzed and *A. alkaniphila* is reassigned to the genus *Vallitalea* as a new strain of *V. guaymasensis*, designated *V. guaymasensis* strain L81. The 6.4 Mbp genome contained 5651 protein encoding genes, whereof 4043 were given a functional prediction. Pathways for fermentation of mono-saccharides, di-saccharides, peptides, and amino acids were identified whereas a complete pathway for the fermentation of *n*-alkanes was not found. Growth on carbohydrates and proteinous compounds supported methane production in co-cultures with *Methanoplanus limicola*. Multiple confurcating hydrogen-producing hydrogenases, a putative bifurcating electron-transferring flavoprotein—butyryl-CoA dehydrogenase complex, and a Rnf-complex form a basis for the observed hydrogen-production and a putative reverse electron-transport in *V. guaymasensis* strain L81. Combined with the observation that *n*-alkanes did not support growth in co-cultures with *M. limicola*, it seemed more plausible that the previously observed degradation patterns of crude-oil in strain L81 are explained by unspecific activation and may represent a detoxification mechanism, representing an interesting ecological function. Genes encoding a capacity for polyketide synthesis, prophages, and resistance to antibiotics shows interactions with the co-occurring microorganisms. This study enlightens the function of the fermentative microorganisms from hydrothermal vents systems and adds valuable information on the bioprospecting potential emerging in deep-sea hydrothermal systems.

Keywords: *Vallitalea guaymasensis*; hydrothermal vent; syntrophy; whole-genome sequence

1. Introduction

In hydrothermal vent systems water-rock reactions produce microbial nutrients such as H_2S, CH_4 and H_2, which create the basis for chemosynthetic food-webs and allow hot spots for biological activity to form in the deep ocean. This primary production results in a steady supply of organic matter that may be utilized by heterotrophic microorganisms [1–3]. Recently, the mesophilic heterotrophic bacterium *Abyssivirga alkaniphila* L81^T (=DSM 29592T = JCM 30920T) was isolated from a biofilm growing on a hydrothermal chimney at Loki's Castle hydrothermal vent field (LCVF) [4]. The LCVF is located on the Arctic Mid-Ocean Ridge (AMOR) at 73°30′ N and 8° E, where the Mohns Ridge migrates into the Knipovich Ridge, at a depth of 2400 m [5,6]. The field is discharging black smoker fluids of 310–320 °C from four chimneys, located at two mounds, roughly 150 m apart [5–7]. Low temperature venting occurs at the eastern flank of the mound, where a field of small barite chimneys is found [5,8]. The high-temperature vent fluids have high concentrations of CH_4, H_2, and CO_2, with CH_4 values of 15.5 mmol kg^{-1}, among the highest reported for a bare-rock hosted field [5]. The vent fluids are further characterized by a pH of 5.5, end-member H_2S content up to 4.7 mmol kg^{-1}, and very high NH_4 concentrations. The high values for CH_4 combined with the NH_4 values, points to the influence of hydrothermal alteration of buried sedimentary organic matter. C_1 to C_4 hydrocarbons of thermogenic origin have been detected in the venting fluids [7]. The microbial mats growing on the black smokers in LCVF are dominated by chemolitoautotrophic *Epsilonproteobacteria* [3,9] supporting growth of heterotrophic *Bacteroidetes* [10,11].

Recently, an objection to our newly described genus *Abyssivirga* was proposed based on 16S rRNA phylogeny by Postec and coworkers (2017) [12]. They argued that *A. alkaniphila* should be reassigned to the genus *Vallitalea*, possibly representing a novel species, *Vallitalea alkaniphila*, if demonstrated by significant DNA-DNA hybridization and phenotyphic difference [12]. The genus *Vallitalea* belongs to the family *Defluviitaleaceae*, within *Clostridiales*, and comprises Gram negative, motile, non-spore-forming, mesophilic rods with fermentative metabolism [13,14]. So far, all strains of this genus have been isolated from marine hydrothermal systems. The type strain *V. guaymasensis* Ra1766G[T] was isolated from microbial mats situated on sediments at Guaymas Basin [14], an area associated with hydrothermal activity and hydrocarbon seeps [15], whereas *V. pronyensis* FatNI3[T] was isolated from a chimney in the hydrothermal alkaline springs at Prony Bay [13]. Here, we present a comparison of *A. alkaniphila* with *V. guaymasensis* RA1766G1[T] and *V. pronyensis* FatNI3[T] establishing *A. alkaniphila* L81[T] as a new strain of *V. guaymasensis*, named *V. guaymasensis* L81. Moreover, a draft genome sequence of *V. guaymasensis* L81 is analyzed. Altogether, an interesting model system for the investigation of fermentative cooperation from hydrothermal vents systems has been established, along with new genomic input for bioprospecting potential attractive enzymes and new antimicrobial compounds.

2. Materials and Methods

Growth experiments were performed on Met II medium as described by Schouw and co-workers (2016) [4]. Cultures were grown in 30 mL glass vials, sealed with butyl rubber stoppers containing 15 mL of medium, and 15 mL of $N_2:CO_2$ (80:20) gas phase. Acetate, arabinose, butyrate, cellobiose, cellulose, chitin, decane, dextrin, formate, fructose, galactose, glucose, heptane, lactose, maltose, mannitol, mannose, octane, palatinose, pectin, pentane, peptone, propionate, pyruvate, rhamnose, ribose, starch, sucrose, tryptone, yeast extract, xylan, and xylose were tested as growth substrates. Decane, heptane, octane, pentane, pentone, and yeast extract were tested as growth substrates for co-cultures of *V. guaymasensis* and *M. limicola*.

Acetate, arabinose, butyrate, cellobiose, dextrin, formate, fructose, galactose, glucose, lactose, maltose, mannitol, mannose, palatinose, propionate, pyruvate, rhamnose, ribose, sucrose, xylan, and xylose were added to a final concentration of 20 mM. Cellulose and chitin were added to a final concentration of 0.25% w/w, and peptone, yeast extract, pectin, starch, and tryptone to a final concentration of 0.1% w/w. Decane, heptane, octane, pentane, and pentone were added to a final concentration of 1 µL/mL medium.

The production of H_2 and CH_4 in culture headspaces was quantified using a SRI 8610C gas chromatograph (GC), equipped with a packed column (molecular sieve 5 Å) and serially connected thermal conductivity (TCD) and flame-ionization detectors (FID), for H_2/CH_4 and CH_4 detection, respectively. Nitrogen was used as a carrier gas, with a column flow rate of 20 mL/min (70 °C isothermal) and detector temperatures of 250 °C. The GC was calibrated using injected moles of commercial gas standard mixtures.

For each measurement, 0.5 mL subsamples of culture headspace gas were taken with Hamilton gas-tight syringes previously flushed with nitrogen to avoid air contamination of the serum vials. The gas samples were then transferred into a 5 mL loop filled with nitrogen, and after dilution and homogenization were then injected onto the GC column through a valve.

Measured moles of each gas were normalized to the total headspace volume. Although partitioning of gases between headspace and liquid phases is occurring, a consideration of the Henry's Law constants for H_2 and CH_4 (7.8×10^{-6} mol/m^3 Pa and 1.4×10^{-5} mol/m^3 Pa at standard temperature in water [16]), indicates that >90% of these gases is present in the headspace at equilibrium, thus direct measurement from headspace gas subsamples accounts for the near totality of H_2 and CH_4 produced. Analytical uncertainty is estimated to be ±5% (2 s) for both CH_4 and H_2.

Interaction between microbial cells was analyzed with Fluorescence In-Situ Hybridization, as described by Glöckner and co-workers (1996) [17]. From each culture, 500 µL were fixed in 2% formaldehyde at room temperature for two hours. The cells were collected on a 0.2 µm polycarbonate filter, which was then washed twice with excess 1X PBS, air-dried and stained with fluorescently labeled oligonucleotides. *V. guaymasensis* L81 cells were targeted with EUB338 probe (5′-GCT GCC TCC CGT AGG AGT-3′) [18] labeled with Alexa488 fluorochrome, while methanogens were targeted with ARCH917 probe (5′-GTG CTC CCC CGC CAA TTC-3′) [19] labeled with Cy3 fluorochrome. For double hybridization, 5 µL of each probe (50 ng/µL stock concentration) were added to 100 µL of hybridization buffer and approximately 20 µL of the resulting mixture were applied to each filter. The hybridization step was performed in 35% formamide.

Stained filters were mounted with Immersol 518F (Carl Zeiss AG, Oberkochen, Germany) and visualized with Zeiss Axio Imager Z1 microscope (Carl Zeiss Microscopy GmbH, Göttingen, Germany), equipped with filter 38 (Alexa488) and 43 (Cy3).

For Scanning Electron Microscopy (SEM), 100 µL aliquots of each culture were fixed with 2% glutaraldehyde, at room temperature for one hour. Each aliquot was increased to 1 mL final volume using 1X PBS and was applied to a 0.2 µm polycarbonate filter to collect cells. Filters were then dehydrated with serial ethanol washes (50%, 75%, 3 × 100%), air-dried, mounted on an aluminum specimen stud with carbon tape and coated with iridium using a Gatan 682 coater (Gatan Inc., Pleasanton, CA, USA). Microbial cells were visualized by a Zeiss Supra 55VP field emission scanning electron microscope (FE-SEM; Carl Zeiss, Stockholm, Sweden), equipped with a Thermo Noran System SIX energy dispersive spectrometer (EDS) system (Carl Zeiss AS, Oslo, Norway) and an in-lens detector.

The cell wall structures in the *Vallitalea* strains were studied in detail using transmission electron microscopy (TEM). Cell cultures were embedded in LR White medium grade resin (Electron Microscopy Sciences, Pasadena, CA, USA) and later cut to 60 nm slices using a ultramikrotom (Reichert-Jung Ultracut. Leica microsystems GmbH, Wetzlar, Germany). Samples were stained with Reynold's lead citrate solution and examined in a Jeol 1011 TEM (Jeol Ltd., Tokyo, Japan).

DNA-DNA hybridization (DDH) included *A. alkaniphila* L81^T, *V. guaymasensis* Ra1766G1^T (DSM-24848) and *V. pronyensis* FatNI3^T (DSM-25904). All strains were cultivated on Met II medium added 0.1% glucose and 0.05% sucrose. Cells were harvested using centrifugation at 10,000× *g* for 30 min at 4 °C. At Deutsche Sammlung von Microorganismen und Zellkulturen (DSMZ), the cells were disrupted by Constant Systems TS 0.75 KW (IUL Instruments, Barcelona, Spain), and DNA in the crude lysate was purified by chromatography on hydroxyapatite as described by Cashion and co-workers (1977) [20]. DNA-DNA hybridization was carried out as described by De Ley and co-workers (1970) [21], with modifications described by Huss and co-workers (1983) [22], using a model Cary 100 Bio UV/VIS-spectrophotometer equipped with a Peltier-thermostatted 6 × 6 multicell changer and a temperature controller with in situ temperature probe (Varian, Palo Alto, CA, USA).

Cultures grown to mid-exponential or stationary phase were fixed to glass slides and Gram-stained with crystal violet and safranin to investigate possible differences in staining characteristics. Cell wall properties was also investigated by the string test as described by Ryu (1938) [23], where 0.3% KOH was added smear from a cell pellet. Living cultures on microscopy slides were stained for flagella visualization, using the staining solution by Ryu (1937) [24] and method similar to Heimbrook and co-workers (1989) [25].

Genomic DNA for sequencing was extracted from a culture grown on Met II medium added glucose to a final concentration of 2 mM [4]. DNA was extracted by the modified Marmur method [26]. Genome sequencing was performed at the Norwegian Sequencing Centre (www.sequencing.uio.no). A 10 kb library was prepared using Pacific Bioscience 10 kb library preparation protocol and BluePippin (Sage Science, Beverly, MA, USA) for the final size selection. In total, four SMRT cells were used for sequencing the library on a Pacific Bioscience RS II Instrument (Pacific Bioscience, Menlo Park, CA, USA) in combination with the P4-C2 chemistry. The raw reads were filtered prior de novo assembly and polishing using HGAP v3, SMRT Analysis v2.2.0 (Pacific Bioscience, Menlo Park, CA, USA). The genome was annotated using RAST [27–29], IMG-ER [30–32] dbCAN [33] and eggNOG 4.5 [34]. Polyketid biosynthetic clusters were identified with antiSMASH 3.0 [35]. Putative hydrogenses annotated by RAST and IMG-ER were classified using HydDB [36].

This Whole Genome Shotgun project has been deposited at DDBJ/ENA/GenBank under the accession number QMDO00000000. The version described in this paper is version QMDO01000000. The annotated genome is available in the JGI GOLD under Project id: Ga0082380, and in RAST under id number: 6666666.108580.

The raw data have been deposited in the NCBI Sequence Read Archive under Study: PRJNA450742 (SRP151092), Sample: L81 genome (SRS3447209), Experiment: VguL81_WGA (SRX4282565) RUN: cell4_Vguaymasensis_s1_p0.2.bax.h5 (SRR7410928), RUN: cell3_Vguaymasensis _s1_p0.2.bax.h5 (SRR7411504), RUN: cell2_Vguaymasensis_s1_p0.2.bax.h5 (SRR7411505), RUN: cell1_Vguaymasensis_s1_p0.2.bax.h5 (SRR7411506).

3. Results

3.1. Reclassification of Abyssivirga alkaniphila L81^T

Strain L81 was originally classified as *Abyssivirga alkaniphila* L81^T (=DSM 29592T = JCM 30920T) [4]. Later Postec and co-workers (2016) suggested that the 16S rRNA gene sequences of *A. alkaniphila* L81^T and *Valitalia guaymasensis* RA1766G1^T, were so similar that they likely belonged to the same species [12]. To clarify this, a DNA-DNA hybridization was performed of *A. alkaniphila* L81^T with *V. guaymasensis* RA1766G1^T and *V. pronyensis*, respectively. The results revealed a similarity between *A. alkaniphila* L81^T and *V. guaymasensis* of 69.7% and with *V. pronyensis* of 13.35%. With the inherent variability of the method, this places *A. alkaniphila* L81^T and *V. guaymasensis* RA1766G1^T within the 70% DDH similarity recommended as the species cut-off value [37,38], supporting that they are strains of the same species. Based on a polyphasic analysis, including 16S rRNA gene similarities; DNA-DNA hybridization values; optimal temperature for growth; gram staining; flagella type and utilization of substrates, we argue that strain L81 (formerly named *Abyssivirga alkaniphila* L81^T) should be reclassified to *Vallitalea guaymasensis* L81. Supporting data are presented in supplementary data (Supplementary Table S1).

3.2. Emended Description of Vallitalea guaymasensis

Cells are gram positive, motile rods of 0.5 μm × 2–5 μm during exponential growth. Cells are usually single, but can occur in long chains. Stationary phase cells are non-motile with a spherical morphology. Growth is not observed below 15 °C or above 42 °C, and the optimal growth temperature is 37 °C. A minimum of 0.5% NaCl is required for growth, and maximum concentration of 6% NaCl is tolerated. Optimum growth rate occurs at 3% NaCl. The pH range was broad, with an optimum at pH 7.0–8.2. Reduced growth medium is required. Arabinose, cellobiose, dextrin, fructose, galactose, glucose, lactose, maltose, mannose, palatinose, pectin, peptone, ribose, starch, sucrose, tryptone, yeast extract, xylan, and xylose are utilized. Acetate, cellulose, chitin, formate, mannitol, propionate, pyruvate and rhamnose are not. A small amount of yeast extract is required for growth for some substrates, while acetate inhibit growth. Major fermentation products are H_2 and CO_2. Main whole-cell sugar is ribose, and trace amounts of galactose. The G + C content of chromosomal DNA is 31.7 mol%.

The type strain is *Vallitalea guaymasensis* RA1766G1^T (=DSM 24848 = JCM 17997).

3.3. Emended Description of Vallitalea Gen

Cells are motile, mesophilic rods with a fermentative and obligately organoheterotrophic metabolism. Genus is Gram-variable (Gram staining reaction) and includes both spore forming and non-spore forming strains. A minimum of 0.5% NaCl is required for growth. No quinones are detected. The major fatty acids are anteiso-C15:0, iso-C15:0, anteiso-DMA-C15:0 and C16:0.

3.4. Syntrophic Growth

Cultures supplemented with pentane, heptane, octane or decane, respectively, produced the same amount of methane as the negative control cultures in pure Met II medium. This indicated that the added alkanes did not support growth. When the cultures were given additional yeast extract after 11 months incubation, an immediate increase in methane-production was observed, confirming that the cultures were viable and were likely consuming components of the yeast extract only.

3.5. General Genomic Features

The permanent draft assembly of *V. guaymasensis* L81 resulted in 7 contigs with a total of 6.4 Mbp with an average coverage of 75. This places the *V. guaymasensis* L81 genome among the larger sequenced genomes [39,40]. The genome of *V. guaymasensis* L81 represents the first published genome from the genus *Vallitalea*. The GC content of 31.2% is uncharacteristically low for such a large genome [40], with the majority of coding sequences on the leading strand, a common feature in low GC *Firmicutes* [41]. The genome has 6 identical copies of the 16S rRNA gene. There are 17 regions of phage related genes on the genome, and repeated regions in these prophages cause difficulties in genome assembly. Figure 1 shows a circular representation of the genome, illustrating how the contigs are mainly split in phage regions. There are no clear indications of horizontal gene transfer on the genome, however, the low GC content makes transferred regions, tending towards lower GC content than the host genome [42], hard to identify. Table 1 summarizes the major properties of the genome.

Table 1. General genome features.

Category	RAST	IMG-ER
Genome size	6419149 bp	6419149 bp
Contigs	7	7
GC content	31.2%	31.2%
Coding sequences	5628	5651
RNA genes	67 (1.2%)	121 (2.1%)
rRNA genes	11	17
5S rRNA		6
16S rRNA		6
23S rRNA		5
tRNA genes	56	56
Other RNA genes		48
Genes with function prediction	3820 (67.87%)	4043 (70.05%)
Genes without function prediction	1808 (32.13%)	1608 (27.86%)

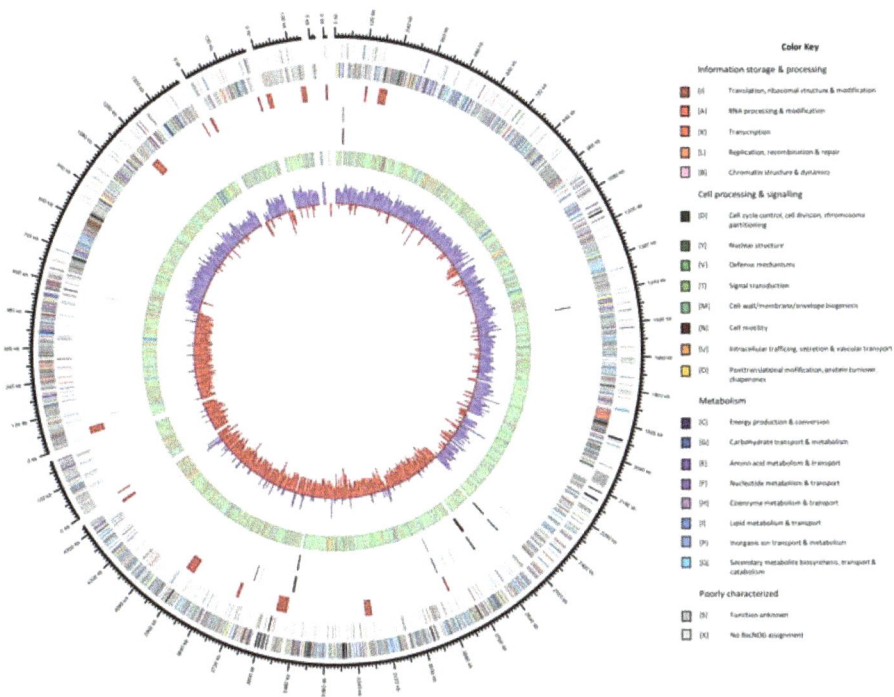

Figure 1. Circular representation of the *V. alkaniphila* L81 genome displaying relevant genome features. Contig order clockwise from top: 0, 5, 2, 6, 3, 7, 1. Circles representing the following (from center to outside): 1, G + C skew [(G − C)/(G + C) using a 2-kbp sliding window] (blue, positive; red, negative); 2, Taxonomy by Uniref90 top hits; *Firmicutes* (green), *Archaea* (red), *Eukaryota* (yellow), other Bacteria (blue), unknown (orange), no hit to Uniref90 (grey); 3, tRNAs (black); 4, rRNA operons (dark red); 5, Prophages; 6, Coding DNA sequence (CDS) on the reverse strand; 7, CDS on the forward strand. Color coding of CDS was based on COG categories. The figure was build using Circos version. 0.67-1.

3.6. Carbohydrate Metabolism and Transport

A high percentage of the annotated genes (23.8% of genes in RAST) can be related to the metabolism of carbohydrates, consistent with growth experiments. Moreover, 85 ABC transporters putatively related to carbohydrate transport were identified (Supplementary Table S1). *V. guaymasensis* L81 was shown to grow on arabinose, fructose, glucose, glycerol, galactose, galacturonate, lactate, pectin, polygalacturonate, and ribose in the original characterization of the strain [4]. Additionally, cellobiose, dextrin, lactose, maltose, mannose, palatinose, starch, sucrose, xylan, and xylose were confirmed to support growth in pure culture in this study.

A complete glycolysis pathway was identified, as were all enzymes involved in gluconeogenesis (Supplementary Figure S1 and Table S2). The Entner–Doudoroff pathway and the oxidative part of the pentose phosphate pathway were incomplete, as observed in other Gram-positive bacteria [43]. In addition to anabolic functions, the partial pentose phosphate pathway may serve as a pathway for isomerization and rearrangement of sugars to fuel the glycolysis in *V. guaymasensis* L81. Starch, isomaltulose, cellobiose, dextrin, galactose, lactose, maltose, mannose, palatinose, and sucrose could all be converted to glucose-6-phosphate for the glycolysis pathway using the enzymes listed in Supplementary Table S2, while glycerol enters the glycolysis pathway as 3-phosphoglycerate (Supplementary Figure S1 and Table S2).

Xylan, arabinoxylan, xylose arabinose, and ribose could enter the glycolysis through the pentose phosphate pathway. The genome also encoded a putative capacity for the degradation of chitin that would enter the glycolysis as fructose-6-phosphate. A putative complete pathway for cellulose degradation to glucose and fructose-6-phosphate was annotated by RAST [27–29] (Supplementary Figure S1 and Table S2). The enzyme annotated as an endoglucanase in the cellulose degradation pathway (peg.3524) appears to be a dubious annotation. No carbohydrate binding modules were identified in the protein by dbCAN [33], and it contained several regions of glycine repeats indicative of a phage origin. This leaves the pathway annotation incomplete, though cultivation experiments suggest that it may be functional.

The pyruvate resulting from glycolysis could fuel the citric acid (TCA) cycle; either as oxaloacetate or acetyl-CoA catalyzed by pyruvate carboxylase and pyruvate ferredoxin oxidoreductase, respectively. The TCA cycle is incomplete due to lacking genes for succinyl-CoA synthetase, and probably function as a biosynthesis pathway for amino acids and tetrapyrroles.

Pyruvate can also be directly fermented to lactate by lactate dehydrogenase, or enter the acetyl-CoA pool by pyruvate:ferredoxin oxidoreductase or pyruvate-formate lyase (Supplementary Figure S1 and Table S2), with a concomitant production of formate by the latter enzyme. The acetyl-CoA could be converted to acetate and ethanol in the conventional mixed acid fermentation pathway, to acetate via acetyl phosphate, or to acetate directly (Supplementary Figure S1 and Table S2). Acetyl-CoA can also be fermented to butyrate via two different pathways where the conversion of acetyl-CoA to butyryl-CoA is identical in both pathways, with acetoacetyl-CoA, 3-hydroxybutanoyl-coA and crotonyl-CoA as intermediates (Supplementary Figure S1 and Table S2). Butyryl-CoA can either be coupled with acetate to form butyrate, a reaction catalyzed by butyrate-acetoacetate CoA transferase; or form butyrate via butyryl phosphate, catalyzed by phosphotransbutyrylase and butyrate kinase. Genes for converting butyryl-CoA to butanol were also present in the genome (Supplementary Table S2). Genes encoding pathways leading to propionate, acetone or 2-propanol were not detected.

3.7. Protein Metabolism and Transport

Vallitalea guaymasensis L81 can grow on peptone and yeast extract as carbon and energy source [4], congruently, the genome encodes proteases and peptidases as well as membrane transport systems for oligopeptides, dipeptides, branched chain amino acids, and polar amino acids (Supplementary Table S2). The genome analysis shows that *V. guaymasensis* L81 is unable to synthesize L-phenylalanine, L-tyrosine, L-histidine, L-arginine, L-isoleucine, L-leucine, L-methionine, L-lysine or L-threonine, and thus rely on acquiring these amino acids from the external environment. Contrarily, genes for the synthesis of L-alanine, L-aspartate, L-glutamate, L-tryptophan, L-glycine, L-aspargine, L-glutamine, L-valine, L-serine, L-cysteine, and L-proline were identified.

Stickland reactions are used by amino acid degrading *Clostridia*, like *Clostridium sticklandii*, and require amino acid pairs, where one amino acid is reduced and the other oxidized [44]. In methanogenic consortia, methanogens can take the role of the reductive part of the Stickland reaction, removing the need for amino acid pairs [45]. A syntrophic relationship is also favorable in terms of thermodynamics. Low hydrogen pressure is favorable for the amino acid degradation reactions, and for some reactions, such as the degradation of alanine to acetate, syntrophy is a requirement [45,46]. The genomic data of *V. guaymasensis* L81 suggests that amino acids are fermented with protons as electron acceptor, and not degraded via Stickland reactions, as no candidate for an amino acid reductase was identified. The observation was supported by our laboratory experiments where low amounts of hydrogen were produced by pure cultures, and methane by consortia grown on yeast extract and peptone. This is similar to the *Acetoanaerobium pronyense* that was isolated from the Prony hydrothermal vent field [47].

The degradation pathways present for L-alanine, L-threonine, L-glycine, and L-serine all lead to pyruvate. The fructoselysine and L-lysine degradation pathway identified, is identical to that previously described for *Intestimonas* strain AF211 [48], where fructoselysine is phosphorylated to fructoselycine-6-phosphate, and subsequently split into L-lysine and glucose-6-phosphate. Glucose-6-phosphate enters the glycolysis, while L-lysine is fermented to acetate and butyrate.

There are no complete annotated ATP-yielding pathways for histidine and glutamate degradation in this genome, however, for the putative methylaspartate pathway [46,49], only the citramalate lyase was missing. It is possible that an enzyme different from those previously described could catalyze this reaction.

3.8. Alkane Activation and Degradation

Vallitalea guaymasensis L81 encodes two putative alkylsuccinate synthases for activation of *n*-alkanes by addition of fumarate [4] (Supplementary Table S2). A complete putative pathway for alkane degradation was however not identified on the genome. The enzymes involved in the carbon skeleton rearrangement and decarboxylation of the methylalkylsuccinic acids resulting from fumarate addition are not conclusively described [50]. An acyl-CoA synthetase has been suggested as a possible candidate for the addition of S-CoA in *D. alkenivorans* AK01 [50]. One putative acyl-CoA synthetase (peg.3802), and two long chain fatty acyl-CoA synthetases (peg.2630, peg.3316) were identified in the *V. guaymasensis* L81 genome. It has been suggested that an enzyme analogous to methylmalonyl-CoA mutase is responsible for the carbon skeleton rearrangement [50,51]. No homolog for a methylmalonyl-CoA mutase was identified on the *V. guaymasensis* L81 genome. Two acetyl-CoA carboxyl transferases (peg.444, peg.4125) that could be involved in decarboxylation of the fatty acyl-CoA prior to β-oxidation were also identified.

A putative complete pathway for β-oxidation of the fatty acids was identified (Supplementary Figure S1 and Table S2), where three putative acyl-CoA dehydrogenases were identified, of which two were located directly upstream of electron transfer flavoprotein (Etf) complexes (peg.3682, 1149). Moreover, three putative enoyl-CoA hydratases, two 3-hydroxyacyl-CoA dehydrogenases and two 3-ketoacyl-CoA thiolases were identified.

Regeneration of fumarate is required for this mode of alkane activation and has been shown to occur via the methylmalonyl-CoA pathway in *D. alkenivorans* AK-01 [50]. Of the required enzymes for this pathway, we could not identify homologs for succinyl-CoA synthetase or succinate dehydrogenase in *V. guaymasensis* L81.

3.9. Redox-Components

Three putative bifurcating/confurcating, multimeric [FeFe] hydrogenases (two trimeric and one tetrameric) were identified in *V. guaymasensis* L81 (Figure 2). The three multimeric [FeFe] hydrogenases have conserved regions similar to the trimeric bifurcating/confurcating hydrogenase from *Thermotoga maritima* [52]. The α-subunits (HydA) of all three enzymes contained the H-cluster binding domain and the conserved upstream FeS clusters similar to the *T. maritima* α-subunits [53,54]. However, all three α-subunits of the *V. guaymasensis* L81 multimeric hydrogenases are lacking the C-terminal [2Fe-2S] binding domain found in *T. maritima*. Since the fourth subunit (HydD) (peg.2899) of the tetrameric [FeFe] hydrogenases of *V. guaymasensis* strain L81carries a [2Fe-2S]-cluster binding domain similar to the C-terminal [2Fe-2S]-cluster domain in the *T. maritima* α-subunit, this isoenzyme appears most similar in architecture to the *T. maritima* trimeric enzyme. Moreover, a γ-subunit (HydC) with a 2Fe-2S domain was part of each hydrogenase. The β-subunits (HydB) in *V. guaymasensis* L81 carry the FMN/NADH-binding domain and the [2Fe-2S]-cluster binding domain as identified in the *T. maritima* enzyme. However, the two-[4Fe-4S] binding domain is only conserved in the tetrameric enzyme (peg.2898, 2899, 2900, 2901). This domain is absent in the first trimeric hydrogenase (peg.2066, 2067, 2068), and substituted by an enlarged β-subunit (peg.3895) with a GltD binding domain after the FMN/NADH-binding domain in the second trimeric hydrogenase (peg.3894, 3895, 3896), suggesting a second NAD(P)H binding site.

One monomeric [FeFe] hydrogenase (peg.4378) was classified by HydDB [36] as belonging to group C1, linked to a sensor histidine kinase (peg.4376). Group C1 hydrogenases are postulated to be involved in hydrogen sensing, and regulation of transcription of other hydrogenases [55,56].

Two [FeFe] hydrogenases (peg.96 and 98) were classified by HydDB as belonging to group C3 and group B hydrogenases respectively. The function of group C3 hydrogenases remain undescribed, but the members of this group are co-transcribed with hydrogen evolving hydrogenases associated with fermentation, and appear to regulate the transcription of these hydrogenases [55–59]. The function of group B hydrogenases remains unconfirmed, but indirect evidence suggests they are involved in hydrogenogenic fermentation, coupling the reoxidation of reduced ferredoxin to hydrogen evolution [55,57,60].

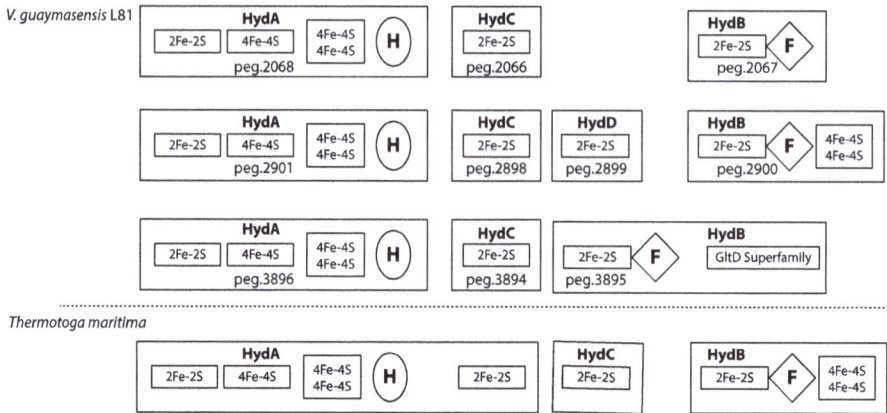

Figure 2. Domain organization of multimeric Fe-Fe hydrogenases. Large boxes represent hydrogenase subunits. The trimeric bifurcating/confurcating hydrogenase from *T. maritima* is shown for comparison. H, H-cluster; 2Fe-2S, cluster binding site; 4Fe-4S, cluster binding site; F, FMN and NAD$^+$ binding site. *T. maritima* data from Soboh and co-workers (2004) [61].

Elements of a putative [NiFe] hydrogenase were identified, consisting of HypE (peg.4591), HypD (peg.4592), HypC (peg.4593), HypF (peg.4594), HyaD (peg.4596), a [NiFe] hydrogenase large subunit (peg.4597), and a [NiFe] hydrogenase small subunit (peg.4599). The large and small subunits were annotated by RAST and HydDB as belonging to the [NiFe] group 1 uptake hydrogenases.

The two electron-transfer flavoprotein (Etf) complexes encoded in the genome, are directly linked to the butyryl-CoA/acyl-CoA dehydrogenase complexes. One of the Etf β-subunits (peg.3683) contains the NADH and FAD binding site marker sequences for a bifurcating Etf [62,63].

The genome encodes both a V-type ATPase and a F-type ATPase. F-ATPases generate ATP via proton translocation, while V-ATPases utilize ATP to translocate protons across membranes [64]. Contrary to F-ATPases, the function of V-ATPases is not reversible, which renders them unable to generate ATP. The function of V-ATPases is thus solely to produce a proton motive force [64], and the presence of a V-ATPase makes energy generation by electron transport phosphorylation probable in *V. guaymasensis* L81.

The genome also encodes a proton/sodium ion-translocating Rnf (*Rhodobacter* nitrogen fixation) complex (peg.4370–4375), catalyzing the reversible oxidation of reduced ferredoxin with NAD$^+$ [9] (Figure 3).

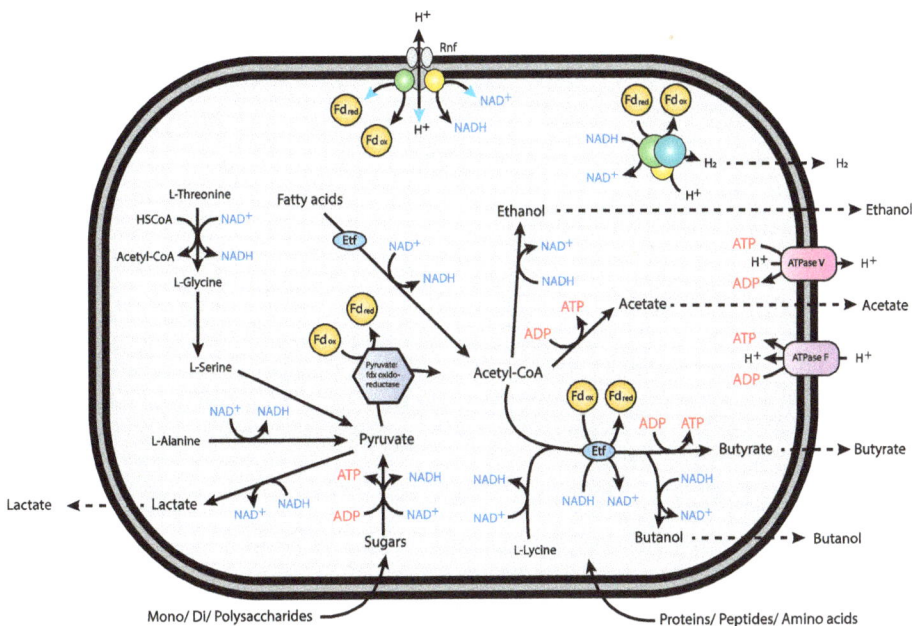

Figure 3. Overview of the main metabolic pathways and energy production in *V. guaymasensis* L81. The overall scheme for energy conservation in *V. guaymasensis* L81 is simple. Proton gradient generation is probably done by Rnf mediated Fd_{red} oxidation. In the absence of a terminal electron acceptor or corresponding electron transport chains, hydrogen generated by cytoplasmic hydrogenases is probably produced as a non-respiratory, fermentative mechanism. As Fd_{red} is needed for H_2 generation at higher partial pressure, the bifurcating/confurcating [FeFe] hydrogenases may be capable of conserving energy, coupling oxidation of Fd_{red} and NADH [65].

The genome of *V. guaymasensis* L81 lacks homologues corresponding to the complexes of the mitochondrial respiratory chain, including genes encoding a NADH:ubiquinone oxidoreductase (Nuo) complex. In addition, genes corresponding to respiratory cytochromes are absent.

3.10. Polyketid Synthesis

The antiSMASH analysis revealed 9 different gene clusters potentially involved in the synthesis of polyketids (Supplementary Table S2). Four of the clusters showed similarities with previously described biosynthetic clusters, as listed below, while five clusters contained genes for non-ribosomal peptide synthases and polyketide synthases without showing similarities to previously described polyketide synthesis clusters.

Cluster 3 showed similarity with several previously described biosynthetic clusters: A polyketide synthase (peg.4909) and a 3-hydroxybutyryl-CoA dehydrogenase (peg.4915) showed similarity to genes from zwittermycin A [66], paenilamicin [67], and colibactin [68] biosynthetic clusters. A similar 3-hydroxybutyryl-CoA dehydrogenase is also found in the pellasoren biosynthetic cluster [69], along with a butyryl-CoA dehydrogenase similar to peg.4913 from the *V. guaymasensis* L81 cluster 3. The cluster also contains genes encoding a lanthibiotic transport permease protein (peg.4929), and a lanthibiotic transport ATP-binding protein (peg.4931) similar to those found in the ericin A [70], entianin [71], and subtilin [72] biosynthetic clusters. The lacticin 481 [73,74] biosynthetic cluster contains genes encoding proteins similar to the lanthibiotic transport ATP-binding protein (peg.4931), and a transposase (peg.4926) from cluster 3. The low similarity, in terms of homologous proteins per cluster, with previously described biosynthetic clusters makes it difficult to predict the nature of the polyketide produced by cluster 3.

Cluster 5 showed similarity with several previously described biosynthetic clusters, namely those for nosperin [75], thiomarinol [76] kalimantacin [77], thailandamide [78–80], bacillaene [81–83], elansolid [84], bongkrekic acid [85], calyculin [86], cylindrocyclophane [87], and carbamidocyclophane [88]. Apart from two putative polyketide synthases (peg.1164 and 1168), the similarity lies in a string of 8 small proteins that are represented in various configurations in the mentioned biosynthetic clusters. The hydroxymethylglutaryl-CoA synthase (peg.3083 and 3085) has homologs in all the clusters, while the 3-oxoacyl synthase (peg.3084) has homologs in all clusters apart from the one for calyculin. A methylglutaconyl-CoA hydratase (peg.3082) was homologous with enoyl-CoA hydratases from the thiomarinol, kalimantacin, thailandamide, bacillaene and calyculin clusters, while the enoyl-CoA hydratase (peg.3081) was homologous with a second enoyl-CoA in the kalimantacin, thailandamide, bacillaene, elansolid and calyculin clusters. The enoyl-CoA from cluster 5 was also homologous to enoyl-CoA hydratases in the nosperin and bongkrekic acid clusters. The acyl carrier protein (peg.3086) had homologs in the thiomarinol, kalimantacin thailandamide, bacillaene, and elansolid clusters. A malonyl-CoA acyl carrier protein had homologs in the nosperin, kalimantacin, bacillaene, elansolid, and bongkrekic acid clusters. The only homolog for the acetyl-CoA carboxyl transferase (peg.3090) was found in the thailandamide cluster.

Cluster 6 harbor genes homologous to LanB, LanC and the ABC-transporter, LanT, of the type I lanthibiotic synthesis pathway [89]. The gene cluster was most similar to the penisin biosynthetic gene cluster from *Penibacillus eheimensis* A3 [90]. Homologs for the precursor, Lan A (PenA), and the serine protease, LanP, were not identified, so it is uncertain if lanthibiotics are produced by the strain. LanP is also missing from the *P. eheimensis* A3 penisin cluster, demonstrating that LanP is non-essential for the synthesis of functional lanthibiotics. The *P. eheimensis* A3 penisin cluster contains two additional proteins, PenD and PenR. PenD has been suggested as a dehydratase, and PenR as a possible tanscriptional regulator [90]. The genome of *V. guaymasensis* L81 contains three homologs to PenD (peg.331, 361, 4587) and one homolog to PenR (peg.359), however, neither are encoded in the same cluster as LanB, LanC, and LanT, and thus, uncertain to be involved in polyketide synthesis. Finally, except for two genes similar to the ABC transporter genes eqbK and eqbL found in the equibactin biosynthetic cluster from *Streptococcus equi* [91], cluster 9 showed no similarities with other described biosynthetic clusters.

3.11. Intracellular Compartments

SEM micrographs revealed the presence of internal membrane structures in the cells, however, the function of these structures remains unknown [4]. Interestingly, the genome contains genes for the formation of a combined propanediol utilizing (PDU) and ethanolamine utilizing (EUT) type of bacterial microcompartments (BMC) [92,93]. These processes have to be sequestered in compartments since the intermediate, propionaldehyde from propanediol degradation, is mutagenic, and hence toxic to the cells [94–97]. Furthermore, acetaldehyde from ethanolamine degradation is volatile and needs to be sequestered to not escape the cell [95,96,98]. Present evidence suggests that these PDU/EUT fusion loci produce separate PDU and EUT BMCs [92]. PDU microcompartments have also been shown to be involved in bacterial degradation of rhamnose and fucose [99,100]. The function of BMCs in *V. guaymasensis* L81 is uncertain, as no propanediol oxidoreductase was identified. However, an alpha-L-fucosidase is encoded in the BMC locus in *V. guaymasensis* strain L81, indicating a role in fucose degradation. This would require a different enzyme to perform the part of the missing propanediol oxidoreductase.

4. Discussion

Cultivation experiments and genome analysis have revealed that *V. guaymasensis* L81 has the capacity to fulfill multiple metabolic roles in its environment at Loki's castle hydrothermal vent field.

Fermentation of mono-, di- and polysaccharides occurs via the glycolysis pathway and may lead to a formation of acetate and H_2, which in co-cultures supports the growth of methanogenic partners and subsequent CH_4-production. *Vallitalea guaymasensis* strain L81 also encodes a capacity for the formation of lactate, ethanol, butyrate, or butanol as fermentation products. Amino acids are predicted to be oxidatively converted to acetate, CO_2 (and NH_4^+) with concomitant interspecies hydrogen transfer to methanogenic partners [101].

During the fermentation of saccharides and amino acids like L-Histidine, L-threonine and L-alanine, Fd_{red} is produced by oxidation of pyruvate to acetyl-CoA and CO_2 in a reaction catalyzed by pyruvate:ferredoxin oxidoreductase, while reduced NADH is formed in multiple redox-reactions (Figure 2, Supplementary Figure S1). The formation of H_2 as an intermediate in these fermentation reactions is predicted to be facilitated by cytoplasmic bifurcating/confurcating hydrogenases. The presence of the GltD-domain on one of the hydrogenase subunits (HydB) may indicate a specific link to the oxidation or assimilation of glutamate. In butyrogenic fermentation, the endergonic reduction of ferredoxin with NADH could be coupled to the exergonic reduction of crotonyl-CoA to butyryl-CoA catalyzed by a putative butyryl-CoA/Etf complex. The two Etf-complexes encoded in the genome, are directly linked to the butyryl-CoA/acyl-CoA dehydrogenase complexes. Moreover, one of the Etf β-subunits contain the marker sequences for a bifurcating Etf [62,63]. This suggests that the *V. guaymasensis* strain L81 can use electron bifurcation via the electron transferring flavoprotein-butyryl-CoA dehydrogenase (Etf-Bcd) complex to generate reduced ferredoxin during acetyl-CoA fermentation to butyrate [62,102–105]. The non-bifurcating Etf complex is directly linked to an acyl-CoA dehydrogenase, and in close proximity to a long-chain fatty acid CoA ligase and a 3-hydroxyacyl-CoA dehydrogenase (Supplementary Table S1). This indicates that it may be involved in the conversion of acyl-CoA to enoyl-CoA in the β-oxidation cycle, as described for *Smithella* spp. [15]. The crotonyl-CoA formed in the fermentation of fructoselysine or lysine may be disproportionated to acetate, butyrate, and H_2, where Fd_{red} for the H_2 formation is, as for the butyrogenic fermentation, generated by electron bifurcation with crotonyl-CoA and NADH. Since *V. guaymasensis* L81 encodes a Rnf complex and a V-type ATPase, re-oxidation of Fd_{red} may also be catalyzed by the encoded Rnf-complex, representing an alternative route for energy conservation. The energy difference between reduced ferredoxin and NAD^+ of about 200 mV is proposed to be used by the Rnf-complex to generate an electrochemical H^+ or Na^+ gradient [103,106]. In *C. ljungdahlii* the Rnf complex plays a crucial role in pumping protons out of the cell membrane for energy conservation during acetogenic, autotrophic growth, but it was also suggested to contribute to ATP synthesis during heterotrophic growth on fructose by generating a proton gradient [107]. The Rnf complex may also contribute to the appropriate $NADH/Fd_{red}$ ratio that will be affected by biosynthesis and the oxidation state of the growth substrates. The direct association of a putative sensory hydrogenase (peg.4378) and a sensory histidine kinase (peg.4376) to the Rnf complex (peg.4370–4375) indicates that the activity of the Rnf complex may be regulated by extracellular H_2 concentrations.

As previously reported, *V. guaymasensis* L81 appears to degrade a wide spectrum of alkanes as assessed by whole-oil gas chromatography [4]. In contrast to a confirmed capacity to perform methanogenic syntrophic growth on protein-rich compounds and saccharides, a complete pathway for utilization of alkanes was not identified in the genome analysis. Growth on alkanes requires enzymes for the activation of alkanes and for β-oxidation of fatty acid intermediates [51]. In the absence of external electron acceptors, such as iron, nitrate or sulfate, this process is an obligate syntrophic reaction, and hydrogenases for H_2 production are required [108–110]. Interestingly, *V. guaymasensis* L81 appears to have the capacity for reverse electron transport driven H_2-production that would be required for syntrophic fermentation of hydrocarbons to methane. ATP may be formed via substrate level phosphorylation in the conversion of acetyl-CoA intermediates to acetate during the oxidation of fatty acids. This ATP could then be used for creating a proton gradient by pumping protons to the periplasmic space. The resulting proton-gradient could again support Rnf-driven ferredoxin reduction coupled to reoxidation of NADH formed in the oxidation of 3-hydroxybutanoyl to acetoacetyl-CoA. This Fd_{red} could subsequently support H_2-production by one or more of the putative multimeric bifuricating [FeFe] hydrogenases in *V. guaymasensis* L81. Moreover, this Fd_{red} could feed the bifuricating Bcd-Etf complex, coupling the endergonic oxidation of butyryl-CoA to crotonyl_CoA with the exergonic reduction of NAD^+ with Fd_{red}. In the fermentation pathway of *n*-alkanes an enzyme homologous to methylmalonyl-CoA mutase has been proposed to be responsible for carbon skeleton rearrangement of the methylalkylsuccinates resulting from fumarate activation [50,51]. As no enzyme homologous to methylmalonyl-CoA mutase was identified in the *V. guaymasensis* L81 genome, we cannot say if this reaction is plausible in this organism. Altogether, there is potentially a metabolic capacity for syntrophic methanogenic growth on alkanes in *V. guaymasensis* L81. However, significant methane-production was not observed during growth with a selection of *n*-alkanes as substrate, and further work is needed to confirm this metabolism. Another possibility is that the degradation of crude oil-components observed in *V. guaymasensis* L81 [4] is a result of an unspecific activation, and not a full hydrocarbon metabolism directly yielding energy for the cell. It has previously been demonstrated that the hydrocarbon activation enzymes show a relaxed specificity, particularly those for *n*-alkanes, and activate a broader range of hydrocarbons than can be fully metabolized by the cell [111,112]. One possible function of this broad-spectrum activation is detoxification [111]. Hydrocarbons are toxic to microorganisms and can diffuse over the cell membrane [111]. Jarling and co-workers (2015) proposed that the transformation of hydrocarbons to di-acids by activating enzymes, reduces the toxic effect, as the two negative charges of the di-acid prevents the molecule from penetrating and disrupting the cell membrane [111]. This potentially allows the hydrocarbon degrading bacteria to grow closer to an oil-water interface than they otherwise could. The ability to detoxify is also beneficial to microorganisms not utilizing hydrocarbons for energy, as it enables them to grow in hydrocarbon-contaminated environments [111]. Hydrocarbons like methane, acetylene, ethylene, ethane, and butane have been detected in venting fluids from Loki's castle [7], demonstrating that this may be a useful function in the system. As the venting fluids at Loki's castle also contain hydrogen concentrations of up to 5.5 mmol kg^{-1} [7], this would shift obligately syntrophic, low energy reactions, such as hydrocarbon degradation and fermentation of amino acids such as alanine, towards being unfavorable. These reactions are thus more probable to occur in surrounding sediments.

Successful syntrophic interactions rely on effective transfer of metabolites, such as hydrogen and formate, between partner organisms [113–118]. The flagellum proteins FliC and FliD have been shown to be important in the adherence of *Pelotomaculum thermopropionicum* to its methanogenic partners *Methanothermobacter thermautotrophicus* and *Methanosaeta thermophila* during syntrophic growth [119, 120], and FliD was also shown to enhance the methanogenic activity of *M. thermautotrophicus* [120]. A complete set of genes for flagellar assembly was observed in *V. guaymasensis* L81 indicating that a similar mechanism could occur in this organism. When studied using light microscopy, cells of *V. guyamasensis* L81 and *M. limicola* were observed both free-living and clumped together in aggregates of cells and precipitated iron sulfides. A potential for biofilm formation was observed using SEM and FISH analyses in the stationary phase of co-cultures grown on glucose, but no flagella or pili were observed (Figure 4). The SEM investigation indicates the formation of an extracellular substance that likely aids cell aggregation (Figure 4). Cells generally showed a higher tendency to aggregate during growth on unfavorable substrates, and with low substrate concentrations. There are several other features revealed through the genome analysis that point to interactions with other organisms in the ecosystem. The genome of *V. guaymasensis* L81 encodes genes for the synthesis of various polyketides that could be utilized as means of keeping competitors at bay. Genes for vancomycin resistance and aminoglycoside resistance were also identified. These may be useful as protection from antibiotics produced by the cell itself, or as defense against competitors. The prophages present on the genome reflect the constant impact of viruses on the microbial community.

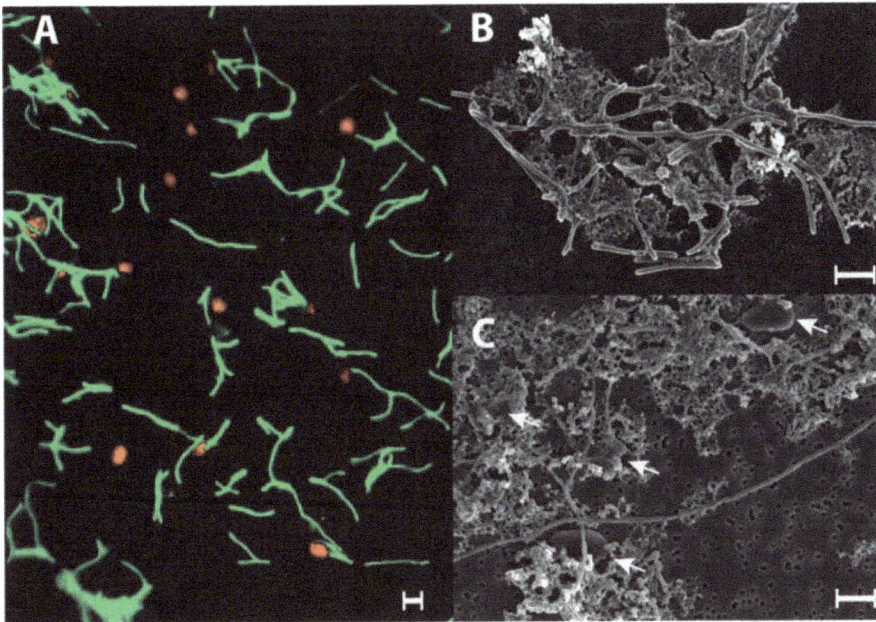

Figure 4. (**A**) Fluorescent in situ hybridization image of a stationary phase co-culture of *V. guaymasensis* L81 (labeled in green) and *M. limicola* (labeled in orange), utilizing glucose as substrate for growth; (**B**) Scanning electron micrograph of a stationary phase co-culture of *V. guaymasensis* L81, and *M. limicola*, given glucose; (**C**) Scanning electron micrograph (SEM) of a stationary phase co-culture of *V. guaymasensis* L81, and *M. limicola*, given crude oil. Arrows mark cells of *M. limicola*. Scale bars: 2 µm.

The varied metabolic toolbox, coupled with polyketid biosynthesis clusters, and multiple prophages, makes *V. guaymasensis* L81 an interesting subject for further studies.

5. Conclusions

V. Guaymasensis L81 is a versatile organism, with the ability to utilize a wide range of carbohydrates and peptides, both in pure culture, and in co-cultures with a methanogenic partner. The observed hydrocarbon degradation facilitated by the strain is proposed to be unspecific activation, possibly as a detoxification mechanism, rather than energy metabolism. The genome infers an ability to degrade complex polymers such as chitin and xylan. Along with 9 putative polyketide synthesis clusters, this makes the strain interesting from an industrial perspective.

Supplementary Materials: The following are available online at http://www.mdpi.com/2076-2607/6/3/63/s1, Figure S1: Core metabolic reactions, Table S1: Phenotypic properties important for reclassification, Table S2: Annotations related to Figure S1.

Author Contributions: Conceptualization, I.H.S. and G.B.; Methodology, I.H.S., G.B.; R.S.; A.S. and E.R.; Validation, A.S.; I.R. and F.V.; Formal Analysis, R.S.; A.S. and I.R.; Investigation, A.S.; I.R. and F.V.; Resources, I.H.S.; G.B., R.S. and E.R.; Data Curation, R.S.; Writing, Original Draft Preparation, A.S.; I.H.S.; I.R. and W.H.; Writing, Review & Editing, I.H.S.; A.S.; G.B.; R.S.; E.R.; F.V.; Visualization, A.S. and R.S.; Supervision, I.H.S.; G.B.; R.S.; Project Administration, I.H.S.; Funding Acquisition, I.H.S.

Funding: This work was supported by the Norwegian Research Council (projects 208491 and 179560) and from the European Union's Horizon 2020 research and innovation program (Blue Growth: Unlocking the potential of Seas and Oceans) through the Project 'INMARE' under grant agreement No. 634486.

Conflicts of Interest: The authors declare no conflict of interest.

References

1. Rogers, K.L.; Amend, J.P. Energetics of potential heterotrophic metabolisms in the marine hydrothermal system of Vulcano Island, Italy. *Geochim. Cosmochim. Acta* **2006**, *70*, 6180–6200. [CrossRef]
2. Slobodkina, G.B.; Kolganova, T.V.; Tourova, T.P.; Kostrikina, N.A.; Jeanthon, C.; Bonch-Osmolovskaya, E.A.; Slobodkin, A.I. *Clostridium tepidiprofundi* sp. nov.; a moderately thermophilic bacterium from a deep-sea hydrothermal vent. *Int. J. Syst. Evolut. Microbiol.* **2008**, *58 Pt 4*, 852–855. [CrossRef] [PubMed]
3. Stokke, R.; Dahle, H.; Roalkvam, I.; Wissuwa, J.; Daae, F.L.; Tooming-Klunderud, A.; Thorseth, I.H.; Pedersen, R.B.; Steen, I.H. Functional interactions among filamentous *Epsilonproteobacteria* and *Bacteroidetes* in a deep-sea hydrothermal vent biofilm. *Environ. Microbiol.* **2015**, *17*, 4063–4077. [CrossRef] [PubMed]
4. Schouw, A.; Eide, T.L.; Stokke, R.; Pedersen, R.B.; Steen, I.H.; Bødtker, G. *Abyssivirga alkaniphila* gen. nov.; sp. nov.; an alkane-degrading, anaerobic bacterium from a deep-sea hydrothermal vent system, and emended descriptions of *Natranaerovirga pectinivora* and *Natranaerovirga hydrolytica*. *Int. J. Syst. Evolut. Microbiol.* **2016**, *66*, 1724–1734. [CrossRef] [PubMed]
5. Pedersen, R.B.; Rapp, H.T.; Thorseth, I.H.; Lilley, M.D.; Barriga, F.J.; Baumberger, T.; Flesland, K.; Fonseca, R.; Früh-Green, G.L.; Jorgensen, S.L. Discovery of a black smoker vent field and vent fauna at the Arctic Mid-Ocean Ridge. *Nat. Commun.* **2010**, *1*, 126. [CrossRef] [PubMed]
6. Pedersen, R.B.; Thorseth, I.H.; Nygård, T.; Lilley, M.D.; Kelley, D.S. Diversity of hydrothermal systems on slow spreading ocean ridges. *Geophys. Monogr. Ser.* **2010**, *188*. [CrossRef]
7. Baumberger, T.; Früh-Green, G.L.; Thorseth, I.H.; Lilley, M.D.; Hamelin, C.; Bernasconi, S.M.; Okland, I.E.; Pedersen, R.B. Fluid composition of the sediment-influenced Loki's Castle vent field at the ultra-slow spreading Arctic Mid-Ocean Ridge. *Geochim. Cosmochim. Acta* **2016**, *187*, 156–178. [CrossRef]
8. Eickmann, B.; Thorseth, I.H.; Peters, M.; Strauss, H.; Bröcker, M.; Pedersen, R.B. Barite in hydrothermal environments as a recorder of subseafloor processes: A multiple-isotope study from the Loki's Castle vent field. *Geobiology* **2014**. [CrossRef] [PubMed]
9. Dahle, H.; Roalkvam, I.; Thorseth, I.H.; Pedersen, R.B.; Steen, I.H. The versatile in situ gene expression of an *Epsilonproteobacteria*-dominated biofilm from a hydrothermal chimney. *Environ. Microbiol. Rep.* **2013**, *5*, 282–290. [CrossRef] [PubMed]
10. Bauer, S.L.; Roalkvam, I.; Steen, I.H.; Dahle, H. *Lutibacter profundi* sp. nov.; isolated from a deep-sea hydrothermal system on the Arctic Mid-Ocean Ridge and emended description of the genus Lutibacter. *Int. J. Syst. Evolut. Microbiol.* **2016**, *66*, 2671–2677. [CrossRef] [PubMed]
11. Wissuwa, J.; Bauer, S.L.; Steen, I.H.; Stokke, R. Complete genome sequence of *Lutibacter profundi* LP1T isolated from an arctic deep-sea hydrothermal vent system. *Stand. Genom. Sci.* **2017**, *12*, 5. [CrossRef] [PubMed]
12. Postec, A.; Olivier, B.; Fardeau, M.-L. Objection to the proposition of the novel genus *Abyssivirga*. *Int. J. Syst. Evolut. Microbiol.* **2016**, *67*, 174. [CrossRef]
13. Ben Aissa, F.; Postec, A.; Erauso, G.; Payri, C.; Pelletier, B.; Hamdi, M.; Ollivier, B.; Fardeau, M.-L. *Vallitalea pronyensis* sp. nov.; isolated from a marine alkaline hydrothermal chimney. *Int. J. Syst. Evolut. Microbiol.* **2014**, *64 Pt 4*, 1160–1165. [CrossRef] [PubMed]
14. Lakhal, R.; Pradel, N.; Postec, A.; Hamdi, M.; Ollivier, B.; Godfroy, A.; Fardeau, M.-L. *Vallitalea guaymasensis* gen. nov.; sp. nov.; isolated from marine sediment. *Int. J. Syst. Evolut. Microbiol.* **2013**, *63*, 3019–3023. [CrossRef] [PubMed]
15. Teske, A.; Callaghan, A.V.; LaRowe, D.E. Biosphere frontiers of subsurface life in the sedimented hydrothermal system of Guaymas Basin. *Front. Microbiol.* **2014**, *5*, 362. [CrossRef] [PubMed]
16. Sander, R. Compilation of Henry's law constants (version 4.0) for water as solvent. *Atmos. Chem. Phys.* **2015**, *15*, 4399–4981. [CrossRef]
17. Glöckner, F.; Amann, R.; Alfreider, A.; Pernthaler, J.; Psenner, R.; Trebesius, K.; Schleifer, K.-H. An in-situ hybridization protocol for detection and identification of planktonic bacteria. *Syst. Appl. Microbiol.* **1996**, *19*, 403–406. [CrossRef]
18. Amann, R.I.; Binder, B.J.; Olson, R.J.; Chisholm, S.W.; Devereux, R.; Stahl, D.A. Combination of 16S rRNA-targeted oligonucleotide probes with flow cytometry for analyzing mixed microbial populations. *Appl. Environ. Microbiol.* **1990**, *56*, 1919–1925. [PubMed]

19. Loy, A.; Lehner, A.; Lee, N.; Adamczyk, J.; Meier, H.; Ernst, J.; Schleifer, K.-H.H.; Wagner, M. Oligonucleotide microarray for 16S rRNA gene-based detection of all recognized lineages of sulfate-reducing prokaryotes in the environment. *Appl. Environ. Microbiol.* **2002**, *68*, 5064–5081. [CrossRef] [PubMed]

20. Cashion, P.; Holder-Franklin, M.; McCully, J.; Franklin, M. A rapid method for the base ratio determination of bacterial DNA. *Anal. Biochem.* **1977**, *81*, 461–466. [CrossRef]

21. De Ley, J.; Cattoir, H.; Reynaerts, A. The quantitative measurement of DNA hybridization from renaturation rates. *Eur. J. Biochem.* **1970**, *12*, 133–142. [CrossRef]

22. Huss, V.A.R.; Festl, H.; Schleifer, K.H. Studies on the spectrophotometric determination of DNA hybridization from renaturation rates. *Syst. Appl. Microbiol.* **1983**, *4*, 184–192. [CrossRef]

23. Ryu, E. On the Gram-differentiation of bacteria by the simplest method. *J. Jpn. Soc. Vet. Sci.* **1938**, *15*, 205–207. [CrossRef]

24. Ryu, E. A simple method of staining bacterial flagella. *Kitasato Arch. Exp. Med.* **1937**, *14*, 218–219.

25. Heimbrook, M.E.; Wang, W.L.; Campbell, G. Staining bacterial flagella easily. *J. Clin. Microbiol.* **1989**, *27*, 2612–2615. [PubMed]

26. Roalkvam, I.; Drønen, K.; Stokke, R.; Daae, F.-L.; Dahle, H.; Steen, I.H. Physiological and genomic characterization of *Arcobacter anaerophilus* IR-1 reveals new metabolic features in *Epsilonproteobacteria*. *Front. Microbiol.* **2015**, *6*, 987. [CrossRef] [PubMed]

27. Aziz, R.K.; Bartels, D.; Best, A.A.; DeJongh, M.; Disz, T.; Edwards, R.A.; Formsma, K.; Gerdes, S.; Glass, E.M.; Kubal, M.; et al. The RAST Server: Rapid annotations using subsystems technology. *BMC Genom.* **2008**, *9*, 75. [CrossRef] [PubMed]

28. Brettin, T.; Davis, J.J.; Disz, T.; Edwards, R.A.; Gerdes, S.; Olsen, G.J.; Olson, R.; Overbeek, R.A.; Parrello, B.; Pusch, G.D.; et al. RASTtk: A modular and extensible implementation of the RAST algorithm for building custom annotation pipelines and annotating batches of genomes. *Sci. Rep.* **2015**, *5*, 8365. [CrossRef] [PubMed]

29. Overbeek, R.A.; Olson, R.; Pusch, G.D.; Olsen, G.J.; Davis, J.J.; Disz, T.; Edwards, R.A.; Gerdes, S.; Parrello, B.; Shukla, M.; et al. The SEED and the rapid annotation of microbial genomes using subsystems technology (RAST). *Nucleic Acids Res.* **2014**, *42*, D206–D214. [CrossRef] [PubMed]

30. Chen, I.-M.A.; Markowitz, V.M.; Chu, K.; Palaniappan, K.; Szeto, E.; Pillay, M.; Ratner, A.; Huang, J.; Andersen, E.; Huntemann, M.; et al. IMG/M: Integrated genome and metagenome comparative data analysis system. *Nucleic Acids Res.* **2017**, *45*, D507–D516. [CrossRef] [PubMed]

31. Markowitz, V.M.; Chen, I.-M.A.; Chu, K.; Szeto, E.; Palaniappan, K.; Pillay, M.; Ratner, A.; Huang, J.; Pagani, I.; Tringe, S.; et al. IMG/M 4 version of the integrated metagenome comparative analysis system. *Nucleic Acids Res.* **2014**, *42*, D568–D573. [CrossRef] [PubMed]

32. Markowitz, V.M.; Chen, I.-M.A.; Palaniappan, K.; Chu, K.; Szeto, E.; Pillay, M.; Ratner, A.; Huang, J.; Woyke, T.; Huntemann, M.; et al. IMG 4 version of the integrated microbial genomes comparative analysis system. *Nucleic Acids Res.* **2014**, *42*, D560–D567. [CrossRef] [PubMed]

33. Yin, Y.; Mao, X.; Yang, J.; Chen, X.; Mao, F.; Xu, Y. dbCAN: A web resource for automated carbohydrate-active enzyme annotation. *Nucleic Acids Res.* **2012**, W445–W451. [CrossRef] [PubMed]

34. Huerta-Cepas, J.; Szklarczyk, D.; Forslund, K.; Cook, H.; Heller, D.; Walter, M.C.; Rattei, T.; Mende, D.R.; Sunagawa, S.; Kuhn, M.; et al. eggNOG 4.5: A hierarchical orthology framework with improved functional annotations for eukaryotic, prokaryotic and viral sequences. *Nucleic Acids Res.* **2016**, *44*, D286–D293. [CrossRef] [PubMed]

35. Weber, T.; Blin, K.; Duddela, S.; Krug, D.; Kim, H.; Bruccoleri, R.; Lee, S.; Fischbach, M.A.; Müller, R.; Wohlleben, W.; et al. antiSMASH 3.0-a comprehensive resource for the genome mining of biosynthetic gene clusters. *Nucleic Acids Res.* **2015**, *43*, W237–W243. [CrossRef] [PubMed]

36. Søndergaard, D.; Pedersen, C.; Greening, C. HydDB: A web tool for hydrogenase classification and analysis. *Sci. Rep.* **2016**, *6*, 34212. [CrossRef] [PubMed]

37. Stackebrandt, E.; Ebers, J. Taxonomic parameters revisited: Tarnished gold standards. *Microbiol. Today* **2006**, *33*, 152–155.

38. Wayne, L.G.; Brenner, D.J.; Colwell, R.R.; Grimont, P.A.; Kandler, O.; Krichevsky, M.I.; Moore, L.H.; Moore, W.E.; Murray, R.; Stackebrandt, E.S. Report of the ad hoc committee on reconciliation of approaches to bacterial systematics. *Int. J. Syst. Bacteriol.* **1987**, *37*, 463–464. [CrossRef]

39. Koonin, E.V.; Wolf, Y.I. Genomics of bacteria and archaea: The emerging dynamic view of the prokaryotic world. *Nucleic Acids Res.* **2008**, *36*, 6688–6719. [CrossRef] [PubMed]

40. Land, M.; Hauser, L.; Jun, S.-R.; Nookaew, I.; Leuze, M.R.; Ahn, T.-H.; Karpinets, T.; Lund, O.; Kora, G.; Wassenaar, T.; et al. Insights from 20 years of bacterial genome sequencing. *Funct. Integr. Genom.* **2015**, *15*, 141–161. [CrossRef] [PubMed]
41. Rocha, E. Is there a role for replication fork asymmetry in the distribution of genes in bacterial genomes? *Trends Microbiol.* **2002**, *10*, 393–395. [CrossRef]
42. Daubin, V.; Lerat, E.; Perrière, G. The source of laterally transferred genes in bacterial genomes. *Genome Biol.* **2003**, *4*, 1–12. [CrossRef] [PubMed]
43. Richardson, A.; Somerville, G.; Sonenshein, A. Regulating the intersection of metabolism and pathogenesis in gram-positive bacteria. In *Metabolism and Bacterial Pathogenesis*, 1st ed.; Department of Health and Human Services: Washington, DC, USA, 2015; Volume 1, pp. 129–165.
44. Mead, G. The amino acid-fermenting clostridia. *J. Gen. Microbiol.* **1971**, *67*, 47–56. [CrossRef] [PubMed]
45. Schink, B.; Stams, A.J. Syntrophism among prokaryotes. In *The Procaryotes*; Dworkin, M., Falkow, S., Rosenberg, E., Schleifer, K.H., Stackebrandt, E., Eds.; Springer: Berlin, Germany, 2006; Volume 2, pp. 309–335. ISBN 978-0-387-30740-4.
46. Buckel, W. Unusual enzymes involved in five pathways of glutamate fermentation. *Appl. Microbiol. Biotechnol.* **2001**, *57*, 263–273. [CrossRef] [PubMed]
47. Bes, M.; Merrouch, M.; Joseph, M.; Quéméneur, M.; Payri, C.; Pelletier, B.; Ollivier, B.; Fardeau, M.-L.; Erauso, G.; Postec, A. *Acetoanaerobium pronyense* sp. nov.; an anaerobic alkaliphilic bacterium isolated from a carbonate chimney of the Prony Hydrothermal Field (New Caledonia). *Int. J. Syst. Evolut. Microbiol.* **2015**, *65*, 2574–2580. [CrossRef] [PubMed]
48. Bui, T.; Ritari, J.; Boeren, S.; de Waard, P.; Plugge, C.M.; de Vos, W.M. Production of butyrate from lysine and the Amadori product fructoselysine by a human gut commensal. *Nat. Commun.* **2015**, *6*, 10062. [CrossRef] [PubMed]
49. Barker, H. Explorations of bacterial metabolism. *Annu. Rev. Biochem.* **1978**, *47*, 1–33. [CrossRef] [PubMed]
50. Callaghan, A.V.; Morris, B.E.L.; Pereira, I.A.C.; McInerney, M.J.; Austin, R.N.; Groves, J.T.; Kukor, J.J.; Suflita, J.M.; Young, L.Y.; Zylstra, G.J.; et al. The genome sequence of *Desulfatibacillum alkenivorans* AK-01: A blueprint for anaerobic alkane oxidation. *Environ. Microbiol.* **2012**, *14*. [CrossRef] [PubMed]
51. Wilkes, H.; Rabus, R.; Fischer, T.; Armstroff, A.; Behrends, A.; Widdel, F. Anaerobic degradation of n-hexane in a denitrifying bacterium: Further degradation of the initial intermediate (1-methylpentyl) succinate via C-skeleton rearrangement. *Arch. Microbiol.* **2002**, *177*, 235–243. [CrossRef] [PubMed]
52. Schut, G.J.; Adams, M.W. The iron-hydrogenase of *Thermotoga maritima* utilizes ferredoxin and NADH synergistically: A new perspective on anaerobic hydrogen production. *J. Bacteriol.* **2009**, *191*, 4451–4457. [CrossRef] [PubMed]
53. Mulder, D.W.; Shepard, E.M.; Meuser, J.E.; Joshi, N.; King, P.W.; Posewitz, M.C.; Broderick, J.B.; Peters, J.W. Insights into [FeFe]-hydrogenase structure, mechanism, and maturation. *Structure* **2011**. [CrossRef] [PubMed]
54. Vignais, P.; Billoud, B.; Meyer, J. Classification and phylogeny of hydrogenases. *FEMS Microbiol. Rev.* **2001**, *25*, 455–501. [CrossRef] [PubMed]
55. Greening, C.; Biswas, A.; Carere, C.R.; Jackson, C.J.; Taylor, M.C.; Stott, M.B.; Cook, G.M.; Morales, S.E. Genomic and metagenomic surveys of hydrogenase distribution indicate H_2 is a widely utilised energy source for microbial growth and survival. *ISME J.* **2016**, *10*, 761–777. [CrossRef] [PubMed]
56. Vignais, P.; Billoud, B. Occurrence, classification, and biological function of hydrogenases: An overview. *Chem. Rev.* **2007**, *107*, 4206–4272. [CrossRef] [PubMed]
57. Calusinska, C.; Happe, T.; Joris, B.; Wilmotte, A. The surprising diversity of *clostridial* hydrogenases: A comparative genomic perspective. *Microbiology* **2010**, *156*, 1575–1588. [CrossRef] [PubMed]
58. Shaw, A.; Hogsett, D.; Lynd, L. Identification of the [FeFe]-hydrogenase responsible for hydrogen generation in *Thermoanaerobacterium saccharolyticum* and demonstration of increased ethanol yield via hydrogenase knockout. *J. Bacteriol.* **2009**, *191*, 6457–6464. [CrossRef] [PubMed]
59. Zheng, Y.; Kahnt, J.; Kwon, I.; Mackie, R.; Thauer, R. Hydrogen formation and its regulation in *Ruminococcus albus*: Involvement of an electron-bifurcating [FeFe]-hydrogenase, of a non-electron-bifurcating [FeFe]-Hydrogenase, and of a Putative Hydrogen-Sensing [FeFe]-Hydrogenase. *J. Bacteriol.* **2014**, *196*, 3840–3852. [CrossRef] [PubMed]

60. Wolf, P.; Biswas, A.; Morales, S.; Greening, C.; Gaskins, R. H_2 metabolism is widespread and diverse among human colonic microbes. *Gut Microbes* **2016**, *7*, 235–245. [CrossRef] [PubMed]

61. Soboh, B.; Linder, D.; Hedderich, R. A multisubunit membrane-bound [NiFe] hydrogenase and an NADH-dependent Fe-only hydrogenase in the fermenting bacterium *Thermoanaerobacter tengcongensis*. *Microbiology* **2004**, *150*, 2451–2463. [CrossRef] [PubMed]

62. Chowdhury, N.P.; Kahnt, J.; Buckel, W. Reduction of ferredoxin or oxygen by flavin-based electron bifurcation in *Megasphaera elsdenii*. *FEBS J.* **2015**, *282*, 3149–3160. [CrossRef] [PubMed]

63. Costas, A.M.; Poudel, S.; Miller, A.-F.; Schut, G.J.; Ledbetter, R.N.; Fixen, K.R.; Seefeldt, L.C.; Adams, M.W.W.; Harwood, C.S.; Boyd, E.S.; et al. Defining electron bifurcation in the electron-transferring flavoprotein family. *J. Bacteriol.* **2017**, *199*, e00440-17. [CrossRef] [PubMed]

64. Perzov, N.; Padler-Karavani, V.; Nelson, H.; Nelson, N. Features of V-ATPases that distinguish them from F-ATPases. *FEBS Lett.* **2001**, *504*, 223–228. [CrossRef]

65. Visser, M.; Worm, P.; Muyzer, G.; Pereira, I.; Schaap, P.J.; Plugge, C.M.; Kuever, J.; Parshina, S.N.; Nazina, T.N.; Ivanova, A.E.; et al. Genome analysis of *Desulfotomaculum kuznetsovii* strain 17T reveals a physiological similarity with *Pelotomaculum thermopropionicum* strain SIT. *Stand. Genom. Sci.* **2013**, *8*, 69–87. [CrossRef] [PubMed]

66. Kevany, B.M.; Rasko, D.A.; Thomas, M.G. Characterization of the complete Zwittermicin A biosynthesis gene cluster from *Bacillus cereus*. *Appl. Environ. Microbiol.* **2009**, *75*, 1144–1155. [CrossRef] [PubMed]

67. Garcia-Gonzalez, E.; Müller, S.; Hertlein, G.; Heid, N.; Süssmuth, R.D.; Genersch, E. Biological effects of paenilamicin, a secondary metabolite antibiotic produced by the honey bee pathogenic bacterium *Paenibacillus larvae*. *MicrobiologyOpen* **2014**, *3*, 642–656. [CrossRef] [PubMed]

68. Homburg, S.; Oswald, E.; Hacker, J.; Dobrindt, U. Expression analysis of the colibactin gene cluster coding for a novel polyketide in *Escherichia coli*. *FEMS Microbiol. Lett.* **2007**, *275*, 255–262. [CrossRef] [PubMed]

69. Jahns, C.; Hoffmann, T.; Müller, S.; Gerth, K.; Washausen, P.; Höfle, G.; Reichenbach, H.; Kalesse, M.; Müller, R. Pellasoren: Structure elucidation, biosynthesis, and total synthesis of a cytotoxic secondary metabolite from *Sorangium cellulosum*. *Angew. Chem.* **2012**, *51*, 5239–5243. [CrossRef] [PubMed]

70. Stein, T.; Borchert, S.; Conrad, B.; Feesche, J.; Hofemeister, B.; Hofemeister, J.; Entian, K.-D. Two different lantibiotic-like peptides originate from the ericin gene cluster of *Bacillus subtilis* A1/3. *J. Bacteriol.* **2002**, *184*, 1703–1711. [CrossRef] [PubMed]

71. Fuchs, S.W.; Jaskolla, T.W.; Bochmann, S.; Kötter, P.; Wichelhaus, T.; Karas, M.; Stein, T.; Entian, K.-D. Entianin, a novel subtilin-like lantibiotic from *Bacillus subtilis* subsp. *spizizenii* DSM 15029T with high antimicrobial activity. *Appl. Environ. Microbiol.* **2011**, *77*, 1698–1707. [CrossRef] [PubMed]

72. Klein, C.; Kaletta, C.; Schnell, N.; Entian, K.-D. Analysis of genes involved in biosynthesis of the lantibiotic subtilin. *Appl. Environ. Microbiol.* **1992**, *58*, 132–142. [PubMed]

73. Rince, A.; Dufour, A.; Le Pogam, S.; Thuault, D.; Bourgeois, C.M.; Le Pennec, J.P. Cloning, expression, and nucleotide sequence of genes involved in production of lactococcin DR, a bacteriocin from *Lactococcus lactis* subsp. *lactis*. *Appl. Environ. Microbiol.* **1994**, *60*, 1652–1657. [PubMed]

74. Rincé, A.; Dufour, A.; Uguen, P.; Le Pennec, J.P.; Haras, D. Characterization of the lacticin 481 operon: The *Lactococcus lactis* genes lctF, lctE, and lctG encode a putative ABC transporter involved in bacteriocin immunity. *Appl. Environ. Microbiol.* **1997**, *63*, 4252–4260. [PubMed]

75. Kampa, A.; Gagunashvili, A.N.; Gulder, T.A.M.; Morinaka, B.I.; Daolio, C.; Godejohann, M.; Miao, V.P.W.; Piel, J.; Andrésson, O.S. Metagenomic natural product discovery in lichen provides evidence for a family of biosynthetic pathways in diverse symbioses. *Proc. Natl. Acad. Sci. USA* **2013**, *110*, E3129–E3137. [CrossRef] [PubMed]

76. Fukuda, D.; Haines, A.S.; Song, Z.; Murphy, A.C.; Hothersall, J.; Stephens, E.R.; Gurney, R.; Cox, R.J.; Crosby, J.; Willis, C.L.; et al. A natural plasmid uniquely encodes two biosynthetic pathways creating a potent anti-MRSA antibiotic. *PLoS ONE* **2011**, *6*, e18031. [CrossRef] [PubMed]

77. Mattheus, W.; Gao, L.-J.J.; Herdewijn, P.; Landuyt, B.; Verhaegen, I.; Masschelein, I.; Volckaert, G.; Lavigne, R. Isolation and purification of a new kalimantacin/batumin-related polyketide antibiotic and elucidation of its biosynthesis gene cluster. *Chem. Biol.* **2010**, *17*, 149–159. [CrossRef] [PubMed]

78. Ishida, K.; Lincke, T.; Behnken, S.; Hertweck, C. Induced biosynthesis of cryptic polyketide metabolites in a *Burkholderia thailandensis* quorum sensing mutant. *J. Am. Chem. Soc.* **2010**, *132*, 13966–13968. [CrossRef] [PubMed]

79. Ishida, K.; Lincke, T.; Hertweck, C. Assembly and absolute configuration of short-lived polyketides from *Burkholderia thailandensis. Angew. Chem.* **2012**, *51*, 5470–5474. [CrossRef] [PubMed]
80. Nguyen, T.A.; Ishida, K.; Jenke-Kodama, H.; Dittmann, E.; Gurgui, C.; Hochmuth, T.; Taudien, S.; Platzer, M.; Hertweck, C.; Piel, J. Exploiting the mosaic structure of trans-acyltransferase polyketide synthases for natural product discovery and pathway dissection. *Nat. Biotechnol.* **2008**, *26*, 225–233. [CrossRef] [PubMed]
81. Chen, X.; Koumoutsi, A.; Scholz, R.; Eisenreich, A.; Schneider, K.; Heinemeyer, I.; Morgenstern, B.; Voss, B.; Hess, W.R.; Reva, O.; et al. Comparative analysis of the complete genome sequence of the plant growth–promoting bacterium *Bacillus amyloliquefaciens* FZB42. *Nat. Biotechnol.* **2007**, *25*, 1007–1014. [CrossRef] [PubMed]
82. Chen, X.-H.; Vater, J.; Piel, J.; Franke, P.; Scholz, R.; Schneider, K.; Koumoutsi, A.; Hitzeroth, G.; Grammel, N.; Strittmatter, A.W.; et al. Structural and Functional Characterization of three polyketide synthase gene clusters in *Bacillus amyloliquefaciens* FZB 42. *J. Bacteriol.* **2006**, *188*, 4024–4036. [CrossRef] [PubMed]
83. Moldenhauer, J.; Chen, X.-H.; Borriss, R.; Piel, J. Biosynthesis of the antibiotic bacillaene, the product of a giant polyketide synthase complex of the trans-AT Family. *Angew. Chem.* **2007**, *46*, 8195–8197. [CrossRef] [PubMed]
84. Dehn, R.; Katsuyama, Y.; Weber, A.; Gerth, K.; Jansen, R.; Steinmetz, H.; Höfle, G.; Müller, R.; Kirschning, A. Molecular basis of elansolid biosynthesis: Evidence for an unprecedented quinone methide initiated intramolecular Diels–Alder cycloaddition/macrolactonization. *Angew. Chem.* **2011**, *50*, 3882–3887. [CrossRef] [PubMed]
85. Moebius, N.; Ross, C.; Scherlach, K.; Rohm, B.; Roth, M.; Hertweck, C. Biosynthesis of the respiratory toxin bongkrekic acid in the pathogenic bacterium *Burkholderia gladioli. Chem. Biol.* **2012**, *19*, 1164–1174. [CrossRef] [PubMed]
86. Wakimoto, T.; Egami, Y.; Nakashima, Y.; Wakimoto, Y.; Mori, T.; Awakawa, T.; Ito, T.; Kenmoku, H.; Asakawa, Y.; Piel, J.; et al. Calyculin biogenesis from a pyrophosphate protoxin produced by a sponge symbiont. *Nat. Chem. Biol.* **2014**, *10*. [CrossRef] [PubMed]
87. Nakamura, H.; Hamer, H.A.; Sirasani, G.; Balskus, E.P. Cylindrocyclophane biosynthesis involves functionalization of an unactivated carbon center. *J. Am. Chem. Soc.* **2012**, *134*, 18518–18521. [CrossRef] [PubMed]
88. Preisitsch, M.; Heiden, S.E.; Beerbaum, M.; Niedermeyer, T.H.J.; Schneefeld, M.; Herrmann, J.; Kumpfmüller, J.; Thürmer, A.; Neidhardt, I.; Wiesner, C.; et al. Effects of halide ions on the carbamidocyclophane biosynthesis in *Nostoc* sp. CAVN2. *Mar. Drugs* **2016**, *14*, 21. [CrossRef] [PubMed]
89. Pag, U.; Sahl, H.-G.G. Multiple activities in lantibiotics—Models for the design of novel antibiotics? *Curr. Pharm. Des.* **2002**, *8*, 815–833. [CrossRef] [PubMed]
90. Baindara, P.; Chaudhry, V.; Mittal, G.; Liao, L.M.; Matos, C.O.; Khatri, N.; Franco, O.L.; Patil, P.B.; Korpole, S. Characterization of the antimicrobial peptide penisin, a class Ia novel lantibiotic from *Paenibacillus* sp. Strain A3. *Antimicrob. Agents Chemother.* **2016**, *60*, 580–591. [CrossRef] [PubMed]
91. Heather, Z.; Holden, M.T.; Steward, K.F.; Parkhill, J.; Song, L.; Challis, G.L.; Robinson, C.; Davis-Poynter, N.; Waller, A.S. A novel streptococcal integrative conjugative element involved in iron acquisition. *Mol. Microbiol.* **2008**, *70*, 1274–1292. [CrossRef] [PubMed]
92. Axen, S.D.; Erbilgin, O.; Kerfeld, C.A. A Taxonomy of Bacterial Microcompartment Loci Constructed by a Novel Scoring Method. *PLoS Comput. Biol.* **2014**, *10*, e1003898. [CrossRef] [PubMed]
93. Buchrieser, C.; Rusniok, C.; Kunst, F.; Cossart, P.; Glaser, P. The *Listeria* Consortium. Comparison of the genome sequences of *Listeria monocytogenes* and *Listeria innocua*: Clues for evolution and pathogenicity. *FEMS Immunol. Med. Microbiol.* **2003**, *35*, 207–213. [CrossRef]
94. Bobik, T.A.; Havemann, G.D.; Busch, R.J.; Williams, D.S.; Aldrich, H.C. The propanediol utilization (pdu) operon of *Salmonella enterica serovar Typhimurium* LT2 includes genes necessary for formation of polyhedral organelles involved in coenzyme B(12)-dependent 1, 2-propanediol degradation. *J. Bacteriol.* **1999**, *181*, 5967–5975. [PubMed]
95. Bobik, T.; Lehman, B.; Yeates, T. Bacterial microcompartments: Widespread prokaryotic organelles for isolation and optimization of metabolic pathways. *Mol. Microbiol.* **2015**, *98*, 193–207. [CrossRef] [PubMed]
96. Kerfeld, C.; Heinhorst, S.; Cannon, G. Bacterial microcompartments. *Annu. Rev. Microbiol.* **2010**, *64*, 391–408. [CrossRef] [PubMed]

97. Rondon, M.R.; Kazmierczak, R.; Escalante-Semerena, J.C. Glutathione is required for maximal transcription of the cobalamin biosynthetic and 1,2-propanediol utilization (cob/pdu) regulon and for the catabolism of ethanolamine, 1,2-propanediol, and propionate in *Salmonella typhimurium* LT2. *J. Bacteriol.* **1995**, *177*, 5434–5439. [CrossRef] [PubMed]

98. Stojiljkovic, I.; Bäumler, A.J.; Heffron, F. Ethanolamine utilization in *Salmonella typhimurium*: Nucleotide sequence, protein expression, and mutational analysis of the cchA cchB eutE eutJ eutG eutH gene cluster. *J. Bacteriol.* **1995**, *177*, 1357–1366. [CrossRef] [PubMed]

99. Chowdhury, C.; Sinha, S.; Chun, S.; Yeates, T.O.; Bobik, T.A. Diverse Bacterial microcompartment organelles. *Microbiol. Mol. Biol. Rev.* **2014**, *78*, 438–468. [CrossRef] [PubMed]

100. Petit, E.; LaTouf, G.W.; Coppi, M.V.; Warnick, T.A.; Currie, D.; Romashko, I.; Deshpande, S.; Haas, K.; Alvelo-Maurosa, J.G.; Wardman, C.; et al. Involvement of a bacterial microcompartment in the metabolism of fucose and rhamnose by *Clostridium phytofermentans*. *PLoS ONE* **2013**, *8*, e54337. [CrossRef] [PubMed]

101. Schink, B. Energetics of syntrophic cooperation in methanogenic degradation. *Microbiol. Mol. Biol. Rev. MMBR* **1997**, *61*, 262–280. [PubMed]

102. Buckel, W.; Thauer, R.K. Energy conservation via electron bifurcating ferredoxin reduction and proton/Na+ translocating ferredoxin oxidation. *Biochim. Biophys. Acta BBA Bioenerg.* **2013**, *1827*, 94–113. [CrossRef] [PubMed]

103. Chowdhury, N.P.; Klomann, K.; Seubert, A.; Buckel, W. Reduction of flavodoxin by electron bifurcation and sodium ion-dependent reoxidation by NAD+ catalyzed by ferredoxin-NAD+ reductase (Rnf). *J. Biol. Chem.* **2016**, *291*, 11993–12002. [CrossRef] [PubMed]

104. Chowdhury, N.P.; Mowafy, A.M.; Demmer, J.K.; Upadhyay, V.; Koelzer, S.; Jayamani, E.; Kahnt, J.; Hornung, M.; Demmer, U.; Ermler, U.; et al. Studies on the mechanism of electron bifurcation catalyzed by electron transferring flavoprotein (Etf) and butyryl-CoA dehydrogenase (Bcd) of *Acidaminococcus fermentans*. *J. Biol. Chem.* **2014**, *289*, 5145–5157. [CrossRef] [PubMed]

105. Herrmann, G.; Jayamani, E.; Mai, G.; Buckel, W. Energy Conservation via electron-transferring flavoprotein in anaerobic bacteria. *J. Bacteriol.* **2008**, *190*, 784–791. [CrossRef] [PubMed]

106. Biegel, E.; Schmidt, S.; Gonzalez, J.M.M.; Müller, V. Biochemistry, evolution and physiological function of the Rnf complex, a novel ion-motive electron transport complex in prokaryotes. *Cell. Mol. Life Sci. CMLS* **2011**, *68*, 613–634. [CrossRef] [PubMed]

107. Tremblay, P.-L.; Zhang, T.; Dar, S.A.; Leang, C.; Lovley, D.B. The Rnf complex of *Clostridium ljungdahlii* is a proton-translocating Ferredoxin: NAD+ oxidoreductase essential for autotrophic growth. *MBio* **2013**, *4*, e00406-12. [CrossRef]

108. Dolfing, J.; Larter, S.R.; Head, I.M. Thermodynamic constraints on methanogenic crude oil biodegradation. *ISME J.* **2008**, *2*, 442–452. [CrossRef] [PubMed]

109. Jones, D.M.; Head, I.M.; Gray, N.D.; Adams, J.J.; Rowan, A.K.; Aitken, C.M.; Bennett, B.; Huang, H.; Brown, A.; Bowler, B.F.; et al. Crude-oil biodegradation via methanogenesis in subsurface petroleum reservoirs. *Nature* **2008**, *451*, 176–180. [CrossRef] [PubMed]

110. Zengler, K.; Richnow, H.H.; Rosselló-Mora, R.; Michaelis, W.; Widdel, F. Methane formation from long-chain alkanes by anaerobic microorganisms. *Nature* **1999**, *401*, 266–269. [CrossRef] [PubMed]

111. Jarling, R.; Kühner, S.; Janke, E.B.; Gruner, A.; Drozdowska, M.; Golding, B.T.; Rabus, R.; Wilkes, H. Versatile transformations of hydrocarbons in anaerobic bacteria: Substrate ranges and regio- and stereo-chemistry of activation reactions. *Front. Microbiol.* **2015**, *6*, 880. [CrossRef] [PubMed]

112. Wilkes, H.; Buckel, W.; Golding, B.T.; Rabus, R. Metabolism of hydrocarbons in n-Alkane-utilizing anaerobic bacteria. *J. Mol. Microbiol. Biotechnol.* **2016**, *26*, 138–151. [CrossRef] [PubMed]

113. De Bok, F.; Plugge, C.; Stams, A.J. Interspecies electron transfer in methanogenic propionate degrading consortia. *Water Res.* **2004**, *38*, 1368–1375. [CrossRef] [PubMed]

114. Ishii, S.; Kosaka, T.; Hori, K.; Hotta, Y.; Watanabe, K. Coaggregation facilitates interspecies hydrogen transfer between *Pelotomaculum thermopropionicum* and *Methanothermobacter thermautotrophicus*. *Appl. Environ. Microbiol.* **2005**, *71*, 7838–7845. [CrossRef] [PubMed]

115. Ishii, S.; Kosaka, T.; Hotta, Y.; Watanabe, K. Simulating the contribution of coaggregation to interspecies hydrogen fluxes in syntrophic methanogenic consortia. *Appl. Environ. Microbiol.* **2006**, *72*, 5093–5096. [CrossRef] [PubMed]

116. Kato, S.; Hashimoto, K.; Watanabe, K. Microbial interspecies electron transfer via electric currents through conductive minerals. *Proc. Natl. Acad. Sci. USA* **2012**, *109*, 10042–10046. [CrossRef] [PubMed]
117. Kouzuma, A.; Kato, S.; Watanabe, K. Microbial interspecies interactions: Recent findings in syntrophic consortia. *Front. Microbiol.* **2015**, *6*, 477. [CrossRef] [PubMed]
118. Stams, A.J. Metabolic interactions between anaerobic bacteria in methanogenic environments. *Antonie van Leeuwenhoek* **1994**, *66*, 271–294. [CrossRef] [PubMed]
119. Kato, S.; Watanabe, K. Ecological and evolutionary interactions in syntrophic methanogenic consortia. *Microbes Environ.* **2010**, *25*, 145–151. [CrossRef] [PubMed]
120. Shimoyama, T.; Kato, S.; Ishii, S.; Watanabe, K. Flagellum mediates symbiosis. *Science* **2009**, *323*, 1574. [CrossRef] [PubMed]

microorganisms

MDPI

Article

Molecular Tunnels in Enzymes and Thermophily: A Case Study on the Relationship to Growth Temperature

Juan Miguel Gonzalez[ORCID]

Instituto de Recursos Naturales y Agrobiología, Consejo Superior de Investigaciones Científicas, IRNAS-CSIC, Avda. Reina Mercedes 10, 41702 Sevilla, Spain; jmgrau@irnase.csic.es; Tel.: +34-95-462-4711; Fax: +34-95-462-4002

Received: 29 August 2018; Accepted: 16 October 2018; Published: 20 October 2018

Abstract: Developments in protein expression, analysis and computational capabilities are decisively contributing to a better understanding of the structure of proteins and their relationship to function. Proteins are known to be adapted to the growth rate of microorganisms and some microorganisms (named (hyper)thermophiles) thrive optimally at high temperatures, even above 100 °C. Nevertheless, some biomolecules show great instability at high temperatures and some of them are universal and required substrates and cofactors in multiple enzymatic reactions for all (both mesophiles and thermophiles) living cells. Only a few possibilities have been pointed out to explain the mechanisms that thermophiles use to successfully thrive under high temperatures. As one of these alternatives, the role of molecular tunnels or channels in enzymes has been suggested but remains to be elucidated. This study presents an analysis of channels in proteins (i.e., substrate tunnels), comparing two different protein types, glutamate dehydrogenase and glutamine phosphoribosylpyrophosphate amidotransferase, which are supposed to present a different strategy on the requirement for substrate tunnels with low and high needs for tunneling, respectively. The search and comparison of molecular tunnels in these proteins from microorganisms thriving optimally from 15 °C to 100 °C suggested that those tunnels in (hyper)thermophiles are required and optimized to specific dimensions at high temperatures for the enzyme glutamine phosphoribosylpyrophosphate amidotransferase. For the enzyme glutamate dehydrogenase, a reduction of empty spaces within the protein could explain the optimization at increasing temperatures. This analysis provides further evidence on molecular channeling as a feasible mechanism in hyperthermophiles with multiple relevant consequences contributing to better understand how they live under those extreme conditions.

Keywords: molecular tunnels; enzyme structure; enzyme activity; temperature; thermophile; thermophily; enzyme thermostability

1. Introduction

Biomacromolecular channels (tunnels and pores) present critical structural importance which is intimately related to the biological function and structural stability of macromolecules [1–3]. The access of substrates to the active sites of enzymes occurs in many cases through tunnels that connect the exterior to the internal protein spaces and in a reverse way for the products of biological reactions. In addition, tunnels have been shown to form internal passages between active sites and protein subunits leading to a quick interconnection between proteins. In this way, specific substrates can be transferred, increasing stability and accelerating reaction times [4]. Channels (at least 15 Å long) have been reported in 64% of enzymes with known crystal structures [5], suggesting that they are common features for the structure of many enzymes and relevant to their functionality.

Molecular channeling has been proposed to have great importance in various aspects of protein structure and enzyme activity [6,7]. For instance, molecular channeling reduces the time required for the transit of reaction intermediates between active centers of different enzymes, or within multifunctional enzymes, connecting one active site to another. The mechanism of substrate tunneling contributes to minimize the loss of reaction intermediates by diffusion contributing to increase enzyme efficiency. Another advantage of tunneling is to protect labile reaction intermediates that otherwise would be rapidly degraded by exposure to solvents or high temperatures. In addition, the isolation through molecular channels of substrates, products or intermediates could avoid potential inhibitory effects to the cell and to other enzymes. Additionally, using tunnels, substrates, and reaction intermediates, can be preserved from their use in other distinct metabolic pathways, diminishing competition and avoiding a potential unfavorable equilibrium. On top of these numerous features, molecular tunneling could limit the use of substrates, intermediates and products to a specific multi-step chain of related enzymes within a pathway, in order to increase metabolic efficiency and accelerate the stepping-stone progress of a sequential catalytic transformation. Although the potential importance of molecular tunnels has been shown [5–7], the relevance of this process on the ability of (hyper)thermophiles to thrive under high temperatures remains clearly understudied.

There are numerous studies explaining the possibilities and different alternatives so that nucleic acids, proteins and lipids can maintain their stability in cells living at high temperatures [8–10]. Nevertheless, a gap in knowledge remains on how thermophiles can thrive under high temperatures using mostly the same low-molecular weight, thermolabile, and biomolecules that all living cells possess. Examples of thermolabile biomolecules are ATP, NAD/H, NADP/H, and pyridoxal phosphate, among others, many of which are universal and essential for the central metabolism of all cells. Reports early this century [11,12] pointed out that several low-molecular weight biomolecules are thermolabile and their turnover rate at the growth temperatures of (hyper)thermophiles was not permissive as an efficient living alternative. This is an unresolved question, arising since the discovery of hyperthermophiles during the last half of the previous century [13,14]. Recently, Cuecas et al. [15] presented a simple mechanism that could allow thermophiles to maintain the thermal stability of some low molecular weight biomolecules such as NADH. Cuecas et al. [15] proposed that by maintaining the intracellular viscosity, those thermolabile biomolecules could be greatly stabilized. Nevertheless, this mechanism appeared to be inefficient at >80 °C, i.e., for hyperthermophiles, because of the great reduction of viscosity in aqueous solutions at temperatures close to the boiling point of water. Other complementary possibilities for the maintenance of the thermal stability of low-molecular weight biomolecules have been suggested [11,12,15] but the most likely alternative is the occurrence of molecular channeling, so that the low-molecular weight thermolabile biomolecules could be transferred directly between enzymes and maintained within molecular channels or tunnels, such that their thermal stability is preserved much longer than we would typically believe from experimentation in standard diluted aqueous solutions. However, technical and experimental difficulties have limited demonstrations of the relevance of molecular channeling and substrate tunnels in hyperthermophiles. A first step to provide evidence on the relevance of molecular channeling could be to show that the extension and volume of molecular tunnels are somehow governed by the optimum growth temperature of the microorganisms, mainly for hyperthermophiles.

With this aim in mind, this study attempts to show if there is a dependency of molecular tunnels in enzymes as a function of the optimum growth temperature for the microorganisms, with a focus mainly on (hyper)thermophiles. This analysis was comparatively performed on a range of microorganisms with optimum growth temperatures ranging from 15 °C to 100 °C, and on two enzymes, glutamate dehydrogenase (GDH) and glutamine phosphoribosylpyrophosphate amidotransferase (GPAT). GDH is a major constituent of the total protein content in hyperthermophiles (up to 20% of total protein in the hyperthermophilic archaeon *Pyrococcus furiosus* [16]; but it has not been reported to be related to molecular tunnels. GPAT has been long reported as a clear example of substrate channeling in enzymes [1,4,6,7,17]. It is hypothesized that the dimensions of the molecular tunnels in GPAT should be maintained in hyperthermophiles, although this could not be the case for GDH in hyperthermophiles, assuming enzymes need to be optimized to properly function at high temperatures.

2. Materials and Methods

Two case studies were analyzed, an enzyme frequently reported to present substrate tunnels of functional importance, the glutamine phosphoribosylpyrophosphate amidotransferase (GPAT; [1,4,6,7]), and an enzyme of major relevance to cells not reported to present functionally important substrate tunnels, glutamate dehydrogenase (GDH; [16,18]). Amino acid sequences for the studied enzymes covering most of the growth temperature range for microorganisms were obtained from NCBI (National Center for Biotechnology Information; https://www.ncbi.nlm.nih.gov/), and their 3D-structures were predicted by Phyre2 [19]. For GPAT, 129 and 110 3D-structures were analyzed from bacteria (109 sequences) and archaea (21 sequences), respectively. For GDH, a total of 130 3D-structures were analyzed, including bacteria and archaea. These sequences corresponded to prokaryotes with optimum growth temperatures from 15 °C to 100 °C (Supplementary Material). From the predicted structures, the tunnels in each protein were identified and localized by using the software MOLE [20]. According to Petrek et al. [20], MOLE presents a strategy for exploring molecular voids based on Voronoi diagrams, which overcomes some of the limitations (excessive computer demands and errors due to grid extrapolation) from previous software. From MOLE's results, the total tunnel length, lateral surface and volume were calculated. The total distance or length, lateral surface and volume of the tunnels in a protein were the sums of each one of the sections identified and localized in that protein. For each section of the tunnels, lateral surface (S) and volume (V) were calculated assuming the shape of a cylinder or truncated cone using the following formulas, where r_1 and r_2 are the two radii of the truncated cone and h represents its height.

$$V = (1/3) \, \pi h \left(r_1{}^2 + r_1 r_2 + r_2{}^2 \right)$$

$$S = \pi \, (r_1 + r_2) \, \sqrt{(r_1 - r_2)^2 + h^2}$$

Available model structures from GDH and GPAT were included in the analyses for comparative purposes against the results from predicted structures. These model structures were obtained from the Protein Data Bank (PDB; http://www.rcsb.org). For this study, only one subunit was considered so that their results are comparable to the GPAT and GDH predicted structures. A list of the model structures included in this study is available in the supplementary material, together with their PDB accession number, optimum growth temperature and bacterial species. The upper and lower limits of distribution that correspond to maximum and minimum tunnel dimensions at increasing temperatures were estimated by linear regression using the four most extreme data points, either as high or low values and these regression lines were drawn in the figures.

3. Results and Discussion

Tunnel distance, surface and volume estimates for each protein were plotted against the optimum growth temperature reported for the corresponding microorganism. For GDH, results showed that for proteins from microorganisms growing at high temperatures, the tunnel distance, surface and volume tend generally to decrease (Figures 1 and 2). Approximated upper limit lines were drawn resulting in estimates of maximum tunnel distance, surface and volume values at 80 °C of 49 Å, 648 Å^2, and 581 Å^3, respectively, for bacterial GDH. The arqueal GDH data set suggested maximum tunnel distance, surface and volume values at 100 °C of 38 Å, 499 Å^2, and 430 Å^3, respectively. This decrease of channels and open spaces within the GDH molecules suggests that at increasing optimum growth temperatures these proteins get more compact, for example, increasing the number of interactions among closed amino acid residues and reducing the void spaces within the molecule. This is in agreement with previous understanding on the adaptations that proteins experience from hyperthermophiles [10] and specifically in GDH from hyperthermophiles [18,21]. The case of GDH, an enzyme in which molecular channels have not been reported to be functionally relevant, would show a decrease of the empty spaces within the protein 3D-structure to a minimum at high temperature. Thermophily imposes adaptive mechanisms, and so, the microorganisms adapted to grow at high temperatures present GDHs with a trend to minimum molecular channels.

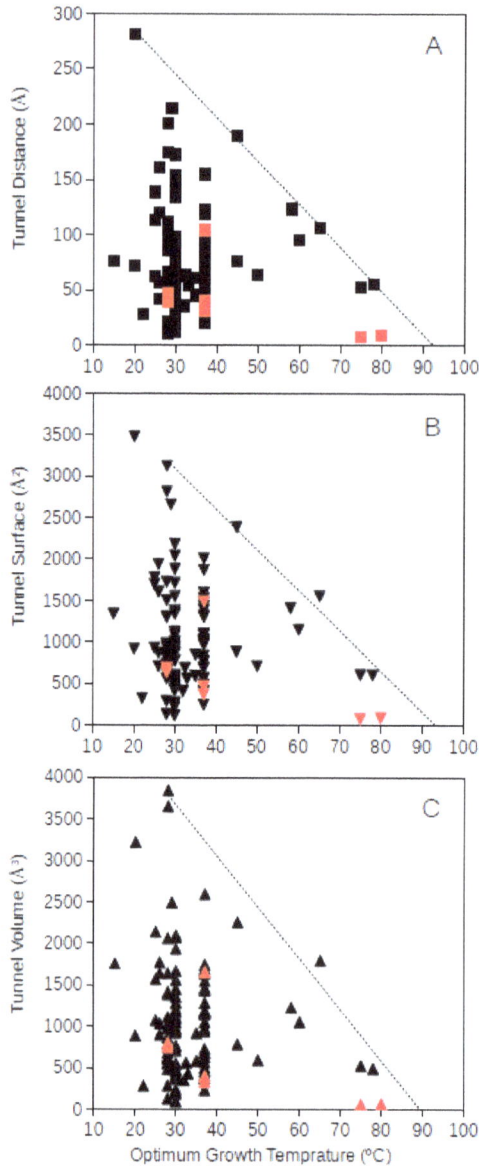

Figure 1. Tunnel distance (**A**), surface (**B**) and volume (**C**) (in Å) dimensions predicted for bacteria GDHs (Glutamate dehydrogenase) ranging in optimum growth temperature from 15 °C to 99 °C. Grey line represents a visualization of the upper limit for these dimensions at increasing temperatures. Black symbols correspond to predicted protein structures; red symbols correspond to resolved model structures from the Protein Data Bank.

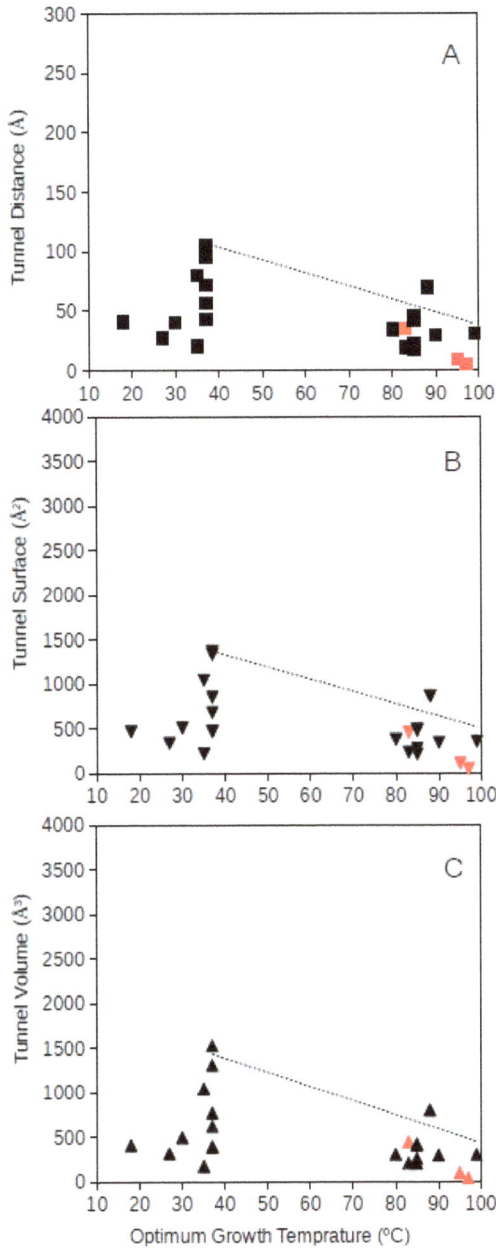

Figure 2. Tunnel distance (**A**), surface (**B**) and volume (**C**) (in Å) dimensions predicted for archaea GDHs ranging in optimum growth temperature from 15 °C to 99 °C. Grey line represents a visualization of the upper limit for these dimensions at increasing temperatures. Black symbols correspond to predicted protein structures; red symbols correspond to resolved model structures from the Protein Data Bank.

Bacteria and archaea GPAT structures where analyzed separately. Bacterial GPAT tunnel distance, surface and volume where plotted against the optimum growth temperature of the corresponding bacteria (Figure 3). The results showed that at increasing optimum growth temperatures, the dimensions of the molecular tunnels points to a highly specific value narrowing down the large variability observed among mesophiles. The estimated values for bacterial GPAT were in the range 112–122 Å, 1368–1503 Å^2, and 1262–1408 Å^3 for the tunnel distance, surface and volume, respectively, (Figure 3) and were calculated at 80 °C from the regression lines drawn from the upper and lower limits of these dimensions. These estimates point towards the prediction of narrow ranges of dimensions for these tunnels in bacterial GPATs for microorganisms growing at high temperatures. These values obtained for bacterial GPAT are much higher than those observed for GDH (see above), suggesting that molecular tunnels are well maintained in GPATs from hyperthermophiles. These results also suggest an optimization of molecular channels within the bacterial GPAT molecules, because at increasing optimum growth temperatures the restrictive conditions push towards minimum unnecessary spaces within the molecule. Nevertheless, GPAT has been reported to require molecular tunnels within the enzyme, and so, the results from Figure 3 suggest that the required tunnels are maintained but superfluous spaces are compacted or filled by interactions among amino acid residues so that the protein is able to maintain its stability and function at the high temperatures required for growth. From this information, one can deduce that bacteria adapt to high temperatures by generating compact proteins, but the requirement for functionally relevant molecular tunnels is maintained. These results do not directly prove that molecular channels are important to maintain the metabolism in (hyper)thermophiles although they represent unimportant evidence on the relevance of these tunnels in hyperthermophiles. Maintaining the molecular tunnel dimensions, and so their functional features, at high optimum growth temperatures indicates that these tunnels are essential for thriving under those extreme conditions.

Archaeal GPAT tunnel distance, surface and volume plotted against the optimum growth temperature for the archaeon corresponding to each of the studied proteins is presented in Figure 4. Archaeal GPATs present apparently more disperse dimension values of molecular tunnels than bacterial GPATs, but a similar phenomenon to bacterial GPATs is observed for archaea, and the range of dimensions for the molecular tunnels in archaeal GPATs are limited for the highest studied optimum growth temperatures (above 80 °C; Figure 3). The upper and lower lines drawing these limits of tunnel dimensions point to the range of values of 85–113 Å, 998–1331 Å^2, and 886–1260 Å^3 for the distance, surface and volume, respectively, of archaeal GPAT molecular tunnels at an optimum growth temperature of 100 °C. This suggests that GPATs in hyperthermophiles also maintain the minimum required molecular tunnel dimensions in agreement to the above, when a scenario with similar dimensions for molecular tunnels for the bacterial GPAT was deduced. Interestingly, the dimensions of molecular tunnels for archaea and bacteria at the highest optimum growth temperatures are in agreement, showing similar values which suggest that these proteins are optimized about the channels and free spaces within the enzymes and reduced to the minimum functional required values. Because these minimum dimensions of molecular tunnels are maintained even at the most restrictive conditions imposed in hyperthermophiles and growth at high temperatures, the results suggest that these molecular tunnels are relevant for the functioning of the enzyme GPAT both in hyperthermophilic bacteria and archaea.

Because microorganisms are adapted to their environmental conditions, their enzymes need to reflect that adaptation. Thus, enzymes are optimized to best perform under the optimum growth temperature. The example of GDH shows a strategy to reduce voids within the protein structure so that it improves its thermostability. The case of GPAT suggests that molecular tunnels are required to maintain its functionality, but the dimensions of these tunnels must be within a narrow range to achieve optimum performance and stability at high temperatures.

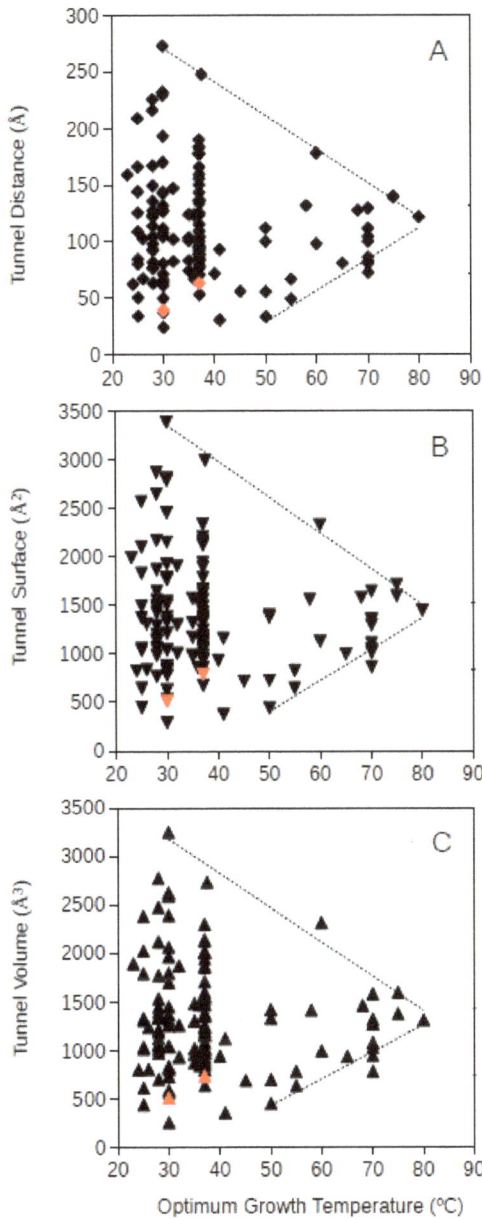

Figure 3. Tunnel distance (**A**), surface (**B**) and volume (**C**) (in Å) dimensions predicted for bacteria GPATs (glutamine phosphoribosylpyrophosphate amidotransferase) ranging in optimum growth temperature from 23 °C to 80 °C. Grey lines represent a visualization of the upper and lower limits for these dimensions at increasing temperatures. Black symbols correspond to predicted protein structures; red symbols correspond to resolved model structures from the Protein Data Bank.

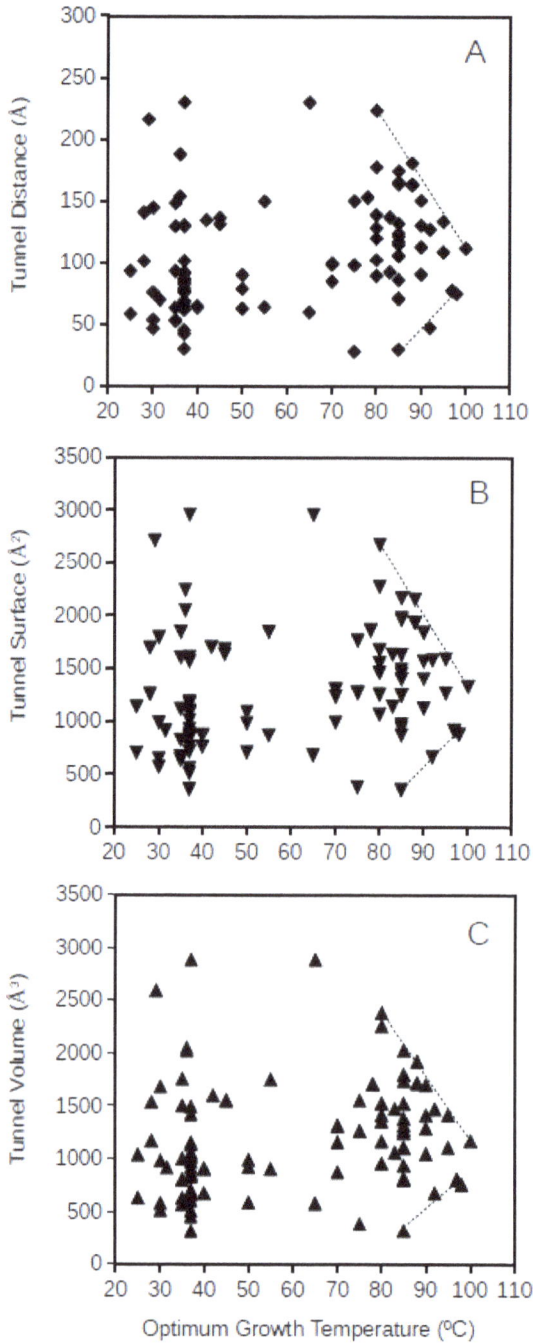

Figure 4. Tunnel distance (**A**), surface (**B**) and volume (**C**) (in Å) dimensions predicted for archaea GPATs ranging in optimum growth temperature from 25 °C to 100 °C. Grey lines represent a visualization of the upper and lower limits for these.

The restrictions of molecular tunnel dimensions required in the microorganisms showing the highest optimum growth temperatures represent a requirement to build functional enzymes such as GPAT in hyperthermophiles. Establishing the requirement of molecular tunnels in specific enzymes of hyperthermophilic archaea and bacteria is a step forward, contributing to understand the mechanisms of thermophily in relationship to protein function and structure. Herein, it has been shown that molecular tunnels are required at high temperatures. Future studies might start to delve deeper into the consequences for this requirement. Although the high thermal stability of proteins (i.e., and enzymes) from hyperthermophiles has been clearly demonstrated [10,22], the capability of hyperthermophiles to maintain the stability of universal, thermolabile low-molecular weight biomolecules (e.g., NADH, NADPH, ATP, pyridoxal phosphate, etc.) remains to be understood. One of the mechanisms most reported for hyperthermophiles to be able to thrive at high temperatures using these thermolabile low-molecular weight biomolecules has been the existence of molecular channels or tunnels that could directly transfer these labile substrates to the active sites of the enzymes maintaining longer stability [11,12,15]. Herein, the requirement of molecular tunnels of specific dimensions is shown for GPAT from hyperthermophiles. This information brings up the potential for the relationship of these molecular tunnels to the maintenance of thermostability and the transference of thermolabile low molecular weight biomolecules to make possible an effective metabolism at high temperatures.

Supplementary Materials: The following are available online at http://www.mdpi.com/2076-2607/6/4/109/s1.

Funding: This research was funded by ERA-IB2 project (ERA-IB-16-049), and Spanish Ministry of Economy and Competitiveness (PCIN-2016-129) and the Regional Government of Andalusia (BIO288). These projects were cofunded by FEDER funds.

Conflicts of Interest: The authors declare no conflict of interest.

References

1. Huang, X.; Holden, H.M.; Raushel, F.M. Channeling of substrates and intermediates in enzyme-catalyzed reactions. *Annu. Rev. Biochem.* **2001**, *70*, 149–180. [CrossRef] [PubMed]
2. Kingsley, L.; Lill, M.A. Substrate tunnels in enzymes: Structure-function relationships and computational methodology. *Proteins* **2015**, *83*, 599–611. [CrossRef] [PubMed]
3. Pravda, L.; Sehnal, D.; Vařeková, R.S.; Navrátilova, V.; Toušek, D.; Berka, K.; Otyepka, M.; Koca, J. ChannelsDB: Database of biomacromolecular tunnels and pores. *Nucleic Acids Res.* **2018**, *46*, D399–D405. [CrossRef] [PubMed]
4. Miles, E.W.; Rhee, S.; Davies, D.R. The molecular basis of substrate channeling. *J. Biol. Chem.* **1999**, *274*, 12193–12196. [CrossRef] [PubMed]
5. Pravda, L.; Berka, K.; Svobodová Vařeková, R.; Sehnal, D.; Banáš, P.; Laskowski, R.A.; Koča, J.; Otyepka, M. Anatomy of enzyme channels. *BMC Bioinform.* **2014**, *15*, 379. [CrossRef] [PubMed]
6. Milani, M.; Pesce, A.; Bolognesi, M.; Bocedi, A.; Ascenzi, P. Substrate channeling. *Biochem. Mol. Biol. Educ.* **2003**, *31*, 228–233. [CrossRef]
7. Raushel, F.M.; Thoden, J.B.; Holden, H.M. Enzymes with molecular tunnels. *Accounts Chem. Res.* **2003**, *36*, 539–548. [CrossRef] [PubMed]
8. Berezovsky, I.N.; Shakhnovich, E.I. Physics and evolution of thermophilic adaptation. *Proc. Natl. Acad. Sci. USA* **2005**, *102*, 12742–12747. [CrossRef] [PubMed]
9. Grogan, D.W. Hyperthermophiles and the problem of DNA instability. *Mol. Microbiol.* **1998**, *28*, 1043–1049. [CrossRef] [PubMed]
10. Vieille, C.; Zeikus, G.J. Hyperthermophilic enzymes: Sources, uses, and molecular mechanisms for thermostability. *Microbiol. Mol. Biol. Rev.* **2000**, *65*, 1–43. [CrossRef] [PubMed]
11. Cowan, D.A. The upper temperature of life—Where do we draw the line? *Trends Microbiol.* **2004**, *12*, 58–60. [CrossRef]
12. Daniel, R.M.; Cowan, D.A. Biomolecular stability and life at high temperatures. *Cell. Mol. Life Sci.* **2000**, *57*, 250–254. [CrossRef] [PubMed]
13. Brock, T.D. Life at high temperatures. *Science* **1967**, *158*, 1012–1019. [CrossRef] [PubMed]

14. Stetter, K.O.; Fiala, G.; Huber, G.; Huber, R.; Segerer, A. Hyperthermophilic microorganisms. *FEMS Microbiol. Rev.* **1990**, *75*, 117–124. [CrossRef]
15. Cuecas, A.; Cruces, J.; Galisteo-López, J.F.; Peng, X.; Gonzalez, J.M. Cellular viscosity in Prokaryotes and thermal stability of low molecular weight biomolecules. *Biophys. J.* **2016**, *111*, 875–882. [CrossRef] [PubMed]
16. Consalvi, V.; Chiaraluce, R.; Politi, L.; Vaccaro, R.; De Rosa, M.; Scandurra, R. Extremely thermostable glutamate dehydrogenase from the hyperthermophilic archaebacterium *Pyrococcus Furiosus*. *Eur. J. Biochem.* **1991**, *202*, 1189–1196. [CrossRef] [PubMed]
17. Mullis, L.S.; Raushel, F.M. Channeling of ammonia through the intermolecular tunnels contained within carbamoyl phosphate synthetase. *J. Am. Chem. Soc.* **1999**, *121*, 3803–3804. [CrossRef]
18. Ma, K.; Robb, F.T.; Adams, M.W. Purification and characterization of NADP-specific alcohol dehydrogenase and glutamate dehydrogenase from the hyperthermophilic archaeon Thermococcus litoralis. *Appl. Environ. Microbiol.* **1994**, *60*, 562–568. [PubMed]
19. Kelley, L.A.; Mezulis, S.; Yates, C.M.; Wass, M.N.; Sternberg, M.J.E. The Phyre2 web portal for protein modeling, prediction and analysis. *Nat. Protoc.* **2015**, *10*, 845–858. [CrossRef] [PubMed]
20. Petřek, M.; Košinová, P.; Koča, J.; Otyepka, M. MOLE: A voronoi diagrama-based explorer of molecular channels, pores, and tunnels. *Structure* **2007**, *15*, 1357–1363. [CrossRef] [PubMed]
21. Britton, K.L.; Yip, K.S.; Sedelnikova, S.E.; Stillman, T.J.; Adams, M.W.; Ma, K.; Maeder, D.L.; Robb, F.T.; Tolliday, N.; Vetriani, C.; et al. Structure determination of the glutamate dehydrogenase from the hyperthermophile *Thermococcus litoralis* and its comparison with that from *Pyrococcus Furiosus*. *J. Mol. Biol.* **1999**, *12*, 1121–1130. [CrossRef] [PubMed]
22. Lee, M.-K.; Gonzalez, J.M.; Robb, F.T. Extremely thermostable glutamate dehydrogenase (GDH) from the freshwater archaeon *Thermococcus waiotapuensis*: Cloning and comparison with two marine hyperthermophilic GDHs. *Extremophiles* **2002**, *6*, 151–159. [CrossRef] [PubMed]

microorganisms

MDPI

Article

Contribution of the Oligomeric State to the Thermostability of Isoenzyme 3 from *Candida rugosa*

María-Efigenia Álvarez-Cao, Roberto González, María A. Pernas and María Luisa Rúa *

Department of Food and Analytical Chemistry, Sciences Faculty of Ourense, University of Vigo, As Lagoas s/n, 32004 Ourense, Spain; mariaealvarez@uvigo.es (M.-E.Á.-C.); robergg@uvigo.es (R.G.); mapernas@mundo-r.com (M.A.P.)

* Correspondence: mlrua@uvigo.es; Tel.: +34-988-387062

Received: 14 September 2018; Accepted: 16 October 2018; Published: 19 October 2018

Abstract: Thermophilic proteins have evolved different strategies to maintain structure and function at high temperatures; they have large, hydrophobic cores, and feature increased electrostatic interactions, with disulfide bonds, salt-bridging, and surface charges. Oligomerization is also recognized as a mechanism for protein stabilization to confer a thermophilic adaptation. Mesophilic proteins are less thermostable than their thermophilic homologs, but oligomerization plays an important role in biological processes on a wide variety of mesophilic enzymes, including thermostabilization. The mesophilic yeast *Candida rugosa* contains a complex family of highly related lipase isoenzymes. Lip3 has been purified and characterized in two oligomeric states, monomer (mLip3) and dimer (dLip3), and crystallized in a dimeric conformation, providing a perfect model for studying the effects of homodimerization on mesophilic enzymes. We studied kinetics and stability at different pHs and temperatures, using the response surface methodology to compare both forms. At the kinetic level, homodimerization expanded Lip3 specificity (serving as a better catalyst on soluble substrates). Indeed, dimerization increased its thermostability by more than 15 °C (maximum temperature for dLip3 was out of the experimental range; >50 °C), and increased the pH stability by nearly one pH unit, demonstrating that oligomerization is a viable strategy for the stabilization of mesophilic enzymes.

Keywords: *Candida rugosa*; lipase; kinetic; interfacial activation; inhibition; dimerization; structure

1. Introduction

Proteins isolated from thermophilic microorganisms, or metagenomes from thermal environments, show high catalytic activity and thermostability at high temperatures. In general, they show high thermostability because they are oligomers that have large hydrophobic cores and increased electrostatic interactions, disulfide bonds, salt-bridging, and surface charges [1]. Oligomerization is recognized as a mechanism for protein stabilization to confer a thermophilic adaptation [2]. Accepting the importance of the oligomers quaternary structure, many authors have studied how to enhance catalytic activity and thermal stability [3]. The strategies followed included site-directed mutagenesis [4–6], insertions [7], or deletions of hydrophobic residues that lead to different molecular aggregation states or improve the thermodynamic stability of the oligomeric assembly [8]. Recently, it was observed that mutations located not only inside but also outside the protein interfaces introduce conformational changes that can alter the oligomeric balance (the transition from monomers to oligomers) by creating new bonds, or through the indirect stabilization of protein dynamics, again suggesting a relation between higher-order oligomeric states and thermostability [9,10]. Furthermore, although mesophilic proteins show less thermostability than their thermophilic homologs, it has been observed that protein-protein interactions allow both the homo-oligomeric [11,12] and hetero-oligomeric [13] organization of a wide variety of mesophilic enzymes, which play an important role in many biological pathways and even in cell-cell adhesion processes. As an example, it was recently demonstrated that the reduction in the oligomeric state, through the substitution of conserved amino acid residues, served to abolish the high thermostability of enzymes from pathogenic microorganisms [14].

Lipases (triacylglycerol acyl hydrolases EC 3.1.1.3.) constitute a wide family of enzymes with the natural catalytic function of hydrolyzing ester bonds in aqueous media, but they can also catalyze ester synthesis in organic media, having thus an enormous biotechnological potential with regard to several industries [15]. *Candida rugosa* (synonym *Candida cylindracea*) lipases are among the most important lipases from a commercial point of view. They are expressed from a complex family of highly related genes, with at least eight of them identified. Several products of these genes (Lip1–Lip5 and LipJ08) have been biochemically obtained, either using crude commercialized extracts [16–19], cultures of the wild-type organism [20–22], or after heterologous expression of synthetic codon-optimized nucleotide sequences [23–33]. Since *C. rugosa* uses a non-universal codon CUG that codes for the amino-acid serine [34], previous attempts to express the native genes were unsuccessful. These purified isoenzymes differ in terms of amino acid sequence, isoelectric point, and glycosylation degree. An excellent recent revision on the subject can be found in Reference [35]. It is accepted that lipases prefer to act on water-insoluble, long-chain triglycerides, often showing interfacial activation phenomena [36]. An explanation of interfacial activation comes from the presence of an amphiphilic movable flap in the structure of most lipases [37]. In the so-called closed or inactive state, the flap covers the active-site region, avoiding its exposure to the solvent, but in the presence of a lipidic substrate, it displaces, leaving a large hydrophobic area exposed around the active site that contributes to the recognition and binding of substrates [38].

The structural basis for the *C. rugosa* lipase activation came from the Lip1 crystals obtained in both its open and closed conformations [39,40], while Lip2 was crystallized in its closed state [41] and Lip3 in an open dimeric conformation [42,43]. Interestingly, isoenzyme Lip3 is so far the only one within the family for which stable dimers have been also purified and biochemically characterized [21,44], providing a perfect model for studying the effect of homodimerization on the thermostability of a mesophilic enzyme. At the kinetic level, the dimerization of *C. rugosa* Lip3 converts this lipase into a very efficient enzyme to hydrolyze water-soluble esters [21,45], but the effect of non-catalytic hydrophobic interphase has not been previously analyzed. We report the effect of homodimerization on both isoenzyme Lip3 thermostabilization and kinetics.

2. Materials and Methods

2.1. Materials

Lipase type VII from *C. rugosa* was obtained from Sigma Chemicals Co. (St Louis, MO, USA), tributyrin and triacetin from Fluka (Deisenhofen, Germany), and sodium deoxycholate from Amresco (Solom, OH, USA). Gels for protein purification, DEAE-Sephacel, Phenyl-Sepharose CL-4B, Sephacryl HR 100, and the Sephacryl S200 column were from GE Healthcare (Piscataway, NJ, USA). All chromatographic steps were performed on a fast protein liquid chromatography (FPLC) system (Pharmacia Biotech, Sweden).

2.2. Lipases Purification

Dimeric Lip3 was purified as described in Reference [19] and monomeric Lip3 as in Reference [18]. Pure dimeric Lip3 was obtained from the commercial crude powder by means of hydrophobic chromatography followed by size exclusion chromatography using, respectively, Phenyl-Sepharose CL-4B and Sephacryl HR100 gels on lab-mounted columns. To obtain monomeric Lip3, an ethanol precipitation step was implemented before the chromatographic steps, namely anionic-exchange chromatography, using DEAE-Sephacel gel, followed by gel filtration with Sephacryl HR100 gel on lab-mounted columns. A scheme of the Lip3 purification procedure (dimer and monomer) was shown previously [45]. During purification, lipase activity was measured in a pH-stat (Methrom, Switzerland) at 30 °C using tributyrin emulsions stabilized with Arabic gum as described in Reference [46]. The reaction was started by adding enzyme aliquots, which caused the release of the fatty acids esterified to the glycerol moiety. pH was kept constant at 7.0 by automatically adding 0.01 M NaOH for at least 10 min. Initial hydrolysis rates were determined from the slope on the linear part of the obtained plots. One unit (U) is the amount of enzyme that liberates 1 µmol of fatty acid per min under the above conditions.

2.3. Size Exclusion Chromatography (SEC)

Purified fractions of dimeric Lip3 were pooled and further purified by SEC using a Sephacryl S200 column equilibrated in 25 mM Tris/HCl buffer (pH 7.5) containing 0.15 M NaCl at 0.3 mL/min. The nature of the interactions responsible for maintaining the homodimers was investigated by SEC as indicated above with the addition of 1% (w/v) sodium cholate in the running buffer. The high molecular weight marker kit from Sigma was used for calibration.

2.4. Enzyme Kinetic Assays

Purified samples of mLip3 and dLip3 were assayed following the initial hydrolysis rate of triacetin in a pH-stat (Methrom, Switzerland). The assays were performed in 5 mM Tris/HCl buffer (pH 7.0) containing 0.1 M CaCl$_2$ at 30 °C, and variable amounts of triacetin (from 35 mM–1.06 M). All assays were done keeping the same stirring speed, while care was taken to avoid the formation of air bubbles in the reaction vessel. The reaction was started with the addition of the enzyme and at least triplicates of each assay were made. One activity unit was defined as the amount of enzyme that released 1 µmol of fatty acids per min. The solubility of triacetin in the reaction conditions was estimated measuring the turbidity as described in Reference [47] and resulted to be 270 mM [21].

2.5. Other Methods

Protein concentration was determined by the Lowry method [48], using bovine serum albumin as standard.

2.6. Effect of Temperature and pH on the Stability of C. rugosa Isoenzyme 3

Optimization of the temperature (T)-pH stability conditions for mLip3 and dLip3 was obtained using the response surface methodology (RSM). To that end, a 2^2-central composite orthogonal was designed. The factors were the independent variables studied at five levels ($-\alpha$, -1, 0, $+1$, $+\alpha$) being 0 the central point and $\alpha = 1.267$ the axial point having for each factor from the center of the experimental domain. The range (pH 5–9 and temperature 30–50 °C) and codification of the variables was as in [22]. The values were coded according to the following equation

$$x_i = (X_i - X_0)/\Delta X_i, \tag{1}$$

where x_i and X_i are the coded and real values of the independent variable i, X_0 is the real value of the independent variable i at the central point, and ΔX_i is the step change value. Table 1 shows real and coded values.

Such a design consisted of a set of 13 experiments where the central point was repeated 5 times to estimate the experimental error. Aliquots of pure lipase solutions were incubated for 30 min (or 1 h) in 0.2 M buffers at different pH values and temperatures. The final protein concentration in the mixtures was 0.25 mg/mL. The following buffers were used: Citrate-phosphate buffer (pHs 5.0 and 5.4), sodium phosphate (pH 7.0), and EPPS (N-[2-hydroxyethyl]piperazine-N'-[3-propane sulfonic acid]) (pHs 8.6 and 9.0). After incubation, the remaining activity was determined in a pH-stat using tributyrin emulsions as described above. Controls consisted of samples of the respective untreated mLip3 and dLip3. The stability of mLip3 and dLip3 was quantified as the percentage of residual activity measured in each condition.

From the experimental data, second-order quadratic models were built to correlate the response with the independent variables or factors, and the optimum point of the empiric model was obtained by solving the regression equation and analyzing the response surface contour plots. The statistical significance of each model was evaluated by analysis of variance (ANOVA) with 95% confidence intervals.

Table 1. Real and coded values of independent variables in the factorial design 2^2-central composite.

Real Values	Coded Values [1]				
	−1.267	−1	0	1	1.267
pH	5	5.4	40	8.6	9
T (°C)	30	32.1	7	47.9	50

^{1} $x_i = (X_i - X_0)/\Delta X_i, i = 1, 2.$

2.7. Statistical Analysis

Statistical data treatment was done using the StatGraphics Centurion XVI (Statpoint Technologies, Inc., Warrenton, OR, USA) package and graphics drawing was performed with the programs GraphPad Prism 5 (La Jolla, CA, USA) and SigmaPlot 12.0 (Systat Software, Inc., San Jose, CA, USA). The significance of the data was tested with a Student's *t*-test and results were considered significant for *p*-values less than or equal to 0.05.

3. Results

3.1. Purification and Molecular Characterisation

Purification of the monomer (mLip3) and dimer (dLip3) of Lip3 *C. rugosa* isoenzyme from a commercial preparation of *C. rugosa* lipase was done following published procedures. The obtained enzymes showed similar specific activities, determined with tributyrin as the substrate, (1030 U/mg for dLip3 and 1500 U/mg for mLip3) and were within the expected range for the pure forms [21,45]. When monomers were re-chromatographed in SEC columns run with buffered solutions (without sodium cholate), only monomers were detected, and the same was found for the dimers.

To verify the nature of the composition of the dLip3, we performed size-exclusion chromatography (SEC), in the absence or presence of the detergent sodium cholate, and the results are shown in Figure 1. By SEC run without detergent, analyses revealed a single protein fraction with an apparent molecular weight of 117 kDa, which correlates with the expected molecular weight of the dimer [21,45]. When pure Lip3 monomers were chromatographed in the same conditions, only monomers were detected. However, if an aliquot of dLip3 was chromatographed, this time the column being equilibrated with 1% (*w/v*) sodium cholate dissolved in equilibration buffer, one high-molecular weight species without activity was eluted in the column void volume, suggesting it could correspond to a high Mw lipase-cholate aggregate. A second protein peak active against tributyrin was eluted in a volume corresponding to 60 kDa (expected for the monomer). In either case, upon 14% SDS-PAGE analysis under reducing conditions, the two tributyrin-active fractions obtained from each run migrated as large homogeneous bands, consistent with the expected Mw of the Lip3 monomer (Figure 2). These results indicate that dimerization of Lip3 molecule is primarily mediated by hydrophobic interactions, which is consistent with the previously reported crystallographic resolution data [21,43]. The hydrophobic nature of the monomer–monomer interaction also explains why the omission of organic solvents and/or detergents during the isolation of the dimer from the commercial impure preparation was crucial [21,45].

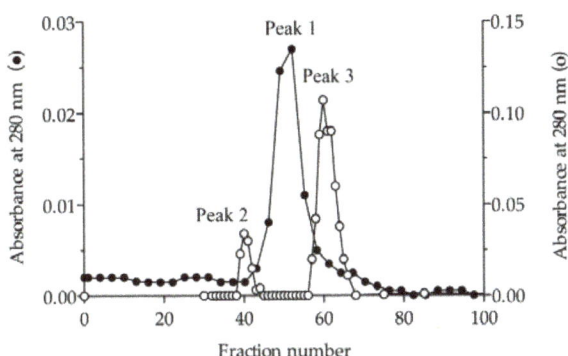

Figure 1. Elution profile in size-exclusion chromatography. Sephacryl S200 column (1.6×60 cm) equilibrated in 25 mM Tris-HCl (pH 7.5) containing 150 mM NaCl in the absence (●) and presence (○) of 1% (*w/v*) sodium cholate. Peak 1 corresponds to dimeric Lip3 (117 kDa). Peak 2 and peak 3 correspond to a high molecular weight lipase-cholate aggregate and monomeric Lip3 (60 kDa), respectively. Loaded sample: Dimeric Lip3. Flow: 0.3 mL/min. Fractions: 3 mL.

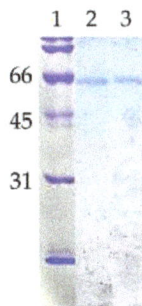

Figure 2. Polyacrylamide gel electrophoresis (SDS-PAGE). Lane 1: Molecular weight standards (kDa); lane 2: mLip 3 (1 μg); lane 3: dLip3 (1 μg). 15% acrylamide gel, Coomassie Blue staining.

3.2. Effect of Oligomerization on Enzyme Kinetics

Previous work on the functional characteristics of *C. rugosa* isoenzymes has shown that monomeric isoenzymes Lip1, Lip2 and Lip3 were kinetically distinct to the Lip3 homodimer. However, substrate specificities with triacylglyceride series [18,22] or cholesteryl oleate [38] were different for each isoenzyme, as monomers all showed proper lipase kinetics with interfacial activation once the solubility limit of triacetin (or tributyrin) was overcome [21,38]. Dimerization allows Lip3 enzyme to hydrolyze soluble triacetin (TA) at a high rate, shifting the kinetic model to a *Michaelis-Menten* type [20,21,45]. Kinetics of mLip3 and dLip3 on triacetin are reproduced in Figure 3a to highlight how differences on specific activity tend to minimize above the solubility limit of TA (270 mM) when a substrate interface is formed, subsequently allowing for mLip3 activation.

In this work, we investigated the effect of including a non-catalytic interface in the reaction media together with TA at different concentrations. To that end, the presence of a hydrophobic organic solvent with a high partition coefficient, such as hexane (log *p* = 3.764), was chosen to be included at a concentration of 25% (*v/v*), enough to form a two-phase system. Figure 3b shows that in the presence of hexane, the activity profiles for mLip3 and dLip3 were almost indistinguishable and compatible with *Michaelis-Menten* type kinetics. For example, at [TA] = 124 mM (well below the solubility of TA), activity was 10.33 ± 0.82 U/mg (mLip3) and 36.30 ± 1.50 U/mg (dLip3) without hexane, but in reaction media supplemented with hexane, both forms showed a specific activity slightly above 90 U/mg

(91.90 ± 1.31 U/mg (mLip3), 99.56 ± 2.15 U/mg (dLip3)). This meant an increase factor of 2.7 for mdLip3, but this was as high as 8.9 for mLip3.

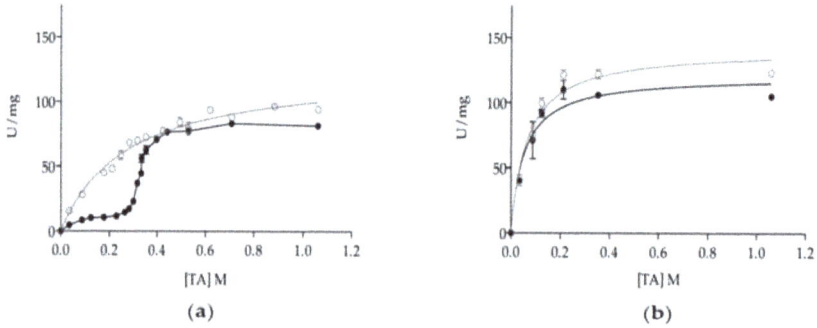

(a)

(b)

Figure 3. Dependence of specific activity of dLip3 (○) and mLip3 (●) on triacetin concentration in the absence (a) and presence (b) of 25% (*v/v*) hexane. The assays were performed at 30 °C and pH 7.0. Triacetin concentration: From 35 mM–1.06 M. Lipase concentration: 80 μM. Triacetin solubility limit was 270 mM. Data from mLip3 series was reproduced with permission from Pernas et al. [20].

3.3. Effect of Dimerization on Enzyme Thermostability

To investigate the relationship between dimerization and thermostability, we next studied the combined effect of pH and temperature on the stability of mLip3 and dLip3 with the response surface methodology (RSM). The design of the experimental matrix and corresponding results for each experiment are given in Table 2. From the results obtained for each enzyme in the experimental matrix, the quadratic polynomial equations to describe stability were:

$$\text{mLip3 (30 min): \% residual activity} = 87 - 31\text{pH} - 12\text{T} - 27\text{pH}^2 - 11\text{pHT} - 6\text{T}^2, \tag{2}$$

$$\text{dLip3 (30 min): \% residual activity} = 89 - 9\text{pH} - 8\text{T} - 11\text{pH}^2 - 7\text{pHT}, \tag{3}$$

$$\text{dLip3 (1 h): \% residual activity} = 89 - 10\text{pH} - 7\text{T} - 15\text{pH}^2 - 10\text{pHT}, \tag{4}$$

Table 2. Experimental design matrix and comparison of observed data and predicted from RSM.

Run	Coded Values		Real Values		Residual Activity (%) [1]			
					mLip3		dLip3	
	x_1	x_2	pH	T	Observed	Predicted	Observed	Predicted
1	1	1	8.6	47.9	1.5	−0.09	40.20	46.08
2	1	−1	8.6	32.1	48.3	47.10	82.50	81.24
3	−1	1	5.4	47.9	83	84.83	83.80	88
4	−1	−1	5.4	32.1	84	86.22	84.30	81.36
5	1.267	0	9	40	2.2	4.51	54.90	51.74
6	−1267	0	5	40	86.2	83.10	78.90	78.39
7	0	1.267	7	50	62.9	62.81	88.70	80.18
8	0	−1.267	7	30	94.3	93.60	95.50	98.24
9	0	0	7	40	86.2	87.42	91.40	89.21
10	0	0	7	40	83.6	87.42	85.60	89.21
11	0	0	7	40	90.4	87.42	87.40	89.21
12	0	0	7	40	87.5	87.42	87.00	89.21
13	0	0	7	40	89.1	87.42	91.10	89.21

[1] The residual activities of mLip3 and dLip3 were measured after 30 min and 1 h of incubation, respectively.

ANOVA tests were performed to validate the quadratic models of the stability of mLip3 and dLip3, testing the statistical significance of each effect by comparing its mean square against an estimated experimental error. As can be seen in the Table 3, linear and quadratic effects for pH and temperature, as well as its interaction (pHT), were statistically significant ($p < 0.05$) to describe mLip3 stability after 30 min of incubation. On the other hand, although the statistically non-significant quadratic effect of temperature (T^2) of both multiple regression analyses applied to dLip3 was eliminated, the model described at 30 min of incubation showed a significant lack of fit (data not shown), while the model shown at 1 h could be validated and adequately used to describe the stability of dLip3 (Table 4). The R^2 statistic explains the 99.56% and 93.81% of the variability of the response attributed to the independent variables for mLip3 and dLip3, respectively, and both models were appropriated under the experimental conditions (lack of fit-test, $p > 0.05$).

Table 3. ANOVA for the response surface quadratic model for the stability of mLip3 after pH and T treatment for 30 min [1].

Factor	SS [2]	DF [3]	MS [4]	F-Value	*p*-Value [5]
pH	6935.61	1	6935.61	998.93	0.0000
T	1063.84	1	1063.84	153.23	0.0002
pH^2	3804.72	1	3804.72	547.99	0.0000
pHT	524.41	1	524.41	75.53	0.0010
T^2	169.87	1	169.87	24.47	0.0078
Lack of fit	27.74	3	9.25	1.33	0.3817
Pure error	27.77	4	6.94	-	-
Total	12,554.0	12	-	-	-

[1] $R^2 = 0.9956$. [2] Sum of squares. [3] Degrees of freedom. [4] Mean square. [5] $p \leq 0.05$ denotes a statistically significant difference at the 5% level.

Table 4. ANOVA for the response surface quadratic model for the stability of dLip3 after pH and T treatment for 1 h [1].

Factor	SS [2]	DF [3]	MS [4]	F-Value	*p*-Value [5]
pH	796.99	1	796.99	117.90	0.0004
T	366.61	1	366.61	54.23	0.0018
pH^2	1166.01	1	1166.01	172.49	0.0002
pHT	436.81	1	436.81	64.62	0.0013
Lack of fit	155.39	4	38.85	5.75	0.0594
Pure error	27.04	4	6.76	-	-
Total	2948.85	12	-	-	-

[1] $R^2 = 0.9381$. [2] Sum of squares. [3] Degrees of freedom. [4] Mean square. [5] $p \leq 0.05$ denotes a statistically significant difference at 5% level.

The surface response 3D-plots allowed us to visualize the interactions of the relative thermostability and the acidophilic character of purified mLip3 and dLip3 (Figure 4). Both monomeric and dimeric forms were more sensitive to alkaline pH values than to high temperatures; dLip3 showed the highest thermostability, but the lack of a "T^2" term in Equation (4) suggests that the optimum for temperature was out of the experimental range (>50 °C). For mLip3, optimum temperature was much lower (35 °C), even though the incubation time was only 30 min compared to 60 min for dLip3. The highest residual activity (maximum stability) was obtained at lower pH values for mLip3 (pH 6.3) than for dLip3 (pH 7.14) (Table 5).

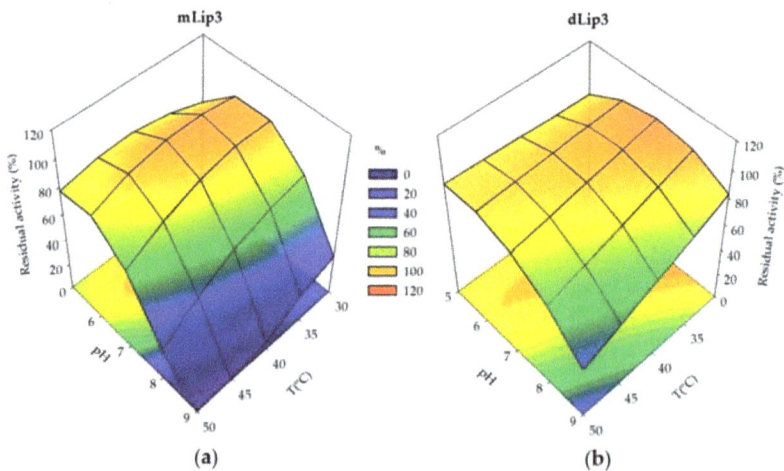

Figure 4. Response surfaces and contour plots generated using experimental design data. (**a**) Residual activity of mLip3 after pH and T treatment for 30 min; (**b**) Residual activity of dLip3 after pH and T treatment for 1 h.

Table 5. Optima pH and temperature values for mLip3 and dLip3 stability.

	mLip3	dLip3
pH	6.3	7.14
Temperature (°C)	35	>50

4. Discussion

Dimers and oligomers are often the functional form of proteins and may have been evolutionarily selected to confer thermostability on them, particularly in thermophilic microorganisms from both *Archaea* and *Bacteria* domains, as these protein subunit associations can result in an extra stabilizations to cope with extreme temperatures [49–51].

In the case of lipases, several have been crystallized in their open conformation without the presence of substrates or inhibitors, suggesting that exposition of hydrophobic areas surrounding the active center occurs in the unbound enzyme [15,52]. The exposed large hydrophobic pocket can promote the association between two open lipases, hence enabling oligomerization [53,54]. Indeed, oligomeric structures appear to be a common feature of the lipases families, and strong correlations between thermostability-aggregation state and catalytic activity-aggregation state have been often described in the literature. For example, non-ionic detergents disaggregate lipases from *Thermosyntropha lipolytica* into less thermostable monomers [55]. The same loss of thermostability was observed for mixtures of dimeric and monomeric forms of lipases in solution under conditions in which disaggregation is promoted (reducing protein concentration and/or including detergents), both in mesophilic and thermophilic microorganisms, such as *Pseudomonas fluorescens* [56], *Alcaligenes* sp. [57], or *Thermomyces lanuginosus* [53]. Mixed results can be found with regard to the effect of aggregation on catalytic activity. For instance, an increase in *P. cepacia* lipase activity was reported as a result of its disaggregation by the addition of 2-propanol [58,59], or *Bacillus thermocatenulatus* lipase (BTL2), treated with some detergents [60], whereas no effect was observed for LipA and LipB from *T. lipolytica* [54]. Still, for *T. lanuginosus*, lipase dimeric forms are less active than monomers [53].

C. rugosa Lip3 isoenzyme has not just been purified to homogeneity as a monomer, but also as a stable dimer [21,44,45] and structure of uncomplexed and linoleate-bound dLip3, known at 1.9 Å and 2.0 Å resolution [42,43]. These gave us the opportunity to study directly the effect of dimerization on catalytic activity and stability [43]. The 3D-structure revealed a dimeric association of two Lip3 monomers with their flaps open (active conformation) and the active-site gorges facing each other, thus shielding the hydrophobic environment of the catalytic triads from the aqueous medium. Four openings are generated at the dimer interface (two independent and two symmetry-related) [43]. In a previous paper, we built a model for TA based on the trilaurin structure available in the Cambridge DataBase (code BTRILA) and, using the program X-PLOR [61], the maximum dimension for the TA model was estimated as 8 Å, smaller than the dimensions of the openings (17 × 9 Å and 13 × 8 Å) [21]. The free access of TA through the openings was also confirmed using the O graphic program [62]. Therefore, the openings were large enough to allow the free entrance of triacetin molecules into the active site, without having to dissociate the dimer [21]. This convincingly explains why dimerization allows the observed shift on the dLip3 kinetic model towards a *Michaelis-Menten* type, as opposed to mLip3, which behaves like a proper lipase showing interfacial activation (Figure 3a). These results pinpoint the importance of a hydrophobic interface in lipases catalysis. To deepen exploration into this aspect of catalysis, we carried out a kinetic experiment, including a non-catalytic hydrophobic interface, into the reaction media, together with the substrate TA. Since polar solvents like isopropanol and acetonitrile often tend to decrease the enzyme activity, hexane (high log *p* value) was used. These results showed how dLip3 and mLip3 kinetics on TA were nearly indistinguishable, proving that monomer and dimer were not catalytically different under those conditions. Thus, the monomer flap opening was triggered by a hydrophobic interface even if this was non-catalytic (hexane).

As indicated above, oligomerization is suggested to be an important mechanism for increasing or maintaining the thermostability of proteins [10]. Indeed, for *C. rugosa* Lip3 isoenzyme, dimerization increases by more than 15 °C its thermostability (maximum T for dLip3 was out of the experimental range; >50 °C), well above what was expected for an enzyme from a mesophile such as *C. rugosa*. In addition, the best pH for stability was nearly one-pH unit higher for the homodimer. Although results are difficult to extrapolate due to the different experimental approaches followed, the stability of *C. rugosa* Lip1 isoenzyme was recently improved by rational design methods, with the increase in optimum temperature around 10 °C and the pH optimum remaining unchanged [63]. The extensive protein-protein contacts generated by Lip3 dimerization have therefore resulted in a substantial stabilization. In addition, enzyme inactivation under acidic and alkaline conditions may be caused by the instability of hydrophilic residues on the molecular surface [63,64]. Thus, increasing the hydrophobic interactions can regulate protein pH tolerance, which is also consistent with our results. Based on the premise that oligomerization is a viable evolutionary strategy for protein stabilization, Fraser et al. [10] conducted a laboratory-directed evolution experiment that allowed the selection of a thermostable variant of the αE7 carboxylesterase from *Lucilia cuprina*, with increased levels of dimeric and tetrameric quaternary structures. As opposed to the *C. rugosa* Lip3 here reported, the greatest thermostability was linked to those oligomers with the lowest catalytic activity.

In summary, our work shows that Lip3 functions as a monomer, but homodimerization expands its specificity (a better catalyst on soluble substrates) and improves pH and T stability. For homodimerization to occur, the key event is the activation of two lipase monomers, which expose a large fraction of hydrophobic and aromatic residues on their interfaces that stabilize the complex [21,43]. In our view, the whole process resembles the so-called domain swapping, an accepted mechanism to explain homodimerization [65,66]. Domain swapping includes opening up of the protein monomeric conformation and the exchange of identical regions between the two monomers [11,67]. A number of studies analyzed the properties of interdomain linker regions since they are flexible and might be responsible for domain swapping. Certain residues (especially Pro) in the interdomain linker region might affect the monomer-dimer equilibrium [11,68,69]. Interestingly, within the *C. rugosa* lipase family, the transition between both open and closed conformers is not restricted to a single flap movement. On the contrary, its secondary structure is refolded and accompanied by a *cis/trans* isomerization of a Pro92 peptide bond that likely increases the energy required for the transition between the two stages [40]. Therefore, our results highlight oligomerization as a viable strategy for protein stabilization of mesophilic enzymes susceptible to be exploited for biotechnological applications.

Author Contributions: M.-E.Á.-C and R.G. contributed equally to this paper, carrying out experimental work, data analysis, and helping with the writing. M.A.P. has contributed in the writing and discussions and M.L.R. designed the experiments and contributed to data analysis and manuscript revision.

Funding: This research was funded by the EU FEDER funds (CIA3-2016).

Conflicts of Interest: The authors declare no conflict of interest. The funders had no role in the design of the study; in the collection, analyses, or interpretation of data; in the writing of the manuscript, and in the decision to publish the results.

References

1. Reed, C.J.; Lewis, H.; Trejo, E.; Winston, V.; Evilia, C. Protein adaptations in archaeal extremophiles. *Archaea* **2013**, *2013*, 1–14. [CrossRef] [PubMed]
2. Giuliani, M.C.; Tron, P.; Leroy, G.; Aubert, C.; Tauc, P.; Giudici-Orticoni, M.T. A new sulfurtransferase from the hyperthermophilic bacterium *Aquifex aeolicus*: Being single is not so simple when temperature gets high. *FEBS J.* **2007**, *274*, 4572–4587. [CrossRef] [PubMed]
3. Devenish, S.R.A.; Gerrard, J.A. The role of quaternary structure in (β/α)8-barrel proteins: Evolutionary happenstance or a higher level of structure-function relationships? *Org. Biomol. Chem.* **2009**, *7*, 833–839. [CrossRef] [PubMed]
4. Linde, M.; Heyn, K.; Merkl, R.; Sterner, R.; Babinger, P. Hexamerization of geranylgeranylglyceryl phosphate synthase ensures structural integrity and catalytic activity at high temperatures. *Biochemistry* **2018**, *57*, 2335–2348. [CrossRef] [PubMed]
5. Loveridge, E.J.; Rodriguez, R.J.; Swanwick, R.S.; Allemann, R.K. Effect of dimerization on the stability and catalytic activity of dihydrofolate reductase from the hyperthermophile *Thermotoga maritima*. *Biochemistry* **2009**, *48*, 5922–5933. [CrossRef] [PubMed]
6. Byun, J.-S.; Rhee, J.-K.; Kim, N.D.; Yoon, J.; Kim, U.; Koh, E.; Oh, J.-W.; Cho, H.-S. Crystal structure of hyperthermophilic esterase EstE1 and the relationship between its dimerization and thermostability properties. *BMC Struct. Biol.* **2007**, *7*, 1–11. [CrossRef] [PubMed]
7. Li, P.-Y.; Chen, X.-L.; Ji, P.; Li, C.-Y.; Wang, P.; Zhang, Y.; Xie, B.-B.; Qin, Q.-L.; Su, H.-N.; Zhou, B.-C.; et al. Interdomain hydrophobic interactions modulate the thermostability of microbial esterases from the Hormone-ensitive Lipase family. *J. Biol. Chem.* **2015**, *290*, 11188–11198. [CrossRef] [PubMed]
8. Singh, M.K.; Shivakumaraswamy, S.; Gummadi, S.N.; Manoj, N. Role of an N-terminal extension in stability and catalytic activity of a hyperthermostable α/β hydrolase fold esterase. *Protein Eng. Des. Sel.* **2017**, *30*, 559–570. [CrossRef] [PubMed]
9. Perica, T.; Kondo, Y.; Tiwari, S.P.; McLaughlin, S.H.; Kemplen, K.R.; Zhang, X.; Steward, A.; Reuter, N.; Clarke, J.; Teichmann, S.A. Evolution of oligomeric state through allosteric pathways that mimic ligand binding. *Science* **2014**, *346*, 1–10. [CrossRef] [PubMed]
10. Fraser, N.J.; Liu, J.; Mabbitt, P.D.; Correy, G.J.; Coppin, C.W.; Lethier, M.; Perugini, M.A.; Murphy, J.M.; Oakeshott, J.G.; Weik, M.; et al. Evolution of protein quaternary structure in response to selective pressure for increased thermostability. *J. Mol. Biol.* **2016**, *428*, 2359–2371. [CrossRef] [PubMed]
11. Nishi, H.; Hashimoto, K.; Madej, T.; Panchenko, A.R. Evolutionary, physicochemical, and functional mechanisms of protein homooligomerization. *Prog. Mol. Biol. Transl. Sci.* **2013**, *117*, 3–24. [CrossRef] [PubMed]
12. Elgharbi, F.; Ben Hlima, H.; Ameri, R.; Bejar, S.; Hmida-sayari, A. A trimeric and thermostable lichenase from *B. pumilus* US570 strain: Biochemical and molecular characterization. *Int. J. Biol. Macromol.* **2017**, *95*, 273–280. [CrossRef] [PubMed]
13. Marsh, J.A.; Teichmann, S.A. Protein flexibility facilitates quaternary structure assembly and evolution. *PLoS Biol.* **2014**, *12*, 1–11. [CrossRef] [PubMed]
14. Anand, S.; Sharma, C. Glycine-rich loop encompassing active site at interface of hexameric *M. tuberculosis* Eis protein contributes to its structural stability and activity. *Int. J. Biol. Macromol.* **2018**, *109*, 124–135. [CrossRef] [PubMed]

15. Aschauer, P.; Rengachari, S.; Lichtenegger, J.; Schittmayer, M.; Padmanabha Das, K.M.; Mayer, N.; Breinbauer, R.; Birner-Gruenberger, R.; Gruber, C.C.; Zimmermann, R.; et al. Crystal structure of the *Saccharomyces cerevisiae* monoglyceride lipase Yju3p. *Biochim. Biophys. Acta-Mol. Cell Biol. Lipids* **2016**, *1861*, 462–470. [CrossRef] [PubMed]

16. Veeraragavan, K.; Gibbs, B.F. Detection and partial purification of two lipases from *Candida rugosa*. *Biotechnol. Lett.* **1989**, *11*, 345–348. [CrossRef]

17. Tomizuka, N.; Ota, Y.; Yamada, K. Studies on lipase from *Candida cylindracea*. *Agric. Biol. Chem.* **1966**, *30*, 576–584. [CrossRef]

18. Rúa, M.L.; Díaz-Mauriño, T.; Fernández, V.M.; Otero, C.; Ballesteros, A. Purification and characterization of two distinct lipases from *Candida cylindracea*. *BBA-Gen. Subj.* **1993**, *1156*, 181–189. [CrossRef]

19. Rúa, M.L.; Ballesteros, A. Rapid purification of two lipase isoenzymes from *Candida rugosa*. *Biotechnol. Tech.* **1994**, *8*, 21–26. [CrossRef]

20. Pernas, M.A.; Pastrana, L.; Fuciños, P.; Rúa, M.L. Regulation of the interfacial activation within the *Candida rugosa* lipase family. *J. Phys. Org. Chem.* **2009**, *22*, 508–514. [CrossRef]

21. Pernas, M.A.; López, C.; Rúa, M.L.; Hermoso, J. Influence of the conformational flexibility on the kinetics and dimerisation process of two *Candida rugosa* lipase isoenzymes. *FEBS Lett.* **2001**, *501*, 87–91. [CrossRef]

22. López, N.; Pernas, M.A.; Pas trana, L.M.; Sánchez, A.; Valero, F.; Rúa, M.L. Reactivity of pure *Candida rugosa* lipase isoenzymes (Lip1, Lip2, and Lip3) in aqueous and organic media. Influence of the isoenzymatic profile on the lipase performance in organic media. *Biotechnol. Prog.* **2004**, *20*, 65–73. [CrossRef] [PubMed]

23. Chang, S.W.; Shieh, C.J.; Lee, G.C.; Shaw, J.F. Multiple mutagenesis of the *Candida rugosa LIP1* gene and optimum production of recombinant *LIP1* expressed in *Pichia pastoris*. *Appl. Microbiol. Biotechnol.* **2005**, *67*, 215–224. [CrossRef] [PubMed]

24. Chang, S.-W.; Lee, G.-C.; Shaw, J.-F. Codon optimization of *Candida rugosa LIP1* gene for improving expression in *Pichia pastoris* and biochemical characterization of the purified recombinant Lip1 lipase. *J. Agric. Food Chem.* **2006**, *54*, 815–822. [CrossRef] [PubMed]

25. Zhao, W.; Wang, J.; Deng, R.; Wang, X. Scale-up fermentation of recombinant *Candida rugosa* lipase expressed in *Pichia pastoris* using the *GAP* promoter. *J. Ind. Microbiol. Biotechnol.* **2008**, *35*, 189–195. [CrossRef] [PubMed]

26. Chang, S.-W.; Lee, G.-C.; Shaw, J.-F. Efficient production of active recombinant *Candida rugosa* Lip3 lipase in *Pichia pastoris* and biochemical characterization of the purified enzyme. *J. Agric. Food Chem.* **2006**, *54*, 5831–5838. [CrossRef] [PubMed]

27. Ferrer, P.; Alarcón, M.; Ramón, R.; Dolors Benaiges, M.; Valero, F. Recombinant *Candida rugosa LIP2* expression in *Pichia pastoris* under the control of the *AOX1* promoter. *Biochem. Eng. J.* **2009**, *46*, 271–277. [CrossRef]

28. Lee, G.-C.; Lee, L.-C.; Sava, V.; Shaw, J.-F. Multiple mutagenesis of non-universal serine codons of the *Candida rugosa LIP2* gene and biochemical characterization of purified recombinant Lip2 lipase overexpressed in *Pichia pastoris*. *Biochem. J.* **2002**, *366*, 603–611. [CrossRef] [PubMed]

29. Lee, L.C.; Chen, Y.T.; Yen, C.C.; Chiang, T.C.-Y.; Tang, S.-J.; Lee, G.-C.; Shaw, J.-F. Altering the substrate specificity of *Candida rugosa* Lip4 by engineering the substrate-binding sites. *J. Agric. Food Chem.* **2007**, *55*, 5103–5108. [CrossRef] [PubMed]

30. Lee, L.-C.; Yen, C.-C.; Malmis, C.C.; Chen, L.-F.; Chen, J.-C.; Lee, G.-C.; Shaw, J.-F. Characterization of codon-optimized recombinant *Candida rugosa* Lipase 5 (Lip5). *J. Agric. Food Chem.* **2011**, *59*, 10693–10698. [CrossRef] [PubMed]

31. Tang, S.-J.; Sun, K.-H.; Sun, G.-H.; Chang, T.-Y.; Lee, G.-C. Recombinant expression of the *Candida rugosa* Lip4 lipase in *Escherichia coli*. *Protein Expr. Purif.* **2000**, *20*, 308–313. [CrossRef] [PubMed]

32. Tang, S.-J.; Shaw, J.-F.; Sun, K.-H.; Sun, G.-H.; Chang, T.-Y.; Lin, C.-K.; Lo, Y.-C.; Lee, G.-C. Recombinant expression and characterization of the *Candida rugosa* Lip4 lipase in *Pichia pastoris*: Comparison of glycosylation, activity, and stability. *Arch. Biochem. Biophys.* **2001**, *387*, 93–98. [CrossRef] [PubMed]

33. Yen, C.-C.; Malmis, C.C.; Lee, G.-C.; Lee, L.-C.; Shaw, J.-F. Site-specific saturation mutagenesis on residues 132 and 450 of *Candida rugosa* Lip2 enhances catalytic efficiency and alters substrate specificity in various chain lengths of triglycerides and esters. *J. Agric. Food Chem.* **2010**, *58*, 10899–10905. [CrossRef] [PubMed]

34. Kawaguchi, Y.; Honda, H.; Taniguchi-Morimura, J.; Iwasaki, S. The codon CUG is read as serine in an asporogenic yeast *Candida cylindracea*. *Nature* **1989**, *341*, 164–166. [CrossRef] [PubMed]

35. Barriuso, J.; Vaquero, M.E.; Prieto, A.; Martínez, M.J. Structural traits and catalytic versatility of the lipases from the *Candida rugosa*-like family: A review. *Biotechnol. Adv.* **2016**, *34*, 874–885. [CrossRef] [PubMed]

36. Verger, R. Enzyme kinetics of lipolysis. In *Methods in Enzymology*; Academic Press, Inc.: Salt Lake City, UT, USA, 1980; Volume 64, pp. 340–392; ISBN 9780121819644.
37. Schrag, J.D.; Cygler, M. Lipases and alpha/beta hydrolase fold. *Methods Enzymol.* **1997**, *284*, 85–107. [CrossRef] [PubMed]
38. Mancheño, J.M.; Pernas, M.A.; Martínez, M.J.; Rúa, M.L.; Hermoso, J.A. Structural insights into the lipase/esterase behavior in the *Candida rugosa* lipases family: Crystal structure of the Lipase 2 isoenzyme at 1.97 Å resolution. *J. Mol. Biol.* **2003**, *332*, 1059–1069. [CrossRef] [PubMed]
39. Grochulski, P.; Li, Y.; Schrag, J.D.; Bouthillier, F.; Smith, P.; Harrison, D.; Rubin, B.; Cygler, M. Insights into interfacial activation from an open structure of *Candida rugosa* lipase. *J. Biol. Chem.* **1993**, *268*, 12843–12847. [PubMed]
40. Grochulski, P.; Li, Y.; Schrag, J.D.; Cygler, M. Two conformational states of *Candida rugosa* lipase. *Protein Sci.* **1994**, *3*, 82–91. [CrossRef] [PubMed]
41. Mancheño, J.M.; Pernas, M.A.; Rúa, M.L.; Hermoso, J.A. Crystallization and preliminary X-ray diffraction studies of two different crystal forms of the Lipase 2 isoform from the yeast *Candida rugosa*. *Acta Crystallogr. Sect. D Biol. Crystallogr.* **2003**, *D59*, 499–501. [CrossRef]
42. Pletnev, V.; Addlagatta, A.; Wawrzak, Z.; Duax, W. Three-dimensional structure of homodimeric cholesterol esterase-ligand complex at 1.4 Å resolution. *Acta Crystallogr.-Sect. D Biol. Crystallogr.* **2003**, *59*, 50–56. [CrossRef]
43. Ghosh, D.; Wawrzak, Z.; Pletnev, V.Z.; Li, N.; Kaiser, R.; Pangborn, W.; Jornvall, H.; Erman, M.; Duax, W.L. Structure of uncomplexed and linoleate-bound *Candida cylindracea* cholesterol esterase. *Structure* **1995**, *3*, 279–288. [CrossRef]
44. Kaiser, R.; Erman, M.; Duax, W.L.; Ghosh, D.; Jörnvall, H. Monomeric and dimeric forms of cholesterol esterase from *Candida cylindracea*. Primary structure, identity in peptide patterns, and additional microheterogeneity. *FEBS Lett.* **1994**, *337*, 123–127. [CrossRef]
45. Pernas, M.; López, C.; Prada, A.; Hermoso, J.; Rúa, M.L. Structural basis for the kinetics of *Candida rugosa* Lip1 and Lip3 isoenzymes. *Colloid Surf. B Biointerfaces* **2002**, *26*, 67–74. [CrossRef]
46. Pernas, M.A.; López, C.; Pastrana, L.; Rúa, M.L. Purification and characterization of Lip2 and Lip3 isoenzymes from a *Candida rugosa* pilot-plant scale fed-batch fermentation. *J. Biotechnol.* **2000**, *84*, 163–174. [CrossRef]
47. Ferrato, F.; Carriere, F.; Sarda, L.; Verger, R. A critical reevaluation of the phenomenon of interfacial activation. *Methods Enzymol.* **1997**, *286*, 327–347. [CrossRef] [PubMed]
48. Randall, R.J.; Lewis, A. Protein measurement with the folin phenol reagent. *Readings* **1951**, *193*, 265–275. [CrossRef]
49. Varejão, N.; De-Andrade, R.A.; Almeida, R.V.; Anobom, C.D.; Foguel, D.; Reverter, D. Structural mechanism for the temperature-dependent activation of the hyperthermophilic Pf2001 esterase. *Structure* **2018**, *26*, 199–208.e3. [CrossRef] [PubMed]
50. Robinson-Rechavi, M.; Alibés, A.; Godzik, A. Contribution of electrostatic interactions, compactness and quaternary structure to protein thermostability: Lessons from structural genomics of *Thermotoga maritima*. *J. Mol. Biol.* **2006**, *356*, 547–557. [CrossRef] [PubMed]
51. Walden, H.; Bell, G.S.; Russell, R.J.M.; Siebers, B.; Hensel, R.; Taylor, G.L. Tiny TIM: A small, tetrameric, hyperthermostable triosephosphate isomerase. *J. Mol. Biol.* **2001**, *306*, 745–757. [CrossRef] [PubMed]
52. Skjold-Jørgensen, J.; Vind, J.; Moroz, O.V.; Blagova, E.; Bhatia, V.K.; Svendsen, A.; Wilson, K.S.; Bjerrum, M.J. Controlled lid-opening in *Thermomyces lanuginosus* lipase–An engineered switch for studying lipase function. *Biochim. Biophys. Acta-Proteins Proteom.* **2017**, *1865*, 20–27. [CrossRef] [PubMed]
53. Gonçalves, K.M.; Barbosa, L.R.S.; Lima, L.M.T.R.; Cortines, J.R.; Kalume, D.E.; Leal, I.C.R.; Mariz e Miranda, L.S.; De Souza, R.O.M.; Cordeiro, Y. Conformational dissection of *Thermomyces lanuginosus* lipase in solution. *Biophys. Chem.* **2014**, *185*, 88–97. [CrossRef] [PubMed]
54. Madsen, J.K.; Sørensen, T.R.; Kaspersen, J.D.; Silow, M.B.; Vind, J.; Pedersen, J.S.; Svendsen, A.; Otzen, D.E. Promoting protein self-association in non-glycosylated *Thermomyces lanuginosus* lipase based on crystal lattice contacts. *Biochim. Biophys. Acta-Proteins Proteom.* **2015**, *1854*, 1914–1921. [CrossRef] [PubMed]
55. Salameh, M.A.; Wiegel, J. Effects of detergents on activity, thermostability and aggregation of two alkalithermophilic lipases from *Thermosyntropha lipolytica*. *Open Biochem. J.* **2010**, *4*, 22–28. [CrossRef] [PubMed]

56. Fernández-Lorente, G.; Palomo, J.M.; Fuentes, M.; Mateo, C.; Guisán, J.M.; Fernández-Lafuente, R. Self-assembly of *Pseudomonas fluorescens* lipase into bimolecular aggregates dramatically affects. *Biotechnol. Bioeng.* **2003**, *82*, 232–237. [CrossRef] [PubMed]

57. Wilson, L.; Palomo, J.M.; Fernández-Lorente, G.; Illanes, A.; Guisán, J.M.; Fernández-Lafuente, R. Effect of lipase–lipase interactions in the activity, stability and specificity of a lipase from *Alcaligenes* sp. *Enzyme Microb. Technol.* **2006**, *39*, 259–264. [CrossRef]

58. Otzen, D.E. Protein unfolding in detergents: Effect of micelle structure, ionic strength, pH, and temperature. *Biophys. J.* **2002**, *83*, 2219–2230. [CrossRef]

59. Dünhaupt, A.; Lang, S.; Wagner, F. Pseudomonas cepacia lipase: Studies on aggregation, purification and on the cleavage of olive oil. *Biotechnol. Lett.* **1992**, *14*, 953–958. [CrossRef]

60. Luisa Rúa, M.; Schmidt-Dannert, C.; Wahl, S.; Sprauer, A.; Schmid, R.D. Thermoalkalophilic lipase of *Bacillus thermocatenulatus*. Large-scale production, purification and properties: Aggregation behaviour and its effect on activity. *J. Biotechnol.* **1997**, *56*, 89–102. [CrossRef]

61. Brunger, A.T. *X-PLOR, version 3.1: A System for X-ray Crystallography and NMR*; Yale University Press: New Haven, CT, USA, 1992.

62. Jones, T.A.; Zou, J.Y.; Cowan, S.W.; Kjeldgaard, M. Improved methods for binding protein models in electron density maps and the location of errors in these models. *Acta Crystallogr. A* **1991**, *47*, 110–119. [CrossRef] [PubMed]

63. Li, G.; Chen, Y.; Fang, X.; Su, F.; Xu, L.; Yan, Y. Identification of a hot-spot to enhance: *Candida rugosa* lipase thermostability by rational design methods. *RSC Adv.* **2018**, *8*, 1948–1957. [CrossRef]

64. Oguchi, Y.; Maeda, H.; Abe, K.; Nakajima, T.; Uchida, T.; Yamagata, Y. Hydrophobic interactions between the secondary structures on the molecular surface reinforce the alkaline stability of serine protease. *Biotechnol. Lett.* **2006**, *28*, 1383–1391. [CrossRef] [PubMed]

65. Gronenborn, A.M. Protein acrobatics in pairs–Dimerization via domain swapping. *Curr. Opin. Struct. Biol.* **2009**, *19*, 39–49. [CrossRef] [PubMed]

66. Bennett, M.J.; Choe, S.; Eisenberg, D. Domain swapping: Entangling alliances between proteins. *Proc. Natl. Acad. Sci. USA* **1994**, *91*, 3127–3131. [CrossRef] [PubMed]

67. Hashimoto, K.; Nishi, H.; Bryant, S.; Panchenko, A.R. Caught in self-interaction: Evolutionary and functional mechanisms of protein homooligomerization. *Phys. Biol.* **2011**, *8*, 1–10. [CrossRef] [PubMed]

68. Hashimoto, K.; Panchenko, A.R. Mechanisms of protein oligomerization, the critical role of insertions and deletions in maintaining different oligomeric states. *Proc. Natl. Acad. Sci. USA* **2010**, *107*, 20352–20357. [CrossRef] [PubMed]

69. Rousseau, F.; Schymkowitz, J.W.H.; Wilkinson, H.R.; Itzhaki, L.S. Three-dimensional domain swapping in p13suc1 occurs in the unfolded state and is controlled by conserved proline residues. *Proc. Natl. Acad. Sci. USA* **2001**, *98*, 5596–5601. [CrossRef] [PubMed]

microorganisms

MDPI

Article

Thermostable Xylanase Production by *Geobacillus* sp. Strain DUSELR13, and Its Application in Ethanol Production with Lignocellulosic Biomass

Mohit Bibra [1], Venkat Reddy Kunreddy [1] and Rajesh K. Sani [1,2,3,4,*]

[1] Department of Chemical and Biological Engineering, South Dakota School of Mines and Technology, Rapid City, SD 57701, USA; mohit.bibra@mines.sdsmt.edu (M.B.); r7sani@gmail.com (V.R.K.)
[2] BuG ReMeDEE Consortium, South Dakota School of Mines and Technology, Rapid City, SD 57701, USA
[3] Composite and Nanocomposite Advanced Manufacturing Centre–Biomaterials (CNAM/Bio), Rapid City, SD 57701, USA
[4] Department of Chemistry and Applied Biological Sciences, South Dakota School of Mines and Technology, Rapid City, SD 57701, USA
*   Correspondence: rajesh.sani@sdsmt.edu; Tel.: +1-605-394-1240

Received: 5 August 2018; Accepted: 31 August 2018; Published: 5 September 2018

Abstract: The aim of the current study was to optimize the production of xylanase, and its application for ethanol production using the lignocellulosic biomass. A highly thermostable crude xylanase was obtained from the *Geobacillus* sp. strain DUSELR13 isolated from the deep biosphere of Homestake gold mine, Lead, SD. *Geobacillus* sp. strain DUSELR13 produced 6 U/mL of the xylanase with the beechwood xylan. The xylanase production was improved following the optimization studies, with one factor at a time approach, from 6 U/mL to 19.8 U/mL with xylan. The statistical optimization with response surface methodology further increased the production to 31 U/mL. The characterization studies revealed that the crude xylanase complex had an optimum pH of 7.0, with a broad pH range of 5.0–9.0, and an optimum temperature of 75 °C. The ~45 kDa xylanase protein was highly thermostable with $t_{1/2}$ of 48, 38, and 13 days at 50, 60, and 70 °C, respectively. The xylanase activity increased with the addition of Cu^{+2}, Zn^{+2}, K^{+} and Fe^{+2} at 1 mM concentration, and Ca^{+2}, Zn^{+2}, Mg^{+2}, and Na^{+} at 10 mM concentration. The comparative analysis of the crude xylanase against its commercial counterpart Novozymes Cellic HTec and Dupont, Accellerase XY, showed that it performed better at higher temperature, hydrolyzing 65.4% of the beechwood at 75 °C. The DUSEL R13 showed the mettle to hydrolyze, and utilize the pretreated, and untreated lignocellulosic biomass: prairie cord grass (PCG), and corn stover (CS) as the substrate, and gave a maximum yield of 20.5 U/mL with the untreated PCG. When grown in co-culture with *Geobacillus thermoglucosidasius*, it produced 3.53 and 3.72 g/L ethanol, respectively with PCG, and CS. With these characteristics the xylanase under study could be an industrial success for the high temperature bioprocesses.

Keywords: Thermostable; xylanase; *Geobacillus*; lignocellulosic biomass; ethanol

---

Highlights:

*   A thermostable xylanase production from a *Geobacillus* sp. strain DUSELR13 with $t_{1/2}$ of 38 days at 60 °C.
*   Use of the lignocellulosic biomass for the xylanase production, and ethanol production.
*   Better xylan hydrolysis by DUSELR13 xylanase than the commercial counterpart at higher temperature.

1. Introduction

The environmental, and economic incentives associated with the utilization of lignocellulosic biomass at industrial scale drive the quest for finding new microorganisms capable of producing enzymes with better characteristics. Using substrates which create direct competition with the food resources had been discouraged at the industrial level. The abundant, economic, and ubiquitously available lignocellulosic biomass (LCB), composed of cellulose and hemicellulose can act as an ideal substrate for the industrial bioprocesses. However, the recalcitrant nature of the LCB due to structural, and chemical hardships require prior hydrolysis of cellulosic and hemicellulosic portions to obtain fermentable sugars [1,2]. Using enzymatic hydrolysis can be environmentally and economically beneficial compared to the other commonly used pretreatment methods [3].

Hemicellulose, which forms about 30–40% of the lignocellulosic biomass (LCB) is mainly constituted by xylan. Xylan is a polysaccharide made up of β-1,4-xylose units or β-1,4-mannose units, and has substitutions of arabinose, methylglucuronic acid, and acetate [4,5]. The complex chemical composition of xylan requires a concerted action of several enzymes, collectively known as hemicellulases, including endo-β-D-xylanases, β-D-xylosidases, α-L-arabinofuranosidases, α-D-glucuronidases, acetylxylan esterases, ferulic and p-coumaric acid esterases. These enzymes act synergistically on the linear chain as well as on the side chains giving xylose and xylooligosaccharides (XOS) as the end products [4,5]. Among the various enzymes required for the effective hydrolysis of xylan component of biomass, endo-1,4-β-xylanase (simply referred to as xylanase), is an essential enzyme that acts on the xylan backbone of the hemicellulose with high specificity, negligible substrate loss, and side products [4] compared to the prevalent chemical methods of hydrolysis.

The branched out application of the xylanases in paper and pulp industry, deinking, pharmaceuticals, cosmetics, food, and feed industry etc. had increased their demand, and significance in the industrial processes [4–9]. The wide operational parameters of these processes require xylanases that can perform well under the extreme conditions of temperatures, pH, salts, osmotic conditions etc. The thermostable enzymes which can operate at high temperatures ranging from 45 to 100 °C have been in demand for a long period in the industrial production, and research bioprocessing set ups. They offer several advantages such as higher mass transfer rates, and low viscosity that may lead to increased solubility of reactants and products; lower risk of contamination from mesophilic microbes, improved hydrolysis performance due to long half-lives at high temperatures, and structural and functional stability at higher temperatures [3,10].

The bioprospecting of several thermophilic bacteria belonging to genus *Geobacillus*, *Bacillus*, *Thermotoga*, *Thermoanerobacterium*, *Anoxybacillus*, and *Acidothermus* etc. had been reported that produce thermostable xylanases [5,6,8,10,11]. The thermostable xylanases obtained from the thermophilic microorganisms are more advantageous in comparison to those obtained from their mesophilic counterparts for industrial bioprocesses [3]. Among several thermophilic organisms studied for xylanase production genus *Geobacillus* had been researched extensively owing to its ability to produced highly thermostable enzymes, utilize various carbon sources [12,13]. The existence of *Geobacillus* spp. in the thermophilic areas has endowed the genus with several species the genome of which encode highly thermostable enzymes with application in several industrial bioprocesses such as the lignocellulosic biomass hydrolysis [4,5], pulp and paper production [6], bioethanol production [14] etc.

Several *Geobacillus* spp. such as *Geobacillus stearothermophilus* [15], *Geobacillus thermodenitrificans* [16], *Geobacillus* sp. strain WSUCF1 [5], and *Geobacillus thermolevorans* [6] etc. had been reported for xylanase production. Although, the *Geobacillus* spp. produce thermostable xylanases, but the duration of thermostability is very small. At 70 °C *Geobacillus thermodinitrificans* strain A333 Xyn had a $t_{1/2}$ of 60 min [4], *Geobacillus* sp. strain TF16 had a $t_{1/2}$ of 10 min [17], and a $t_{1/2}$ of 8 min was observed for xylanase from *Geobacillus thermodinitrificans* strain C5 [18]. Only, the xylanase from *Geobacillus* sp. strain WSUCF1 had an exceptional $t_{1/2}$ of 12 days at 70 °C [5]. However, the recombinant expression of the xylanase from *Geobacillus* sp. strain WSUCF1 in *E. coli* reduced the thermostability from 12 days to only 20 min at 70 °C [19]. Thus the higher thermostability was seen only when the enzyme was produced in the natural host. Several mutagenic studies had been done to increase the thermostability of the xylanases. Irfan and coworkers (2018) carried out site directed mutagenesis in the xylanase from *Geobacillus thermodinitrificans* strain C5 and increased the thermostability from 8 min to 70 min at 70 °C in a triple mutant strain [18]. The thermostability of xylanase from wild type *Geobacillus stearothermophilus* sp. strain XT6 was increased by incorporating substitutions at 13 amino acids. With amino acid substitutions the thermostability increased 52 folds from 3.5 to 182.2 min at 75 °C [20]. In addition to the mutagenic studies, literature studies also reveal several examples of improving thermostability by immobilizing the enzymes. The immobilized enzymes offers several advantages such as improved selectivity, structural stability, reduced sensitivity, and reduced costs due to easy enzyme recovery etc. when compared to the soluble enzymes [17]. With additional benefits of enzyme thermostability the process of immobilization becomes more lucrative. In a study carried out to increase the thermostability of the xylanase enzyme from *Geobacillus* sp. strain TF16 by immobilizing the enzyme with chitosan resulted in enzyme retaining 45% of its activity after 6 h at 55 °C compared to 2 h with the free enzyme [17]. Thus, thermostability of the hydrolytic enzymes is one highly sought characteristic as is evident from the recent works focused on increasing the thermostability of the enzymes [21–23]. However, the anthropogenic measures of increasing thermostability till date are no match for the natural thermostability as observed in the crude xylanase obtained from the wild-type *Geobacillus* sp. strain WSUCF1. Hence, bioprospecting wild type strains that can produce highly thermostable enzymes is very important.

Geobacillus sp. strain DUSELR13 is one such *Geobacillus* sp. which earlier had been reported to produce cellulase with thermostable characteristics [24]. We report here the production of a highly thermostable endoxylanase from *Geobacillus* sp. strain DUSELR13. The production of the xylanase from DUSELR13 was optimized taking into account several physical and biochemical factors. To increase the production further, statistical optimization was carried. A central composite design (CCD) was created in response surface methodology to increase the xylanase production. A comparative analysis of the DUSELR13 xylanase to a commercial enzymatic formulation was further researched to find its suitability on the industrial scale. The xylanase produced was characterized to find the optimum pH, temperature, thermostability, and effect of the metal ions. The xylanase production was further carried with the lignocellulosic biomass to improve the economics of the enzyme production. Finally, a co-culture study was done to study the application of the enzymatic hydrolysis of LCB by *Geobacillus* sp. strain DUSELR13 to produce ethanol with *Geobacillus thermoglucosidasius*.

2. Materials and Methods

2.1. Microorganism and Enzyme Assay

The *Geobacillus* sp. strain DUSELR13, previously known as the *Bacillus* sp. strain DUSELR13., was isolated from the deep biosphere of Homestake gold mine, Lead, South Dakota (44°21′3″ N, and 103°45′57″ W) [24]. It was grown in 90 mL of the minimal media developed previously in our lab [25] with 0.5% (w/v) of xylan as the carbon and energy source in 500-mL Erlenmeyer flasks. The composition of the minimal media per liter is: 0.1 g nitrilotriacetic acid, 0.05 g $CaCl_2 \cdot 2H_2O$, 0.1 g $MgSO_4 \cdot 7H_2O$, 0.01 g NaCl, 0.01 g KCl, 0.3 g NH_4Cl, 0.005 g methionine, 0.2 g yeast extract, 0.01 g casamino acids, 1.8 g of 85% H_3PO_4, 1 mL $FeCl_3$ solution (0.03%), and 1 mL of Nitsch's trace solution. The pH of the medium was kept at 7.0 using 6 M NaOH. The flasks were incubated in a shaker incubator at 60 °C, and 150 rpm for 96 h. A control flask designated as the organism free control was also kept under the similar conditions. After 12 h, 3 mL samples were removed aseptically, and analyzed for growth by measuring OD_{600nm}. The collected samples were centrifuged at 4 °C and $10,000 \times g$ for 10 min, and the supernatant was retained for the endoxylanase activity described below.

2.2. Enzyme Assay

For endoxylanase enzyme assay, performed as per Bailey et al. [26], the reaction mixtures contained 1.8 mL of 1% (w/v) birchwood xylan (Sigma-Aldrich, St. Louis, MO, USA) in phosphate buffer (100 mM, pH 7.0) and 0.2 mL of an appropriate dilution of the supernatant prepared for HPLC analysis containing enzyme. The enzyme-substrate reaction was carried out at 60 °C for 10 min. The reaction was stopped by the addition of 3.0 mL 3,5-Dinitrosalicylic acid (DNSA) solution, boiled for 10 min, and then cooled on ice for color stabilization. The optical absorbance was measured at 540 nm, and the amounts of liberated reducing sugar (xylose equivalents) was estimated against the standard curves for xylose. One unit of xylanase enzyme was defined as the amount of enzyme that releases 1 μmol of xylose per minute under reaction conditions.

2.3. Cell Morphology

The cell morphology was determined using the Scanning electron microscopic (SEM). For sample preparation, 5 mL of culture solution was taken in 15 mL centrifuge, and centrifuged at 10,000 rpm for 10 min. The pellet was washed with PBS (pH 7.2, 100 mM) three times. A mixture of glutaraldehyde and cacodylate buffer 100 mM pH 7.2 was added in a ratio 1:9 (v/v), and mixed gently to fix the samples. Afterwards, the samples were kept on the ice for 45–60 min. The suspension was centrifuged at 6000 rpm for 5–10 min. After centrifugation the supernatant was removed and the pellet was dehydrated using ethyl alcohol. The concentration of the ethyl alcohol was increased in a graduated manner: 30%, 50%, 70%, 90% and 100%. The samples were centrifuged at each step involving addition of ethyl alcohol. With each centrifugation step the supernatant was removed, and after final centrifugation with 100% alcohol the pellet was mounted on an aluminum stub and allowed to dry before microscopic analysis.

2.4. Optimization Studies for Xylanase Production

2.4.1. One Factor at a Time

To find the parameters affecting the xylanase production, different pH's (5.0–9.0 with an increment of 1.0), temperatures (50–70 °C with an increment of 5 °C), xylan concentrations (0.1, 1.0, 2.0, 3.0, and 4.0% (w/v)), and nitrogen sources (yeast extract, urea, tryptone, beef extract, and peptone 0.05% (w/v)) were chosen. The studies were done using one factor at a time (OFAT) approach. In OFAT approach, one factor was optimized at a time, and that value was used to find the optimized value for the other factor. The pH value was selected first and was used to find out the optimal temperature, followed by the xylan concentration and selection of the nitrogen source, for the crude xylanase production. The reaction set for the enzyme optimization studies was prepared as described above.

2.4.2. Response Surface Methodology

Response Surface Methodology (RSM) was used to optimize the fermentation parameters to enhance, and find out the effect of two factor interactions on the extracellular xylanase production, using Design Expert Version 11.0.0 (Stat-Ease Inc., Minneapolis, Minnesota, MN, USA) statistical software. Four variables: temperature, pH, xylan, and tryptone, were chosen for the statistical optimization with all the variables set at a central coded value of zero. Each variable was studied at five different levels ($-\alpha$, -1, 0, $+1$, $+\alpha$) as shown in Table 1. The design included six center points with an alpha value of ± 0.5. All the factorial points, and axial points were studied in triplicates giving a total number of runs 78, as per the combination expression m × 2^k + m × 2 × k + n where 'm' is the number of replicates, 'k' is the number of variables, and 'n' is the number of center points. Quantitative data generated from these experiments shown in Table 2 was subjected to analysis of regression through

RSM to solve multivariate equations. The effects of variables to the response were analyzed by using a second-order polynomial equation:

$$Y = \beta_0 + \beta_1 A + \beta_2 B + \beta_3 C + \beta_4 D + \beta_{11} A^2 + \beta_{22} B^2 + \beta_{33} C^2 + \beta_{44} D^2 + \beta_{12} AB + \beta_{13} AC + \beta_{14} AD + \beta_{23} BC + \beta_{24} BD + \beta_{34} CD \tag{1}$$

where, Y is the predicted response, A, B, C, and D, are the coded levels of the independent parameter, β_0 represents the intercept, β_1, β_2, β_3 and β_4 are linear effect coefficients; β_{11}, β_{22}, β_{33} and β_{44} are the quadratic effect coefficients, β_{12}, β_{13}, β_{14}, β_{23}, β_{24}, and β_{34} are the interaction effect coefficients. The statistical significance of the model was estimated by analysis of variance (ANOVA) with p-value < 0.05 i.e., above 95% confidence level and insignificance of lack of fit test.

The quality of the model developed was evaluated by R-squared values i.e., coefficient of determination: adjusted R^2 and predicted R^2. The fitted polynomial equation was expressed as 3D surface plots to illustrate the relationship between the responses and any two variables to be optimized, keeping the other variables at central positions. Further, numerical optimization method was used for obtaining the optimal solution by keeping the desirability at maximum. The model obtained was validated by running the experiment based on the optimum values obtained for the variables.

Table 1. Experimental range, level, and coded representation of independent variables for the CCD design.

Variables	Code	Range and Levels				
		−1	−0.5	0	+0.5	+1
Temperature (°C)	A	55	57.5	60	62.5	65
pH	B	6	6.25	6.5	6.75	7
Xylan (g/L)	C	5	7.5	10	12.5	15
Tryptone (g/L)	D	1	3.25	5.5	7.75	10

Table 2. Central composite design along with experimental and predicted values of the dependent variable.

Run	A	B	C	D	Xylanase Activity (IU/mL)	
					Experimental	Predicted
1	1.000	1.000	−1.000	1.000	13.60	15.06
2	0.500	0.000	0.000	0.000	26.40	24.92
3	−1.000	−1.000	1.000	1.000	17.60	17.51
4	1.000	1.000	−1.000	−1.000	12.00	11.75
5	0.000	0.000	0.000	0.000	24.70	24.93
6	0.500	0.000	0.000	0.000	16.30	15.80
7	−1.000	−1.000	1.000	−1.000	16.10	15.06
8	−1.000	1.000	1.000	−1.000	23.40	23.24
9	−1.000	1.000	1.000	−1.000	23.10	24.23
10	−0.500	0.000	0.000	0.000	26.20	24.23
11	1.000	1.000	1.000	1.000	18.70	19.83
12	1.000	−1.000	−1.000	1.000	27.50	26.72
13	0.000	0.000	0.000	0.000	20.00	22.51
14	1.000	−1.000	−1.000	−1.000	18.30	17.51
15	1.000	−1.000	1.000	−1.000	26.20	24.92
16	0.000	0.000	0.000	−0.500	15.70	15.80
17	1.000	−1.000	1.000	1.000	15.60	15.16
18	0.000	0.000	0.500	0.000	22.40	24.23
19	−1.000	1.000	−1.000	−1.000	12.30	12.21
20	0.000	0.500	0.000	0.000	19.30	19.48
21	0.000	0.500	0.000	0.000	18.40	19.08
22	0.000	0.000	0.000	−0.500	16.30	16.51
23	0.000	0.000	0.000	0.500	22.60	21.35
24	1.000	1.000	1.000	1.000	23.20	25.89
25	0.000	−0.500	0.000	0.000	18.40	19.83

Table 2. *Cont.*

Run	A	B	C	D	Xylanase Activity (IU/mL)	
					Experimental	Predicted
26	−1.000	1.000	−1.000	1.000	18.60	19.48
27	1.000	−1.000	1.000	1.000	23.80	23.24
28	1.000	1.000	1.000	−1.000	13.30	12.52
29	0.000	0.000	0.000	0.000	15.30	15.06
30	0.000	0.000	0.000	0.000	27.00	24.92
31	0.000	0.000	−0.500	0.000	17.30	16.57
32	−1.000	−1.000	1.000	−1.000	21.20	19.83
33	−1.000	1.000	1.000	1.000	23.10	23.97
34	0.500	0.000	0.000	0.000	23.70	25.89
35	1.000	1.000	1.000	−1.000	22.50	23.40
36	−1.000	1.000	1.000	1.000	21.30	23.24
37	1.000	1.000	−1.000	1.000	16.80	17.85
38	−1.000	1.000	−1.000	−1.000	18.30	17.85
39	−1.000	−1.000	−1.000	−1.000	15.30	15.16
40	1.000	−1.000	−1.000	1.000	14.80	15.16
41	−0.500	0.000	0.000	0.000	19.60	19.48
42	0.000	0.000	0.000	0.500	14.50	13.17
43	−1.000	−1.000	−1.000	−1.000	13.10	12.52
44	0.000	0.000	0.000	−0.500	16.50	16.57
45	1.000	−1.000	−1.000	1.000	15.80	16.51
46	1.000	1.000	−1.000	−1.000	23.40	22.51
47	1.000	−1.000	−1.000	−1.000	26.10	24.92
48	1.000	−1.000	−1.000	−1.000	17.10	16.57
49	0.000	0.000	0.000	0.000	12.10	12.52
50	−1.000	−1.000	−1.000	1.000	12.30	12.21
51	1.000	1.000	1.000	1.000	26.20	24.93
52	0.000	0.000	0.500	0.000	22.00	23.40
53	−1.000	−1.000	1.000	−1.000	24.50	25.89
54	1.000	1.000	−1.000	−1.000	11.20	11.75
55	0.000	0.000	0.000	0.500	28.40	26.72
56	−1.000	−1.000	−1.000	−1.000	13.00	12.44
57	−1.000	−1.000	−1.000	1.000	13.30	13.17
58	0.000	0.500	0.000	0.000	24.50	23.97
59	0.000	0.000	−0.500	0.000	19.00	19.08
60	−0.500	0.000	0.000	0.000	16.00	16.51
61	−1.000	−1.000	−1.000	1.000	17.40	17.85
62	−1.000	1.000	−1.000	−1.000	16.80	17.51
63	−1.000	1.000	1.000	−1.000	20.90	21.35
64	0.000	0.000	0.000	0.000	26.80	24.92
65	0.000	0.000	−0.500	0.000	11.20	11.75
66	−1.000	1.000	1.000	1.000	14.80	15.80
67	−1.000	−1.000	1.000	1.000	12.40	13.17
68	1.000	−1.000	1.000	−1.000	21.30	22.51
69	−1.000	−1.000	1.000	1.000	25.40	24.92
70	−1.000	1.000	−1.000	1.000	21.80	21.35
71	0.000	−0.500	0.000	0.000	24.20	23.97
72	0.000	−0.500	0.000	0.000	13.10	12.44
73	1.000	−1.000	1.000	−1.000	24.50	24.93
74	0.000	0.000	0.500	0.000	12.80	12.21
75	−1.000	1.000	−1.000	1.000	28.30	26.72
76	1.000	1.000	1.000	−1.000	23.60	23.40
77	1.000	−1.000	1.000	1.000	20.40	19.08
78	1.000	1.000	−1.000	1.000	12.20	12.44

2.5. Xylanase Characterization

Sodium dodecyl sulfate-polyacrylamide gel electrophoresis (SDS-PAGE) was performed to find the molecular weight of the protein, as described by Laemmli [27]. Ten milliliters of the supernatant from Erlenmeyer flasks with xylan and PCG was concentrated (5-times) using Amicon

Ultra-15-Millipore (10 kDa cut off). Ten microliters of the concentrated protein was mixed with ten microliters sample buffer (2X). The composition of the sample buffer (4X) per liter was: 250 mL^{-1} M Tris-HCl (pH 6.8), SDS-100 g, 0.1% bromophenol blue (w/v)-80 mL, glycerol-400 mL, 14.3 M β-mercaptoethanol-200 mL, and distilled water to make the volume 1 L. The enzyme extracts containing equal amounts of protein (50 micrograms) were resolved on 10% SDS-PAGE at constant voltage (150 V) till the dye front reached the bottom of the gel. For zymogram, the gel was renatured by washing successively for 30 min with: 20% isopropanol in phosphate buffer saline (PBS, 100 mM, pH 5.9), 8 M urea in PBS, and PBS (pH 5.9) three times. The re-natured gel was placed in sodium phosphate buffer (50 mM, pH 7.0) for 15 min and subsequently, Beechwood xylan (prepared in 50 mM sodium phosphate buffer, pH 7.0), was overlaid on the gel. The gel was incubated at 60 °C for 30 min. This was followed by staining with Congo red (1 mg/mL) for 30 min, and de-staining with 1 M NaCl in PBS until clear bands indicating xylanase activity were visible. The SDS-PAGE gels loaded with the PCG samples were treated with the Silver Stain Plus kit (BioRad, Hercules, CA, USA) as per the manufacturer's instruction.

One percent (w/v) Beechwood xylan was used to determine the relative xylanase activity at various pHs. The pH optimum (pH$_{opt}$) of crude xylanase was estimated by testing enzyme activity in the pH range of 3.0–10.0 using different assay buffers, citrate buffer (100 mM, pH 3–6), phosphate buffer (100 mM, pH 6–7.5), Tris-HCl (100 mM, pH 7.5–9), and glycine-NaOH buffer (100 mM, pH 8.6–10) at 60 °C for 10 min. The enzyme activity obtained at the pH$_{opt}$ was used to calculate the relative enzyme activity at other pHs. The optimum pH of 7.0 was used to determine the optimum temperature for the crude xylanase.

The temperature optimum (T$_{opt}$) for crude xylanase was obtained by performing the enzyme assays at different temperatures. The experiments were carried out at temperatures: 25 °C, 37 °C, and a range of 50–100 °C with an increment of 5 °C under assay conditions as described above. The thermostability of xylanases was assessed by incubating the enzyme at different temperatures 50–100 °C with increments of 10 °C for a period of 80 days. The effect of metal ions on the enzyme activity was determined for Cu, Co, Ca, Mg, Zn, Mn, Na, K, and Fe at 1 mM and 10 mM concentrations. The enzyme solution in 100 mM phosphate buffer (pH—7.0) was doped with the metal salt, and incubated at 60 °C for 1 h. The sampling was done at predetermined time intervals over the period of incubation. The residual activities were determined under optimum pH and temperature conditions using the DNSA method as described above. Throughout the optimization studies, the optimum activity was assumed to be 100%, and the relative enzyme activities were calculated against it. While for the metal ion characterization study, the test study where no metal ion was added was considered as the 100% activity, and all the activities were calculated against it.

2.6. Hydrolysis of Birchwood Xylan

The hydrolysis of Birchwood xylan was carried out in 100 mL conical flask containing 50 mL sodium phosphate buffer (50 mM, pH 7.0) (buffering agent), 1 g xylan, 0.03% (w/v) sodium azide (preservative), sucrose 150 mM (stabilizer) and 20 U crude xylanase/g xylan. The hydrolysis was performed for 48 h at different temperatures (50–75 °C with an increment of 5 °C) with a rotating speed of 150 rpm. Hydrolysis of xylan was also compared using Cellic HTec2 (Novozymes, Franklin, NC, U.S.A.), and Accelerase XY (DuPont, Palo Alto, CA, U.S.A.) with similar enzymatic units as of DUSELR13. A pH of 5.0, and different temperatures in the stable operational range (50–75 °C) of commercial counterparts, was used to compare the hydrolytic potential with the xylanase from *Geobacillus* sp. strain DUSELR13 xylanase. The amount of reducing sugar was measured by HPLC (Shimadzu LC20; Columbia, MD, USA) equipped with a 300 × 7.8 mm Aminex HPX-87H column (Bio-Rad, Torrance, CA, USA). One mL of the samples was removed from the Erlenmeyer flasks, and centrifuged at 10,000 rpm for 10 min. The supernatant obtained was removed and filtered using 0.2 μm pore size membrane filters (Gelman Acrodisc, Sigma Aldrich, St. Louis, MO, USA). The filtered samples were automatically injected onto a heated column (50 °C) and eluted at 0.45 mL/min using 5 mM H$_2$SO$_4$ as the mobile phase in HPLC.

2.7. Enzyme Production with Lignocellulosic Biomass

For enzyme production with lignocellulosic biomass (LCB) 1% (*w/v*) of untreated and mechanically pretreated (Prairie cordgrass (PCG) and corn stover (CS)) were used as the substrate. One hundred ml of the minimal media, supplemented with 10 g/L tryptone, and the pH set to 6.5 using 6 M NaOH, was added to the 500 mL Erlenmeyer flasks. The flasks were autoclaved at 121 °C and 15 psi for 15 min. After autoclaving 1% (*w/v*) of insoluble LCB obtained as described by Bibra et al. (2018) was added to the flasks [28]. The flasks were kept at 59 °C for 96 h. After 12 h 3 mL samples were removed aseptically. The collected samples were centrifuged at 4 °C and 10,000× g for 10 min, and the supernatant was used for measuring the endoxylanase activity (explained above).

2.8. Co-Culture of Geobacillus sp. strain DUSELR13 and Geobacillus thermoglucosidasius for Ethanol Production

The lab strain *Geobacillus* sp. strain DUSELR13 capable of producing thermostable ligninolytic enzymes was used for the ethanol production with ethanol producing strain *Geobacillus thermoglucosidasius* (ATCC 43742). Five percent (*w/v*) of insoluble mechanically treated lignocellulosic biomass (prairie cord grass-PCG, and corn stover-CS) were added to the 100 mL of the minimal media in 500 mL Erlenmeyer flasks. The pH of the media was adjusted to 6.5 using 6 M NaOH, and followed by the addition of 10% (*v/v*) of the actively growing culture of *Geobacillus* sp. strain DUSELR13 to the Erlenmeyer flask. The flasks were kept at 59 °C and 150 rpm for 36 h for enzyme production. After 36 h 20 mL of fresh media (5X) was added to the Erlenmeyer flask for ethanol production, and pH was adjusted to 7.0 using 6 M NaOH. Ten percent (*v/v*) of the actively growing culture of *Geobacillus thermoglucosidasius* was added to the Erlenmeyer flask, and the flasks were capped by polyvinyl stoppers. The amount of ethanol and volatile fatty acids produced were measured with HPLC as explained before [25]. The ethanol yield '$Y_{P/S}$' (g/g) and ethanol productivity 'q_p' (g/L/h) were measured using the equation below:

$$Y_{P/S} \; (g/g) = \frac{\text{Amount of ethanol (g)}}{\text{Amount of susbtrate utilized (g)}} \tag{2}$$

$$q_P \; (g/L/h) = \frac{\text{Amount of ethanol produced (g)}}{\text{Volume (L)} * \text{Time (hours)}} \tag{3}$$

2.9. Material and Energy Balance

The material balance was carried out in terms of mass balance for PCG, and CS. The energy efficiency for conversion of PCG to ethanol was calculated using the following equation:

$$\eta = \frac{\text{xEthanol} * \text{EEthanol} * 100}{\Delta H_{C6,C5LCB} * M_{LCB}} \tag{4}$$

where η represent energy efficiency, xEthanol represents the amount of ethanol produced in moles, EEthanol represents the energy density of ethanol (28.6 MJ/kg), ΔH represents the heat of combustion for hexose and xylose portion of the lignocellulosic biomass, and M_{LCB} represents the amount of lignocellulosic biomass used.

3. Results and Discussion

3.1. Microorganism

The preliminary experiments showed that *Geobacillus* sp. strain DUSELR13 was capable of producing xylanase. The Figure 1 shows the endoxylanase activity, total extracellular protein, and OD_{600nm} with time. *Geobacillus* sp. strain DUSELR13 showed a typical growth curve achieving maximum OD_{600nm} ~0.355 absorption units (A.U.) in 15 h during the exponential phase (Figure 1). After that the OD_{600nm} became constant, signifying a stationary phase. When grown on 0.05% (*w/v*)

Beechwood xylan, DUSELR13 produced 6 U/mL of the thermostable endoxylanase (Xyl) activity and a total extracellular protein of 12.7 mg/mL (Figure 1). The extracellular protein and endoxylanase activity on the other hand still increased after 14 h, thus ruling out any correlation between the growth and the extracellular protein production and the endoxylanase activity. Purohit and coworkers (2017) also observed that the maximum enzyme activity was observed after 24 h when the maximum growth for *Acinetobacter putii* MASK25 was achieved [29]. It was also reported by Shulami and coworkers (2014) that *Geobacillus stearothermophilus* produced xylanase during the late exponential phase, where the cells could survive the nutrient limiting conditions, and produced more xylanase to hydrolyze the xylan and generate more sugar [15].

After 42 h till end, no change in the endoxylanase activity was observed, whereas the extracellular protein still increased which could be due to release of the cell protein after cell lysis during this phase. As the organisms approach towards the end of the stationary phase with nutrient limitation becoming more prominent, they tend to adopt protectionary mechanisms such as the endospores formation. With endospore formation the cell metabolism stops. In the current study, the scanning electron microscopy (SEM) image of the culture during the stationary phase showed rod shaped organism with endospores at the terminal position (Figure 2). This observation regarding formation of endospores during late stationary phase further substantiates the reason for no change in the xylanase activity in the later stages of the bacterial growth.

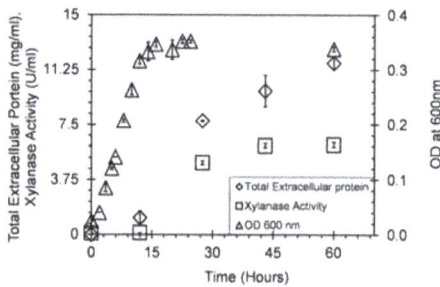

Figure 1. The growth, total extracellular protein, and xylanase activity profile of *Geobacillus* sp. strain DUSELR13 grown on 0.05% (*w/v*) xylan as the substrate at 60 °C. The values are the means of three set of experiments and the error bars represent ± SD of the means with *n* = 3.

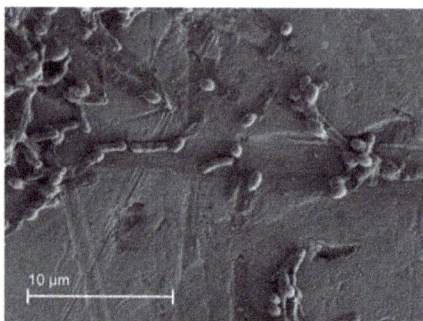

Figure 2. SEM image of *Geobacillus* sp. strain DUSELR13 during the late stationary phase. The majority of the cells show appear as flaccid with the presence of the terminal endospores.

3.2. Optimization of Endoxylanase Production

The secretome obtained during the growth cycle of an organism consists of several proteins, and xylanase might represent a smaller fraction of that secretome. Hence, to increase the production of the endoxylanase several factors were chosen and studied to find the optimal values for xylanase production.

3.2.1. One Factor at a Time

pH

The strain DUSELR13 was grown at different pHs (5–9, with an increment of 1.0). The pH 6.0 (8.54 U/mL-100%) was found to be optimal for the enzyme production. Figure 3A shows the relative percent of the xylanase activity at different pHs. The order of pH for the relative xylanase activity at different pH for the production was 6(100%) > 7(94.14%) > 8(46.3%) > 5(16.05%) > 5(10.12%). The xylanase production decreased as the growth pH deviated from the pH_{opt}. The relative activity of the xylanase decreased by 40–50% at pH's 5, 8, and 9. The pH 6.0 was also found to be optimum for xylanase production by *Geobacillus stearothermophillus* strain KIGBE-IB29 [30] and *Bacillus altitudinis* strain DHN8 [31]. On the contrary, *Geobacillus* sp. strain WSUCF1 exhibited a pH_{opt} of 6.5 for the xylanase production [5]. The external pH affects the transport of the enzymes and chemical products across the membrane [5]. Hence finding a pH value where the extracellular secretion of protein is favored is important.

Temperature

When grown at different temperatures (50–70 °C, with an increment of 5 °C), 60 °C was found to be the T_{opt} for xylanase production (9.23 U/mL-100%) for the strain DUSELR13. The enzyme activity increased from 50 to 60 °C, but decreased from 60 °C to 70 °C, The relative xylanase activity was measured against activity obtained at T = 60 °C, and the order of xylanase activity was 60(100%) > 55(80.13%) > 65(32.7%) > 50(31.12%) > 70(26.44%) °C (Figure 3B). The temperature 60 °C has also been reported as T_{opt} for the endoxylanase production in *Geobacillus* sp. strain WSUCF1 [5] and *Geobacillus steareothermophillus* [15]. Different T_{opt}'s have also been reported for endoxylanase production by several other thermophilic species: *Anoxybacillus flavithermus* Strain TWXYL3 had a T_{opt} of 65 °C [32], whereas *Bacillus amyloliquefaciens* showed T_{opt} of 50 °C for the xylanase production [23]. The production of the enzymes at a high temperature offers advantages such as no contamination and reduced viscosity [3]. A temperature higher than the T_{opt} is more detrimental than a temperature lower than the T_{opt}. High temperatures such as 70 °C can cause the denaturation of several key enzymes required for the cell survival, resulting in reduced growth and lower enzymatic activity. Hence, low xylanase activity was observed at 70 °C. As reported previously, the extracellular protein secretion involves several membrane associated proteins, and is affected by the rate of the folding of such proteins, and the dynamics of complex formation/dissolution within the cell envelope dependent [33]. As the temperature deviates from the T_{opt} production it directly affects all these processes, affecting the extracellular protein secretion, and the enzyme activity.

Xylan Concentration

An optimum substrate concentration ensures to provide the amount of carbon and energy required for biochemical and physiological events. Xylan concentration 0.1, and 1–4% (*w/v*) were used to determine the best possible xylan concentration for xylanase production. It was found that a xylan concentration of 1% (*w/v*) gave maximum xylanase activity (17.4 U/mL). The xylan concentration followed a trend of 1(100%) > 2(76.1%) > 0.1(53%) > 3(40.1%) > 4(9.4%). With increase in the xylan concentration from 1 to 4%, the xylanase activity decreased. A higher concentration of the substrate can cause substrate inhibition, and osmotic effects, resulting in reduced physiological and biochemical activity [25]. On the contrary, a lower concentration of the substrate is unable to provide the required amount of carbon and energy sources, and hence resulted in lower xylanase activity. Kumar and Satyanarayana (2014) also showed reduction in the xylanase activity with increase in the substrate concentration [34]. As the amount of wheat brain increased from 10 g to 40 g, the xylanase activity decreased from 4768 to 3878 U/g dbb [34]. Similar results were also observed by Bibi and coworkers (2014) where with increase in the xylan concentration from 0.5% to 3% (*w/v*) 50% of the xylanase activity was reduced [30].

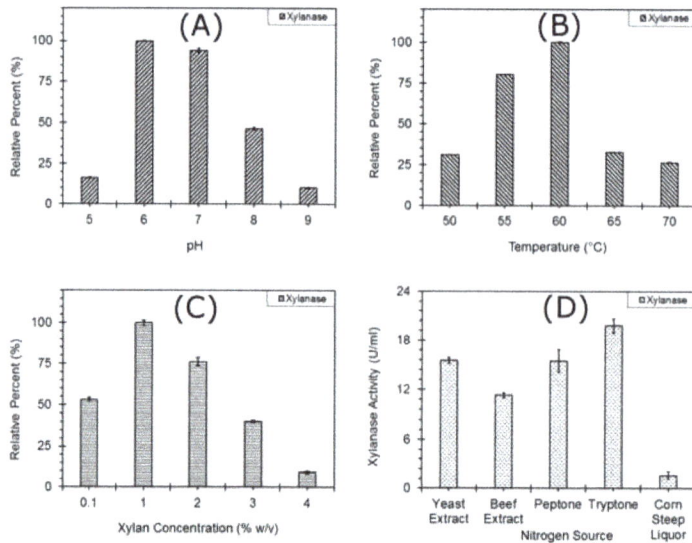

Figure 3. (**A**) The effect of growth pH profile on the xylanase activity of *Geobacillus* sp. strain DUSELR13. The maximum enzyme activity (8.54 U/mL) is taken as 100%; (**B**) the effect of growth temperature profile on the xylanase activity of *Geobacillus* sp. strain DUSELR13. The maximum enzyme activity (9.23 U/mL) is taken as 100%; (**C**) the effect of xylan concentration on the xylanase activity of *Geobacillus* sp. strain DUSELR13. The maximum enzyme activity (17.4 U/mL) is taken as 100%; and (**D**) the effect of nitrogen source on the xylanase activity of *Geobacillus* sp. strain DUSELR13. Values shown were the mean of triplicate experiments and the error bars represent ± SD of the means with *n* = 3.

Nitrogen Source

An optimal source of nitrogen is required for the synthesis of biomolecules e.g., proteins and DNA in a growing cell. Among the different organic nitrogen sources tested, tryptone was found to be having maximum effect on the xylanase production (Figure 3C). The order of xylanase production was tryptone (19.8 U/mL) > yeast Extract (15.5 U/mL) > peptone (15.5 U/mL) > beef extract (11.3 U/mL) > corn steep liquor (1.64 U/mL). The organic nitrogen sources are a good source of amino acids and vitamins and sources such as tryptone and yeast extract have been very commonly used in the fermentation process. Tryptone was also found to be the best organic nitrogen source for xylanase production with *Bacillus subtilis* sp. BS04 [11] and *Geobacillus thermolevarans* [35], however, yeast extract was preferred nitrogen source for xylanase production by *Actinomadura geliboluensis* [36], and beef extract for *Bacillus pumilis* SV-85S [7]. The corn steep liquor, a byproduct from wet corn milling, contains residual sugars from refining, and can be a good source of nitrogen, amino acids, vitamins, and minerals. However, in the present study it showed lower xylanase production in comparison to other nitrogen sources (Figure 3D), the reason for which is unknown. The nitrogen sources which support the bacterial growth better, aid in higher enzyme production. Thus, the presence of an optimal nitrogen source is an essential requirement for the better growth, and hence better enzyme production.

3.2.2. Response Surface Methodology

A quadratic model based on central composite design (CCD) was developed in response surface methodology (RSM) to find the optimum parameter value for the variables mentioned in Table 1. The CCD matrix of the independent variable in coded form is shown in Table 2 with the experimental and predicted values. The analysis of variance of the quadratic model showed that the model was significant based on the p values, and F test (Table 3). The F value for the model was 146.69, with p value < 0.0001 showing high significance of the model. The model values A, B, C, D, AB, AD, A^2, B^2, C^2, and D^2 were considered significant as the p values for these terms in the model was <0.05. A transformation of inverse square root was applied as suggested by the simulation to fit the data in the quadratic design model. The correlation coefficient value was close to 1, indicating high correlation between the experimental and predicted values (Figure 4). A difference of less than 0.2 was obtained between the Adjusted R^2: 0.9636 and Predicted R^2: 0.9538 with an adequate precision of 38.12. A low coefficient of variance value (C.V. %)-2.56 indicated adequate precision and applicability of the model to navigate the design space. The Lack of Fit value -1.61, indicated that the lack of fit was not significant relative to the pure error, and the developed model is fit for predicting the enzyme activity using these variables. The model equation obtained for xylanase activity prediction in coded terms is as follows:

$$\frac{1}{\sqrt{Y}} = 0.2001 + 0.0189\,A + 0.0070\,B + 0.004\,C - 0.0068\,D -$$
$$0.0024\,AB + 0.0006\,AC + 0.0017\,BC + 0.0025\,BD + 0.0016\,CD + \tag{5}$$
$$0.0248A^2 + 0.0231B^2 + 0.0232C^2 - 0.0181\,D^2$$

where A, B, C, D, and Y represent temperature, pH, xylan, tryptone, and xylanase activity respectively.

Table 3. ANOVA for the xylanase activity as a function of independent variables as obtained in the simulation. The main factor effects, and two factor interaction effects having an effect on the xylanase activity are shown.

Source	Sum of Squares	df	Mean Square	F Value	p-Vale	Prob > F
Model	0.0734	14	0.0052	146.69	<0.0001	significant
A-Temperature	0.0177	1	0.0177	496.16	<0.0001	
B-pH	0.0024	1	0.0024	68.47	<0.0001	
C-Xylan	0.0010	1	0.0010	27.96	<0.0001	
D-Tryptone	0.0023	1	0.0023	63.44	<0.0001	
AB	0.0003	1	0.0003	7.64	0.0075	
AC	0.0000	1	0.0000	0.5490	0.4615	
AD	0.0017	1	0.0017	48.19	<0.0001	
BC	0.0001	1	0.0001	3.89	0.0529	
BD	0.0003	1	0.0003	8.63	0.0046	
CD	0.0001	1	0.0001	3.37	0.0711	
A^2	0.0003	1	0.0003	8.57	0.0047	
B^2	0.0003	1	0.0003	7.45	0.0082	
C^2	0.0003	1	0.0003	7.51	0.0080	
D^2	0.0002	1	0.0002	4.55	0.0368	
Residual	0.0023	63	0.0000			
Lack of Fit	0.0005	10	0.0001	1.61	0.1297	Not significant
Pure Error	0.0017	53	0.0000			
Cor Total	0.0756	77				

R^2: 0.9702, adj R^2: 0.9636, predicted R^2: 0.9538, C.V. 2.56% adeq precision 38.13; df = degree of freedom; cor = correlation, Highly significant, $p \le 0.0001$; Significant, $p \le 0.05$; non-significant, $p \ge 0.05$.

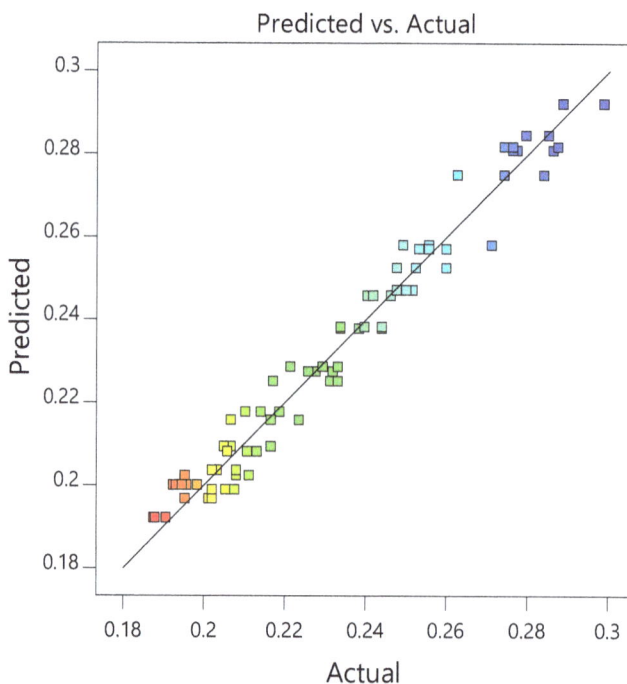

Figure 4. Surface response plot for predicted vs. actual values. The concentration of the data points near the straight line shows high correlation, and adequate precision. The blue color represents the minimum value, whereas the red color represents the maximum value. The rest of the color represents a range between the colors red, and blue. Each square represents the experimental value obtained from the runs.

The Figure 5A shows that a pH from the range 6.25 to 6.5 exhibited higher xylanase activity when the temperature was in the range of 57.5 to 60. Similarly Figure 5B showed that a xylan concentration of 10g/L gave higher enzymatic activity near the optimum temperature of 60 °C. As the temperature increased >60 °C the xylanase activity decreased with other variables. However, interesting results were obtained with tryptone which showed the increase in enzymatic activity at higher and lower concentration i.e., 1g/L, and 10 g/L, while interacting with the other variables (Figure 5C,E,F). A point prediction with numerical optimization showed a xylanase activity of 32.45 U/mL with 8.47 g/L xylan, 10 g/L tryptone, and pH 6.48 at a temperature of 58.9 °C. The post analysis run for the model verification gave an enzymatic activity of 31 U/mL, falling within >95% of the predicted value by the quadratic model depicting the usefulness, and precision of the model. Statistical optimization is a very powerful tool in determining the process parameter values for increasing the desired product yield. Khusro and Coworkers (2016) also obtain 3.7-fold increase in the xylanase production after statistical optimization [37]. In a similar fashion Kumar et al. (2017) increased the xylanase production from 61.09 U/mL to 119.91 U/mL via statistical optimization [38]. They also optimized the selected variables with OFAT approach, before using the statistical optimization later. A combination method of OFAT, factorial design, and/or placket burman designs can help in obtaining the preliminary data which can be used for optimization later. In the present study with statistical optimization, 5.2 times increase in the xylanase activity was observed.

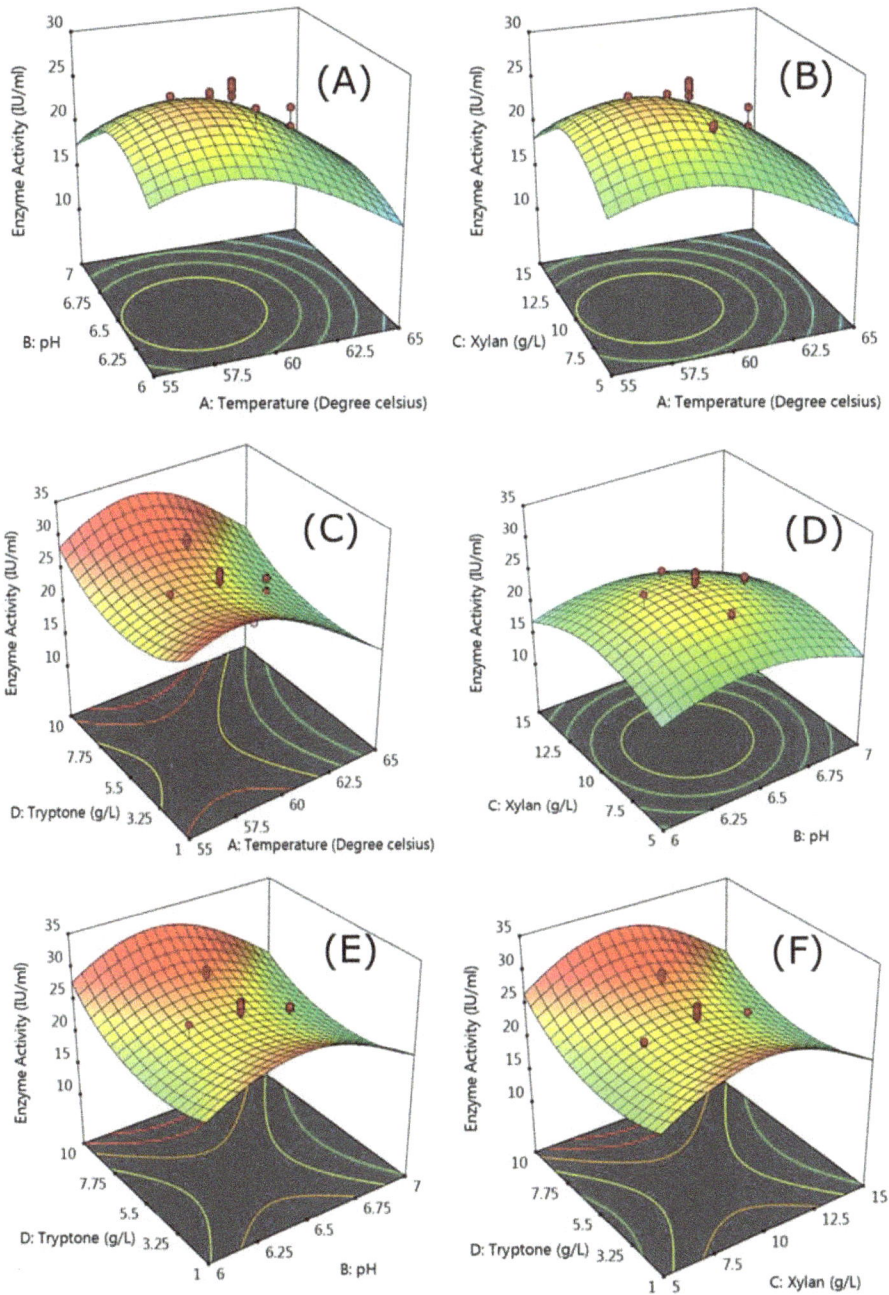

Figure 5. Response surface 3D response plots showing interaction between (**A**) temperature and pH, (**B**) temperature and xylan, (**C**) temperature and tryptone, (**D**) pH and xylan, (**E**) pH and tryptone, and (**F**) xylan and tryptone. The green color represents the minimum value whereas the red color represents the maximum values. The points above or below the response surface area respectively represent the values which are higher or lower than the predicted response value for the enzymatic activity.

3.3. Enzyme Characterization

The characterization studies were carried to find the biochemical properties of the crude enzyme. For SDS-PAGE and zymogram analysis supernatant obtained after the centrifugation of sample was used. The supernatant was concentrated using the 10 kDa centrifuge filter. The SDS-PAGE and zymogram analysis is shown in the Figure 6. The PAGE analysis of total cell proteins showed several bands (Lane A), in comparison to the supernatant collected after the centrifugation where fewer bands were observed (Lane B). A thick prominent band was observed between the molecular weight 37 and 50 kDa in both the lanes A and B. The band position suggested that the molecular weight of the protein was closer to 45 kDa. In addition, a clear single smear against a red background falling in the similar weight category was observed during the zymogram analysis, when the PAGE gels were overlaid with xylan (Lane C). The two analyses confirmed that the band corresponded to the protein xylanase. The past literature studies had shown that several xylanases from the *Geobacillus* spp. and other microorganisms had molecular weight similar to the xylanase under study. The xylanase from *Geobacillus thermodenitrificans* A333 had a molecular weight of 44 kDa [4], and the molecular weight of the xylanase from *Geobacillus thermolevorans* was ~45 kDa [6]. A thermostable xylanase from fungus *Malbranchea pulchella* had a molecular weight of 41.6 kDa [8], while another thermostable xylanase from *Humicola insolens* Y1 [39], had a molecular weight of 44 kDa. However, the molecular weight of a highly thermostable xylanase obtained from the *Geobacillus* sp. strain WSUCF1 was 37 kDa [19]. The SDS-PAGE analysis of the crude xylanase obtained from the growth of DUSEL R13 on PCG showed the presence of several bands of molecular weight ~160, 45, and 20 kDa after silver staining (lane D). The silver staining of the PAGE gels allow to identify even very small concentrations of the proteins because of its greater sensitivity [40]. The presence of bands other than 45 kDa band showed that additional proteins were expressed when DUSELR13 was grown on PCG, which were absent when grown on the xylan. Thus a concerted action of several proteins aid in the growth of the DUSELR13 on the lignocellulosic biomass.

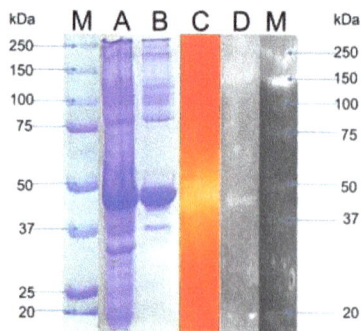

Figure 6. SDS PAGE, and zymogram analysis of the xylanase resolved on the 12.5% polyacrylamide gel. Lane M: Biorad precision plus standard protein marker; lane A: Crude enzyme; lane B Concentrated supernatant; lane C: Zymogram for the xylanase; lane D: SDS PAGE analysis of the xylanase grown on PCG with silver staining.

Further characterization of the crude xylanase revealed that the maximum xylanase activity was observed in the 100 mM phosphate buffer at pH 7.0, with activity in broad pH range from 5.0–9.0. The xylanase from DUSELR13 retained >45% activity at a pH range of 4.0–9.0 and >82% activity in a pH range of 6.0–8.0 9 (Figure 7A). Several other *Geobacillus* spp. had a variable pH_{opt} and a wide pH range for the thermostable xylanases. *Geobacillus* sp. strain WSUCF1 had a pH_{opt} of 6.5 and range: 4.5–8.5 [5]; *Geobacillus* sp. strain TC-W7: pH_{opt}: 8.2 and range: 5.2–10.2 [41]; *Geobacillus thermonitirificans* strain A333: pH_{opt}: 7.5, and range: 7.5–10.0 [4]; and *Geobacillus thermodenitrificans* strain T12: pH$_{opt}$: 6.0, and range: 3.0–9.0 [16] respectively. The broad pH range applicability of the xylanase under current study can be very advantageous for biofuel, paper, and pulp, xylitol, and many other important industries which require direct use of xylanase under different pH conditions [3,42]. Table 4 shows the comparison of characteristics of several thermostable xylanases.

To find the optimum temperature for the xylanase from DUSELR13, the activity was assayed at different temperatures (25, 37, 50–100 °C with an increment of 5 °C). The characterization study for the temperature revealed that 75 °C was the T$_{opt}$ for the xylanase from DUSELR13 (Figure 7B). The xylanase retained more than 60% of the enzyme activity at temperatures between 50 and 85 °C. Around 50% of the tested points fell between temperature ranges of 60–85 °C and accounted for more than 75% of the relative xylanase activity. The lowest relative xylanase activity was observed at 100 °C followed by 25 > 95 > 37 °C suggesting that 25 and 37 °C were too low for the enzyme activity, while 95 and 100 °C were too high temperatures for the enzyme activity. Several thermostable xylanases have T$_{opt}$ similar or closer to strain DUSELR13 e.g., *Geobacillus* sp. strain TC-W7 had a T$_{opt}$ of 75 °C [41]; *Geobacillus thermonitirificans* strain A333, *Geobacillus* sp. strain WSUCF 1 showed a T$_{opt}$ of 70 °C [4], and *Malbranchea pulchela* had a T$_{opt}$ of 80 °C [8]. The results support the ability of the DUSELR13 xyalanse to perform in the industrial process which operate at high temperatures.

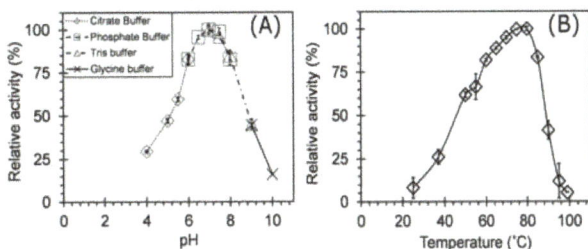

Figure 7. The effect of pH (**A**), and temperature (**B**) on the xylanase activity of *Geobacillus* sp. strain DUSELR13. The maximum enzyme activity is taken as 100%. The values shown are the mean of triplicate experiments and the error bars represent ± SD of the means with $n = 3$.

The thermostability of the DUSELR13 xylanase was determined by incubating it at different temperatures. The xylanase retained >50% of the activity at temperatures 50, 60, and 70 °C for 48, 38, and 13 days respectively (Figure 8A,B). The xylanase was highly thermostable and retained >23.5% of activity at 50 °C after 75 days. At 60 and 70 °C respectively, it retained 21 and 6% activity after 75 and 44 days. However, when the xylanase was incubated at 80, 90, and 100 °C respectively, resulted in the loss of 50% of its activity in 110, 18, and 5 min. The thermostable xylanase from *Geobacillus* sp. strain WSUCF1 retained 70% of its activity after 19 days at 50 °C [5]. Interestingly, other thermostable xylanases did not retain their activity for long time periods. The xylanase from *Geobacillus thermodenitrificans* strain A333 retained 50% of the activity for 60 min at 70 °C [4], *Geobacillus* sp. strain TC-W7 retained 80% of the activity for 90 min at 70 °C [41], and *Geobacillus thermodenitrificans* strain T12 retained only 18% of the initial activity for 60 min at 70 °C [16]. The xylanases XynA1 and XynA2 from *Geobacillus thermodinitrificans* strain NG80-2 reported by Huang and coworkers showed >50% activity at 65 °C for 24 h [43]. However, as the temperature was increased to 75 °C XynA1 and XynA2 respectively showed only 15% activity after 4 h and 15 min [43]. The other xylanase with such high thermostability had been reported from *Geobacillus* sp. strain WSUCF1, which retained 50% of enzymatic activity for 19 and 12 days respectively at 60 and 70 °C [5]. Thus, the xylanase from *Geobacillus* sp. strain DUSELR13 is highly thermostable among the known xylanases and can be of promising use in high temperature bioprocesses that run for longer durations.

Table 4. The characteristics of various thermostable xylanases.

Organism	Type	Enzyme Activity (U/mL)	M.Wt. [#] (kDa)	T_{opt} (°C)	pH_{opt}	Thermostability	Reference
Geobacillus sp. strain DUSELR13	Wild type	31.0	~45	75	7.0	$t_{1/2}$ = 13 days at 70 °C	This Study
Geobacillus sp. strain WSUCF1	Wild type	23.8	37	70	6.5	$t_{1/2}$ = 12 days at 70 °C	[5]
Geobacillus thermoleovorans	Wild type	26.52	~45	80	8.5	$t_{1/2}$ = 50–55 min at 80 °C	[6]
Geobacillus thermodinitrificans strain A333	Wild type	0.02	44	70	7.5	$t_{1/2}$ = 60 min at 70 °C	[4]
Geobacillus thermodinitrificans strain NG80-2 XynA1	Recombinant	40.4	N.A.*	70	7.6	$t_{1/2}$ = 28 h at 65 °C	[43]
Geobacillus thermodinitrificans strain NG80-2 XynA2	Recombinant	36.8	N.A.	70	6.5	$t_{1/2}$ = 26 h at 65 °C	[43]
Geobacillus sp. strain TF16	Recombinant	7.92	38.9	55	8.5	$t_{1/2}$ = 10 min at 70 °C	[17]
Bacillus amyloliquefaciens	Wild type	48.5	~50	50	9.0	$t_{1/2}$ = 45 min at 70 °C	[23]
Bacillus pumilus SV-85S	Wild type	2995	N.A.	50	6.0	$t_{1/2}$ = 25 min at 70 °C	[7]
Cladosporium oxysporum GQ-3	Wild type	55.92	N.A.	50	8.0	$t_{1/2}$ = 30 min at 70 °C	[44]
Malbranchea pulchela	Wild type	3.0	49	80	5.5	$t_{1/2}$ = 260 min at 70 °C	[8]

[#] M.Wt. Molecular weight. * N.A. Not available.

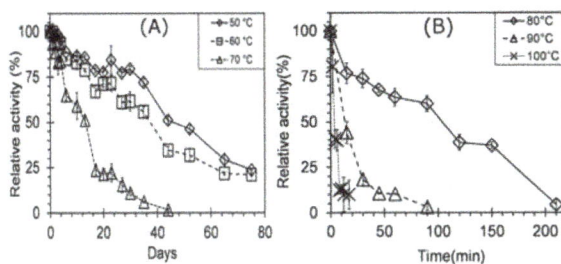

Figure 8. Thermal stability profiling of the xylanase produced by *Geobacillus* sp. strain DUSELR13 at different temperatures (**A**) 50–70 °C and (**B**) 80–100 °C. The enzyme activity is expressed as the percentage of the initial activity that is taken as 100%. The values are mean of triplicate experiments and the error bars represent ± SD of the means with $n = 3$.

The enzyme activity is often affected by the presence of metal ions in the complex substrate (e.g., food waste, agricultural and forestry residues), or the medium. The metal ions either serve as the catalytic center of the enzymes, acts as the bridge to bring the enzyme and substrate in close proximity or maintain other physiological functions [45]. Hence, the effect of different metal ions on the xylanase activity was determined at 1mM and 10mM concentrations (Figure 9). The xylanase activity on the crude xylanase grown with xylan as the substrate without any external metal ion addition was taken as 100%, and all the relative activities were measured against it. At 1 mM concentration the addition of Cu^{+2}, Zn^{+2}, K^+, and Fe^{+2} showed an increase in the endoxylanase activity, but Na^+, Mg^{+2}, and Ca^{+2} resulted in the decrease of xylanase activity. However, at 10mM Ca^{+2}, Zn^{+2}, Na^+, and Mg^{+2} had a positive effect on the xylanase activity, but Cu^{+2}, K^+, and Fe^{+2} reduced the activity of the xylanase. Co^{+2} and Mn^{+2} decreased the xylanase activity at both the concentrations, but the observed decrease was more with Mn^{+2} as the concentration increased from 1 mM to 10 mM. Maximum increase in the xylanase activity, with the increase in the metal ion concentration from 1mM to 10mM, was observed with addition of $CaCl_2$, where the relative activity increased from 94.6% to 105.7%. On the contrary, maximum decrease was observed with the addition of $CuSO_4$, where it decreased from 101.4% to 70.7%.

Several literature studies have shown varying effects of metal ions on the xylanase activity at different concentrations. For xylanase produced by *Geobacillus thermodenitrificans* strain A333 with increase in the concentration from 1mM to 10mM, the addition of Na^+, Cu^{+2}, Mg^{+2}, and Fe^{+2} respectively decreased the xylanase activity from 122% to106%, 79% to 76%, 72% to 70% and 80% to 3%; whereas addition of K^+ increased the xylanase activity from 142% to 145%, and Ca^{+2} and Zn^{+2} had no effect [4]. Two xylanases XynA1, and XynA2 obtained from *Geobacillus thermodenitrificans* strain NG80-2 showed different behavior towards the Zn^{+2} metal ion. In the presence of 5 mM Zn^{+2} the activity of XynA1 increased, whereas with XynA2 for it decreased [43]. The activity of xylanase produced by *Cladosporium oxysporum* increased after the addition of Mg^{+2} and Ca^{+2} (1 mM), but was inhibited after adding Cu^{+2} (1 mM) [44]. Similarly, increased enzymatic activity was observed by the addition of Ca^{+2}, Cu^{+2}, Co^{+2}, and Mn^{+2} (1 mM), but the activity was inhibited by the addition of Cd^{+2}, Hg^{+2}, Ba^{+2}, Mg^{+2}, Fe^{+2} (1 mM) in case of xylanase produced by *Caldiprobacter algeriensis* sp. no. strain TH7C1 [46]. Thus the different concentrations of metal ions have variable effect on the xylanase activity, even for enzymes obtained from similar organism, and hence the metal ion concentration needs to be tightly regulated.

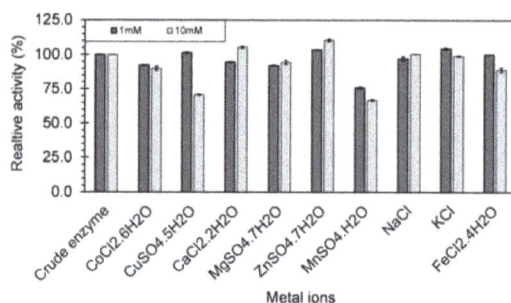

Figure 9. The effect of metal ion on the xylanase activity of *Geobacillus* sp. strain DUSELR13. The activity of the crude xylanase is taken as 100% (30.6 U/mL) and effect of the metal ions on the xylanase activity is expressed as the percentage of the initial activity. The values are mean of triplicate experiments and the error bars represent ± SD of the means with $n = 3$.

3.4. Hydrolysis of the Beechwood Xylan

The thermostable enzymes have been considered with potential to improve the current high temperature fermentative industrial processes. The hydrolytic potential of the xylanase from DUSELR13 was compared to its commercial counterpart Cellic Htec2 [47] and Accellerase XY at their operationally stable conditions. Accellerase XY has an operational stability in the range of 50–75 °C, and pH 4.5 and 7.0 [48], and Cellic HTec 2 has also showed activity at a temperature range of 50–75 °C, and pH 4.5 and 6.75. Figure 10 shows the hydrolysis of the beechwood xylan by the DUSELR13 crude xylanase, and its commercial counterparts. The percentage conversions were calculated on the basis of reducing sugars released from 1 g beechwood xylan. The maximum amount of sugar that can be obtained with xylan hydrolysis under ideal conditions where complete hydrolysis occurs was taken as 100%. The xylanase from DUSELR13 converted more xylan into sugars at high temperature whereas the commercial counterparts showed higher percentage of xylan hydrolysis at relatively lower temperatures. At 75 °C, DUSEL R13 xylanase hydrolyzed 65.4% of the beechwood xylan, whereas Cellic HTec2, and Accellerase XY respectively showed only 41.8%, and 20% hydrolysis. However, at lower temperature the hydrolysis by DUSELR13 was lower in comparison to that obtained with Cellic HTec2, and Accellerase XY. At 50 °C, the xylanase from DUSELR13 showed only 30% xylan hydrolysis in comparison to 74%, and 55.8% respectively by Cellic HTec2 and Accellerase XY. The hydrolysis rate for DUSELR13 increased significantly from 55 to 60 °C, but the change was not that significant from 60 to 75 °C. This was also observed during the T_{opt} for the xylanase activity in Figure 5B, where the increase in the xylanase activity showed similar trend. On the other hand, the xylanase activity for Cellic HTec2 reduced significantly from 55 to 75 °C. With increase in temperature to 75 °C, the xylan hydrolysis by DUSELR13 increased 1.52 times to 65.4%, however, for Cellic HTec2, and Accellerase XY respectively it reduced 1.77 times to 41.8%, and 2.79 times to 20%.

Bhalla and coworkers (2015) also reported that the highly thermostable xylanase from *Geobacillus* sp. strain WSUCF1 performed better than the commercial counterparts at higher temperatures. At 70 °C, WSUCF1 xylanase yielded higher conversions of 68.9% as compared to Cellic HTec2 (49.4%) and Accellerase XY (28.92%). However, the commercial counterparts also outperformed WSUCF1 xylanase at lower temperatures (T = 50 °C), suggesting that the xylanase from WSUCF1 also performed better at high temperature processes [5]. Similarly, the thermostable xylanase obtained from *Thermotoga themarum* hydrolyzed 87% of the beechwood xylan at 85 °C after 3 h of incubation [10]. The present hydrolysis results show that the xylanase from DUSELR13 was better than the commercial counterpart at high temperatures. The high temperature activity, and thermostability of the crude enzyme cocktail from DUSELR13 will be suitable for use in several high temperature industrial processes that involve direct lignocellulosic biomass utilization.

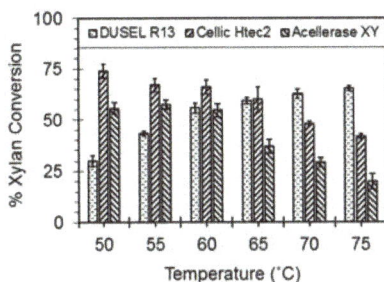

Figure 10. Hydrolysis of the beechwood xylan by DUSELR13, and comparative analysis of the hydrolytic activity with Cellic HTec2 and Accellerase XY. The percentage xylan conversion is calculated on the basis of the sugar released from 1 g of xylan, with complete hydrolysis of xylan considered as 100%. The values are mean of triplicate experiments and the error bars represent ± SD of the means with *n* = 3.

3.5. Enzyme Production with Lignocellulosic Biomass

The use of lignocellulosic biomass for enzyme production is economically advantageous due to its cheap and ubiquitous availability. DUSELR13, was able to use both untreated and mechanically pretreated lignocellulosic biomass for xylanase production. The order of the endoxylanase activity with different carbon sources as the substrate was: xylan (100%: 30.2 U/mL) > pretreated prairie cordgrass—PPCG (67.9%) > untreated prairie cordgrass—PCG (59.5%) > pretreated corn stover—PCS (56%) > untreated corn stover—CS (48.6%) (Figure 11). DUSELR13 showed more enzyme activity when it grew on the mechanically pretreated lignocellulosic biomass, and the xylan, as these sources provide easily accessible carbon for growth in comparison to the untreated lignocellulosic biomass. However, its capability to use the untreated substrates substantiates its industrial potential for enzyme production, using the cheaper carbon sources without pretreatment. The xylanase production using lignocellulosic biomass had been done previously also with *Streptomyces termitum* using bagasse, straw sugarcane, and cocoa pod husk [9] for xylanase production; *Geobacillus* sp. stain WSUCF1 utilizing corn stover and prairie cordgrass for the same [5]; and *Caldicellulosiruptor owensensis* using corn cob [49]. Corn stover and prairie cordgrass used in this study are inexpensive lignocellulosic biomasses that do not compete with food resources for human. CS had been extensively used in several bioprocesses: biobutanol, bioflocculant, bioethanol; and lipid [50]. Prairie cordgrass also had been previously reported in the production of enzymes, and biofuels [5,25]. The availability of a cheaper resource viz. lignocellulosic biomass that can be used in the industrial bioprocesses for providing fermentable sugars is very important. The organisms that can utilize such substrates hold significance, and DUSELR13 is one such strain that can be put into use for this purpose.

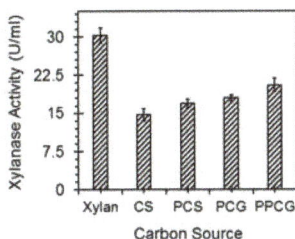

Figure 11. Enzyme production with untreated and mechanically treated corn stover and prairie cordgrass with comparison to xylan under optimized conditions. The values are mean of triplicate experiments and the error bars represent ± SD of the means with *n* = 3.

4. Ethanol Production with Lignocellulosic Biomass

Five percent (w/v) of thermo-mechanically treated insoluble corn stover and prairie cordgrass [25] were used for ethanol production in 500-mL Erlenmeyer flasks. The flasks were inoculated with *Geobacillus* sp. strain DUSELR13 at time 0 h, and with *Geobacillus thermoglucosidasius* at time 36 h. An ethanol yield of 3.53 g/L (0.21 g/g PCG utilized) and 3.72 g/L (0.2 g/g CS utilized), respectively were obtained with prairie cordgrass and corn stover (Figure 12). The material balance showed that 90.5 and 91.6% of the mass was recovered during the process, converting 62.4 and 57.4% energy respectively in prairie cordgrass and corn stover to ethanol (Table 5). The ethanol productivity obtained was 0.059 (g/L/h) and 0.62 (g/L/h) with Corn Stover and prairie cordgrass, respectively.

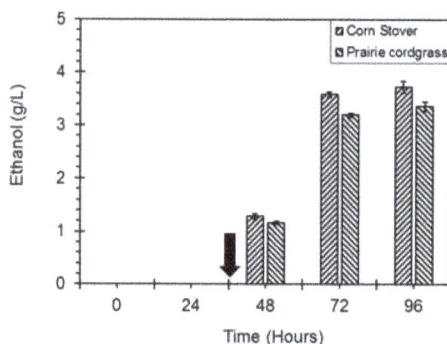

Figure 12. Ethanol production with co-culture of *Geobacillus* sp. strain DUSELR13, and *Geobacillus thermoglucosidasius*. The black arrow represents the point of addition of *Geobacillus thermoglucosidasius*. The values are mean of triplicate experiments and the error bars represent ± SD of the means with $n = 3$.

Table 5. Mass balance for ethanol production using PCG and CS.

Substrate/Metabolite	Mass Balance			
	PCG	%	CS	%
	Mass		Mass *	
Substrate Utilized	13.8	100	16.7 [#]	100
Biomass	-	-	-	-
Acetate	1.46	10.57971	1.68	10.05988
Lactate	4.12	29.85507	5.66	33.89222
Propionate	0.37	2.681159	0.64	3.832335
Ethanol	3.35	24.27536	3.72	22.27545
CO_2	3.2	23.18841	3.6	21.55689
Total	12.5		11.7	
Recovery		90.57971		91.61677

* g/L. [#] Amount of substrate utilized.

Geobacillus sp. strain DUSELR13 had been previously known for producing thermostable cellulase [24], and this study presents the production of a thermostable endoxylanase by this organism. The presence of cellulase and hemicellulases can aid in the hydrolysis of the lignocellulosic biomass, producing the sugars required for fermentation. After the addition of *Geobacillus thermoglucosidasius*, the ethanol production was observed. Along with ethanol, acetate, lactate, and propionate were also produced as the main fermentation products showing that the fermentation was mixed acid type. The ethanol production observed with corn stover was higher than that observed with prairie cordgrass. This can be attributed to the compositional difference among the two substrates. Corn stover has lower lignin content (16%) as compared to prairie cordgrass (21%) [51]. *Geobacillus thermoglucosidasius* known for ethanol production lacks lignocellulosic biomass hydrolyzing enzymes [13], and hence cannot hydrolyze the lignocellulosic biomass, whereas on the other hand *Geobacillus* sp. strain DUSELR13 is unable to produce ethanol. Hence, the ethanol production was observed only when the two cultures were grown together. None of the monocultures showed ethanol production with lignocellulosic biomass. Park et al. (2012) also observed that the co-culture of cellulase producing *Acremonium cellulolyticus* and ethanologen *Saccharomyces cerevisiae* produced 8.7 g/L with 5% (w/v) steam solka-floc as susbtrate [52]. Similar results were also observed by Miyazaki et al. (2008) where co-cultures of aerobic cellulolytic *Geobacillus* sp. strain kpuB3 aided in ethanol production by anaerobic hemicellulolytic *Thermoanerobacterium* sp. strain kpu04 using bean curd refuse [53]. The co-culture of *Geobacillus* sp. strain kpuB3 and *Thermoanerobacterium* sp. strain kpu04 produced a total of 1.26 g/L ethanol whereas the monoculture of *Thermoanaerobacterium* sp. strain kpu04 produced only 0.37 g/L ethanol.

5. Conclusions

A wild-type thermophilic microorganism, like *Geobacillus* sp. strain DUSELR13 which can utilize the lignocellulosic biomass and produce thermostable enzymes, was studied in the present work. The DUSELR13 produced 6 U/mL of a highly thermostable xylanase, which after optimization was increased to 31 U/mL. The xylanase showed better hydrolytic potential than the commercial counterparts with xylan at higher temperatures. The enzyme was active over a wide range of pH and temperatures and had the potential to utilize and produce the xylanase using lignocellulosic biomass (e.g., corn stover and prairie cordgrass). The hydrolytic potential of the xylanase with lignocellulosic biomass helped to hone its characteristics for bioethanol production in co-culture studies with *Geobacillus thermoglucosidasius*. The DUSELR13 strain has already been researched for cellulase, and now this study further substantiates its use for the lignocellulosic biomass hydrolysis at higher temperatures for future biotechnological improvements.

Author Contributions: M.B. and V.R.K. performed the experimental and analysis work. M.B. conducted the statistical optimization, data curation and validation. M.B. wrote the paper. R.K.S. contributed in writing, reviewing, editing, and supervision.

Funding: This research was supported by the US Air Force under Biological Waste to Energy Project (FA4819-14-C-0004). Mohit Bibra also acknowledges the financial support in the form of "Proof of Concept" provided by the South Dakota Governor's Office of Economic Development. The authors gratefully acknowledge the financial support provided by National Aeronautics and Space Administration, Established Program to Stimulate Competitive Research under award No. NNX13AB25A. The support from the Department of Chemical and Biological Engineering at the South Dakota School of Mines and Technology is gratefully acknowledged.

Conflicts of Interest: The authors declare no conflict of interest.

References

1. Maurya, D.P.; Singla, A.; Negi, S. An overview of key pretreatment processes for biological conversion of lignocellulosic biomass to bioethanol. *3 Biotech* **2015**, *5*, 597–609. [CrossRef] [PubMed]
2. Kumar, S.; Bhalla, A.; Bibra, M.; Wang, J.; Morisette, K.; Subramanian, M.R.; Salem, D.; Sani, R.K. Thermophilic biohydrogen production: Challenges at the industrial scale. In *Bioenergy: Opportunities and Challenges*; Krishnaraj, R.N., Yu, J.S., Eds.; CRC Press, Taylor and Francis: Boca Raton, FL, USA, 2015; p. 382.
3. Bhalla, A.; Bansal, N.; Kumar, S.; Bischoff, K.M.; Sani, R.K. Improved lignocellulose conversion to biofuels with thermophilic bacteria and thermostable enzymes. *Bioresour. Technol.* **2013**, *128*, 751–759. [CrossRef] [PubMed]
4. Marcolongo, L.; La Cara, F.; Morana, A.; Di Salle, A.; Del Monaco, G.; Paixão, S.M.; Alves, L.; Ionata, E. Properties of an alkali-thermo stable xylanase from *Geobacillus thermodenitrificans* A333 and applicability in xylooligosaccharides generation. *World J. Microbiol. Biotechnol.* **2015**, *31*, 633–648. [CrossRef] [PubMed]

5. Bhalla, A.; Bischoff, K.M.; Sani, R.K. Highly thermostable xylanase production from A thermophilic *Geobacillus* sp. Strain WSUCF1 utilizing lignocellulosic biomass. *Front. Bioeng. Biotechnol.* **2015**, *3*, 84. [CrossRef] [PubMed]

6. Verma, D.; Satyanarayana, T. Cloning, expression and applicability of thermo-alkali-stable xylanase of *Geobacillus thermoleovorans* in generating xylooligosaccharides from agro-residues. *Bioresour. Technol.* **2012**, *107*, 333–338. [CrossRef] [PubMed]

7. Nagar, S.; Gupta, V.K.; Kumar, D.; Kumar, L.; Kuhad, R.C. Production and optimization of cellulase-free, alkali-stable xylanase by *Bacillus pumilus* SV-85S in submerged fermentation. *J. Ind. Microbiol. Biotechnol.* **2010**, *37*, 71–83. [PubMed]

8. Ribeiro, L.F.; De Lucas, R.C.; Vitcosque, G.L.; Ribeiro, L.F.; Ward, R.J.; Rubio, M.V.; Damásio, A.R.; Squina, F.M.; Gregory, R.C.; Walton, P.H. A novel thermostable xylanase GH10 from *Malbranchea pulchella* expressed in *Aspergillus nidulans* with potential applications in biotechnology. *Biotechnol. Biofuels* **2014**, *7*, 115. [CrossRef] [PubMed]

9. de Sales, A.N.; de Souza, A.C.; Moutta, R.d.O.; Ferreira-Leitão, V.S.; Schwan, R.F.; Dias, D.R. Use of lignocellulose biomass for endoxylanase production by *Streptomyces termitum*. *Prep. Biochem. Biotechnol.* **2017**, *47*, 1–8. [CrossRef] [PubMed]

10. Shi, H.; Zhang, Y.; Li, X.; Huang, Y.; Wang, L.; Wang, Y.; Ding, H.; Wang, F. A novel highly thermostable xylanase stimulated by Ca^{2+} from *Thermotoga thermarum*: cloning, expression and characterization. *Biotechnol. Biofuels* **2013**, *6*, 26. [PubMed]

11. Irfan, M.; Asghar, U.; Nadeem, M.; Nelofer, R.; Syed, Q. Optimization of process parameters for xylanase production by *Bacillus* sp. in submerged fermentation. *J. Radiat. Res. Appl. Sci.* **2016**, *9*, 139–147. [CrossRef]

12. Carlson, C.; Singh, N.K.; Bibra, M.; Sani, R.K.; Venkateswaran, K. Pervasiveness of UVC 254-resistant *Geobacillus* strains in extreme environments. *Appl. Microbiol. Biotechnol.* **2018**, *102*, 1869–1887. [CrossRef] [PubMed]

13. Cripps, R.; Eley, K.; Leak, D.J.; Rudd, B.; Taylor, M.; Todd, M.; Boakes, S.; Martin, S.; Atkinson, T. Metabolic engineering of *Geobacillus thermoglucosidasius* for high yield ethanol production. *Metab. Eng.* **2009**, *11*, 398–408. [CrossRef] [PubMed]

14. Raita, M.; Ibenegbu, C.; Champreda, V.; Leak, D.J. Production of ethanol by thermophilic oligosaccharide utilising *Geobacillus thermoglucosidasius* TM242 using palm kernel cake as a renewable feedstock. *Biomass Bioenergy* **2016**, *95*, 45–54. [CrossRef]

15. Shulami, S.; Shenker, O.; Langut, Y.; Lavid, N.; Gat, O.; Zaide, G.; Zehavi, A.; Sonenshein, A.L.; Shoham, Y. Multiple regulatory mechanisms control the expression of the *Geobacillus stearothermophilus* gene for extracellular xylanase. *J. Biolog. Chem.* **2014**, *289*, 25957–25975. [CrossRef] [PubMed]

16. Daas, M.J.; Martínez, P.M.; van de Weijer, A.H.; van der Oost, J.; de Vos, W.M.; Kabel, M.A.; van Kranenburg, R. Biochemical characterization of the xylan hydrolysis profile of the extracellular endo-xylanase from *Geobacillus thermodenitrificans* T12. *BMC Biotechnol.* **2017**, *17*, 44. [CrossRef] [PubMed]

17. Cakmak, U.; Ertunga, N.S. Gene cloning, expression, immobilization and characterization of endo-xylanase from *Geobacillus* sp. TF16 and investigation of its industrial applications. *J. Mol. Catal. B Enzym.* **2017**, *133*, 288–298. [CrossRef]

18. Irfan, M.; Gonzalez, C.F.; Raza, S.; Rafiq, M.; Hasan, F.; Khan, S.; Shah, A.A. Improvement in thermostability of xylanase from *Geobacillus thermodenitrificans* C5 by site directed mutagenesis. *Enzyme Microb. Technol.* **2018**, *111*, 38–47. [CrossRef] [PubMed]

19. Bhalla, A.; Bischoff, K.M.; Uppugundla, N.; Balan, V.; Sani, R.K. Novel thermostable endo-xylanase cloned and expressed from bacterium *Geobacillus* sp. WSUCF1. *Bioresour. Technol.* **2014**, *165*, 314–318. [CrossRef] [PubMed]

20. Zhang, Z.-G.; Yi, Z.-L.; Pei, X.-Q.; Wu, Z.-L. Improving the thermostability of *Geobacillus stearothermophilus* xylanase XT6 by directed evolution and site-directed mutagenesis. *Bioresour. Technol.* **2010**, *101*, 9272–9278. [CrossRef] [PubMed]

21. Yu, H.; Yan, Y.; Zhang, C.; Dalby, P.A. Two strategies to engineer flexible loops for improved enzyme thermostability. *Sci. Rep.* **2017**, *7*, 41212. [CrossRef] [PubMed]

22. Zhang, X.-F.; Yang, G.-Y.; Zhang, Y.; Xie, Y.; Withers, S.G.; Feng, Y. A general and efficient strategy for generating the stable enzymes. *Sci. Rep.* **2016**, *6*, 33797. [CrossRef] [PubMed]

23. Kumar, S.; Haq, I.; Prakash, J.; Singh, S.K.; Mishra, S.; Raj, A. Purification, characterization and thermostability improvement of xylanase from *Bacillus amyloliquefaciens* and its application in pre-bleaching of kraft pulp. *3 Biotech* **2017**, *7*, 20. [CrossRef] [PubMed]

24. Rastogi, G.; Bhalla, A.; Adhikari, A.; Bischoff, K.M.; Hughes, S.R.; Christopher, L.P.; Sani, R.K. Characterization of thermostable cellulases produced by *Bacillus* and *Geobacillus* strains. *Bioresour. Technol.* **2010**, *101*, 8798–8806. [CrossRef] [PubMed]

25. Bibra, M.; Kumar, S.; Wang, J.; Bhalla, A.; Salem, D.R.; Sani, R.K. Single Pot Bioconversion of Prairie Cordgrass into Biohydrogen by Thermophiles. *Bioresour. Technol.* **2018**, *266*, 232–241. [CrossRef] [PubMed]

26. Bailey, M.J.; Biely, P.; Poutanen, K. Interlaboratory testing of methods for assay of xylanase activity. *J. Biotechnol.* **1992**, *23*, 257–270. [CrossRef]

27. Laemmli, U.K. Cleavage of structural proteins during the assembly of the head of bacteriophage T4. *Nature* **1970**, *227*, 680–685. [CrossRef] [PubMed]

28. Wang, J.; Bibra, M.; Venkateswaran, K.; Salem, D.R.; Rathinam, N.K.; Gadhamshetty, V.; Sani, R.K. Biohydrogen production from space crew's waste simulants using thermophilic consolidated bioprocessing. *Bioresour. Technol.* **2018**, *255*, 349–353. [CrossRef] [PubMed]

29. Purohit, A.; Rai, S.K.; Chownk, M.; Sangwan, R.S.; Yadav, S.K. Xylanase from *Acinetobacter pittii* MASK 25 and developed magnetic cross-linked xylanase aggregate produce predominantly xylopentose and xylohexose from agro biomass. *Bioresour. Technol.* **2017**, *244*, 793–799. [CrossRef] [PubMed]

30. Bibi, Z.; Ansari, A.; Zohra, R.R.; Aman, A.; Ul Qader, S.A. Production of xylan degrading endo-1,4-β-xylanase from thermophilic *Geobacillus stearothermophilus* KIBGE-IB29. *J. Radiat. Res. Appl. Sci.* **2014**, *7*, 478–485. [CrossRef]

31. Adhyaru, D.N.; Bhatt, N.S.; Modi, H.A. Enhanced production of cellulase-free, thermo-alkali-solvent-stable xylanase from *Bacillus altitudinis* DHN8, its characterization and application in sorghum straw saccharification. *Biocatal. Agric. Biotechnol.* **2014**, *3*, 182–190. [CrossRef]

32. Ellis, J.T.; Magnuson, T.S. Thermostable and alkalistable xylanases produced by the thermophilic bacterium *Anoxybacillus flavithermus* TWXYL3. *ISRN Microbiol.* **2012**, *2012*, 517524. [CrossRef] [PubMed]

33. Forster, B.M.; Marquis, H. Protein transport across the cell wall of monoderm Gram-positive bacteria. *Mol. Microbiol.* **2012**, *84*, 405–413. [CrossRef] [PubMed]

34. Kumar, V.; Satyanarayana, T. Production of endoxylanase with enhanced thermostability by a novel polyextremophilic *Bacillus halodurans* TSEV1 and its applicability in waste paper deinking. *Process Biochem.* **2014**, *49*, 386–394. [CrossRef]

35. Sharma, A.; Adhikari, S.; Satyanarayana, T. Alkali-thermostable and cellulase-free xylanase production by an extreme thermophile *Geobacillus thermoleovorans*. *World J. Microbiol. Biotechnol.* **2007**, *23*, 483–490. [CrossRef]

36. Adıgüzel, A.O.; Tunçer, M. Production and Characterization of Partially Purified Thermostable Endoxylanase and Endoglucanase from Novel *Actinomadura geliboluensis* and the Biotechnological Applications in the Saccharification of Lignocellulosic Biomass. *BioResources* **2017**, *12*, 2528–2547. [CrossRef]

37. Khusro, A.; Kaliyan, B.K.; Al-Dhabi, N.A.; Arasu, M.V.; Agastian, P. Statistical optimization of thermo-alkali stable xylanase production from *Bacillus tequilensis* strain ARMATI. *Electron. J. Biotechnol.* **2016**, *22*, 16–25. [CrossRef]

38. Kumar, V.; Chhabra, D.; Shukla, P. Xylanase production from *Thermomyces lanuginosus* VAPS-24 using low cost agro-industrial residues via hybrid optimization tools and its potential use for saccharification. *Bioresour. Technol.* **2017**, *243*, 1009–1019. [CrossRef] [PubMed]

39. Du, Y.; Shi, P.; Huang, H.; Zhang, X.; Luo, H.; Wang, Y.; Yao, B. Characterization of three novel thermophilic xylanases from *Humicola insolens* Y1 with application potentials in the brewing industry. *Bioresour. Technol.* **2013**, *130*, 161–167. [CrossRef] [PubMed]

40. Hempelmann, E.; Krafts, K. The mechanism of silver staining of proteins separated by SDS polyacrylamide gel electrophoresis. *Biotech. Histochem.* **2017**, *92*, 79–85. [CrossRef] [PubMed]

41. Bin, L.; Zhang, N.N.; Zhao, C.; Lin, B.X.; Xie, L.H.; Huang, Y.F. haracterization of a Recombinant Thermostable Xylanase from Hot Spring Thermophilic *Geobacillus* sp. TC-W7. *J. Microbiol. Biotechnol.* **2012**, *22*, 1388–1394.

42. Verma, D.; Anand, A.; Satyanarayana, T. Thermostable and alkalistable endoxylanase of the extremely thermophilic bacterium *Geobacillus thermodenitrificans* TSAA1: cloning, expression, characteristics and its applicability in generating xylooligosaccharides and fermentable sugars. *Appl. Biochem. Biotechnol.* **2013**, *170*, 119–130. [CrossRef] [PubMed]

43. Huang, D.; Liu, J.; Qi, Y.; Yang, K.; Xu, Y.; Feng, L. Synergistic hydrolysis of xylan using novel xylanases, β-xylosidases, and an α-L-arabinofuranosidase from *Geobacillus thermodenitrificans* NG80-2. *Appl. Biochem. Biotechnol.* **2017**, *101*, 6023–6037. [CrossRef] [PubMed]

44. Guan, G.-Q.; Zhao, P.-X.; Zhao, J.; Wang, M.-J.; Huo, S.-H.; Cui, F.-J.; Jiang, J.-X. Production and partial characterization of an alkaline xylanase from a novel fungus *Cladosporium oxysporum*. *BioMed Res. Int.* **2016**, *2016*, 4575024.

45. Lehninger, A. Role of metal ions in enzyme systems. *Physiol. Rev.* **1950**, *30*, 393–429. [CrossRef] [PubMed]

46. Amel, B.-D.; Nawel, B.; Khelifa, B.; Mohammed, G.; Manon, J.; Salima, K.-G.; Farida, N.; Hocine, H.; Bernard, O.; Jean-Luc, C. Characterization of a purified thermostable xylanase from *Caldicoprobacter algeriensis* sp. nov. strain TH7C1T. *Carbohydr. Res.* **2016**, *419*, 60–68. [CrossRef] [PubMed]

47. Novozymes, F.E.A. CELLIC Ctec2 and Htec2—Enzymes for Hydrolysis of Lignocellulosic Materials. Novozymes A/S, Luna, p 01. Available online: http://www.shinshu-u.ac.jp/faculty/engineering/chair/chem010/manual/Ctec2.pdf (accessed on 28 August 2018).

48. DuPont Accellerase XY. Available online: accellerase.dupont.com/fileadmin/.../accellerase/.../DUP-00413_ProdSheet_XY_web.page (accessed on 22 July 2018).

49. Peng, X.; Qiao, W.; Mi, S.; Jia, X.; Su, H.; Han, Y. Characterization of hemicellulase and cellulase from the extremely thermophilic bacterium *Caldicellulosiruptor owensensis* and their potential application for bioconversion of lignocellulosic biomass without pretreatment. *Biotechnol. Biofuels* **2015**, *8*, 131. [CrossRef] [PubMed]

50. Balan, V. Current challenges in commercially producing biofuels from lignocellulosic biomass. *ISRN Biotechnol.* **2014**, *2014*, 463074. [CrossRef] [PubMed]

51. Bibra, M.; Wang, J.; Squillace, P.; Pinkelman, R.; Papendick, S.; Schneiderman, S.; Wood, V.; Amar, V.; Kumar, S.; Salem, D. Biofuels and value-added products from extremophiles. In *Advances in Biotechnology*; Nawani, N.N., Khetmalas, M., Razdan, P.N., Pandey, A., Eds.; I K International Publishing House Pvt. Ltd.: New Delhi, India, 2014; p. 268.

52. Park, E.Y.; Naruse, K.; Kato, T. One-pot bioethanol production from cellulose by co-culture of *Acremonium cellulolyticus* and *Saccharomyces cerevisiae*. *Biotechnol. Bbiofuels* **2012**, *5*, 64. [CrossRef] [PubMed]

53. Miyazaki, K.; Irbis, C.; Takada, J.; Matsuura, A. An ability of isolated strains to efficiently cooperate in ethanolic fermentation of agricultural plant refuse under initially aerobic thermophilic conditions: Oxygen deletion process appended to consolidated bioprocessing (CBP). *Bioresour. Technol.* **2008**, *99*, 1768–1775. [CrossRef] [PubMed]

![microorganisms logo] *microorganisms*

MDPI

Article

Production and Characterization of an Extracellular Acid Protease from Thermophilic *Brevibacillus* sp. OA30 Isolated from an Algerian Hot Spring

Mohamed Amine Gomri [1] [iD], **Agustín Rico-Díaz** [2], **Juan-José Escuder-Rodríguez** [2],
Tedj El Moulouk Khaldi [3], **María-Isabel González-Siso** [2],* [iD] **and Karima Kharroub** [1]

[1]   Equipe Métabolites des Extrêmophiles, Laboratoire de Recherche Biotechnologie et Qualité des
      Aliments (BIOQUAL), Institut de la Nutrition, de l'Alimentation et des Technologies Agro
      Alimentaires (INATAA), Université Frères Mentouri Constantine 1 (UFMC1), Route de Ain El Bey,
      25000 Constantine, Algérie; gomrima@umc.edu.dz (M.A.G.); kkharroub@gmail.com (K.K.)
[2]   Grupo EXPRELA, Centro de Investigacións Científicas Avanzadas (CICA), Facultade de Ciencias,
      Universidade da Coruña, 15071 A Coruña, Spain; agustin.rico.diaz@udc.es (A.R.-D.);
      j.escuder@udc.es (J.-J.E.-R.)
[3]   Laboratoire Alimentation, Nutrition et Santé (ALNUTS), Institut de la Nutrition, de l'Alimentation et des
      Technologies Agro Alimentaires (INATAA), Université Frères Mentouri Constantine 1 (UFMC1),
      Route de Ain El Bey, 25000 Constantine, Algérie; moulouk.khaldi@umc.edu.dz
*   Correspondence: migs@udc.es; Tel.: +34-981-167-000

Received: 14 March 2018; Accepted: 10 April 2018; Published: 12 April 2018

Abstract: Proteases have numerous biotechnological applications and the bioprospection for newly-thermostable proteases from the great biodiversity of thermophilic microorganisms inhabiting hot environments, such as geothermal sources, aims to discover more effective enzymes for processes at higher temperatures. We report in this paper the production and the characterization of a purified acid protease from strain OA30, a moderate thermophilic bacterium isolated from an Algerian hot spring. Phenotypic and genotypic study of strain OA30 was followed by the production of the extracellular protease in a physiologically-optimized medium. Strain OA30 showed multiple extracellular proteolytic enzymes and protease 32-F38 was purified by chromatographic methods and its biochemical characteristics were studied. Strain OA30 was affiliated with *Brevibacillus thermoruber* species. Protease 32-F38 had an estimated molecular weight of 64.6 kDa and was optimally active at 50 °C. It showed a great thermostability after 240 min and its optimum pH was 6.0. Protease 32-F38 was highly stable in the presence of different detergents and solvents and was inhibited by metalloprotease inhibitors. The results of this work suggest that protease 32-F38 might have interesting biotechnological applications.

Keywords: *Brevibacillus* sp. OA30; thermophilic; hot spring; Algeria; protease; characterization

---

1. Introduction

Proteases catalyze the hydrolysis of proteinaceous material, and represent the largest worldwide enzyme sales [1]. Due to their characteristic active sites, in combination with their mode of catalytic action, proteases were assigned to groups of aspartic, cysteine, glutamic acid, serine, threonine, or metalloproteases. Moreover, they can be further subdivided based on their pH preferences into acidic, alkaline or neutral proteases [2]. Numerous commercial proteases, especially isolated from microorganisms, are used in various industrial and analytical processes, such as protein analysis, feed and food biotechnology, pharmaceutical and cosmetic preparations, and cleaning processes [3–5]. For example, they have major applications in detergent formulations, cheese-making, baking, meat tenderization, and leather industries [6–8].

Extracellular proteases produced by microorganisms are of great value for industry since they reduce production costs [9]. Thermophilic microorganisms are an important source of biodiversity and thermostable molecules of biotechnological importance and their unique properties at high temperatures justify the search for new proteases, as well as other enzymes of great value [10,11]. Thermostable proteases offer compatibility with processes that function more optimally at higher temperatures (e.g., through reduced viscosity), can have high catalytic efficiencies, and offer resistance from mesophilic microbial contamination [12]. Their robustness, in addition to their broad substrate specificity, makes thermostable proteases promising candidates for various industrial areas [13].

Brevibacillus belongs to the family Paenibacillaceae, a member of the Firmicutes phylum [14]. Among the 14 validated species of this genus, thermophilic *Brevibacillus thermoruber* and *Brevibacillus levickii* were isolated from different geothermal soils and hot springs [15,16]. These organisms have been reported to produce several molecules of biotechnological relevance, such as proteases, chitinases, exopolysaccharides, and bacteriocins, and to have the ability to be used as biocontrol agents and probiotics [17–20].

The aim of this study was to produce and characterize an extracellular protease from the thermophilic *Brevibacillus* sp. strain OA30 isolated from an Algerian hot spring.

2. Materials and Methods

2.1. Isolation of Strain OA30

A water sample was collected from an Algerian hot spring located at Ouled Ali (36°34′ N; 7°23′ E) (54 °C; pH 7.0 ± 0.05). A total of 0.1 mL of the diluted sample was poured on Plate Counting Agar medium, (pH 7.2 ± 0.2) and incubated for 72 h at 55 °C. Strain OA30 was purified and replated on *Thermus* agar medium (% *w/v*: 3 agar; 0.8 peptone; 0.4 yeast extract; 0.2 NaCl; pH 7.2 ± 0.2) [21]. The chemicals used for this study were principally purchased from Sigma Chemical Co. (St. Louis, MO, USA), Merck and Co., Inc. (Kenilworth, NJ, USA), and Fluka Biochemika (Buschs, Switzerland). All media were sterilized at 120 °C for 20 min prior to inoculation.

2.2. Screening for Extracellular Proteolytic Activity

To reveal the extracellular proteolytic activity, strain OA30 was plated on casein agar plates (% *w/v*: 2.5 agar; 1.0 casein; 0.2 peptone; 0.1 yeast extract; 0.2 NaCl; pH 7.2 ± 0.2) and incubated at 55 °C for 48 h. The appearance of clear zones around the colonies confirmed the presence of the enzymatic activity [22].

2.3. Phenotypic Characterization

The phenotypic characterization of the isolate was performed by different tests referring to Bergey's Manual of Determinative Bacteriology and minimal standards for describing new taxa of aerobic, endospore-forming bacteria [23,24]. The colonies' aspect was examined. Cell morphology was observed using a light microscope (1000×, Leica DM 1000 LED (Leica Microsystems, Wetzlar, Germany)) fitted with a digital camera (Leica EC3 camera) after Gram staining of the cells. The presence of endospores was investigated using the Schaeffer-Fulton technique [23,25].

Requirements for NaCl were determined on *Thermus* agar medium at 0, 1, 3, 3.5, 5, 7.5, and 10% (*w/v*) NaCl. Growth was tested on pH values between 5 and 10 and on a temperature range between 30 and 75 °C. Different biochemical and physiological tests were also carried out: catalase and oxidase activities; indole and urease production; ONPG, Methyl Red (MR) and Voges–Proskauer (VP) reactions; fermentation and use as a carbon source of D-glucose, D-fructose, D-galactose, D-maltose, D-saccharose, and D-lactose; and hydrolysis of gelatin, pectin, and starch [26–29].

2.4. Estimation of Growth Rates

Growth rates were estimated at different temperatures, pH, and NaCl concentrations. Only one parameter was changed each time and the two other parameter values were kept constant. Table 1 shows the different value combinations used. To prepare the preculture, approximately 20 mL of *Thermus* liquid medium were inoculated with strain OA30 and incubated overnight at 55 °C. The preculture was then transferred into a sterile 500 mL flask containing 100 mL of the same modified *Thermus* liquid medium to give an initial absorbance at 660 nm of at least 0.1. The culture was incubated in aerobic conditions using a Thermo Scientific MaxQ 4000 Benchtop Orbital Shaker (Thermo Scientific, Waltham, MA, USA) at 120 rpm for approximately 24 h. At different time intervals, the turbidity of the cultures was determined by measuring the increase in optical density at 660 nm with a Synergy H1 hybrid multi-mode microplate reader. At least 10 absorbance measurements were taken into account.

Table 1. Temperature, pH and NaCl concentration values used to estimate growth rates.

Assay \ Parameter	1	2	3	4	5	6	7	8	9	10
T (°C)	50					55				60
pH *	7.0	7.0	6.5	7.0	7.5	8.0	7.0	7.0	7.0	7.0
[NaCl] (% w/v)				0			1	2	3	0

* Phosphate buffer (0.2 M) was used to adjust the pH values.

2.5. Genotypic Characterization

2.5.1. DNA Extraction, 16S rRNA Gene Amplification, and Sequencing

Strain OA was grown aerobically on *Thermus* medium agar (pH 7.2) at 55 °C for 24 h. Genomic DNA was extracted using a modified protocol described previously [30]. The quantity and quality of the genomic DNA was measured using a NanoDrop spectrophotometer (Thermo Scientific). The 16S rRNA gene was amplified by polymerase chain reaction (PCR) with universal bacterial primers E9F (GAGTTTGATCCTGGCTCA) [31] and U1510R (GGTTACCTTGTTACGACTT) [32]. A typical PCR contained (final concentration): 1× DreamTaq buffer, 1% (v/v) BSA (Bovine Serum Albumin), 1.25 U DreamTaq polymerase (Thermo Scientific), 1 µM (each) of primer, 200 µM of each deoxynucleoside triphosphate, and 10 to 100 ng of template DNA in a 50 mL reaction volume. PCR conditions were as follows: 95 °C for 3 min; 30 cycles of 95 °C for 30 s, 52 °C for 30 s, 72 °C for 85 s; and a final incubation at 72 °C for 5 min. PCR products were electrophoresed and visualized on a 1% (w/v) agarose gel. Amplicons were then purified with the GeneJET PCR purification kit (Thermo Scientific). E9F and U1510R primers were used for capillary sequencing at the Central DNA Sequencing Facility, University of Stellenbosch (South Africa).

2.5.2. Phylogenetic Analysis

Identities with described taxa were investigated using the nBLAST tool against the EzBioCloud database of cultured organisms [33]. Multiple sequence alignments were performed using ClustalW [34]. 16S rRNA gene-based phylogenetic tree was constructed based on neighbor-joining [35] and maximum composite likelihood models [36] with 1000 bootstrap replications [37] using the MEGA 7 program package [38]. The sequence of *Sulfobacillus acidophilus* DSM 10332^T was used as the outgroup.

2.6. Enzyme Production

For the production of extracellular proteases, two different media were used: casein medium (M1) (% w/v: 0.8 casein, 0.3 peptone, 0.2 yeast extract, 0.2 glucose, optimum concentration of NaCl, 0.01 CaCl$_2$·2H$_2$O, 0.02 MgSO$_4$·7H$_2$O, 0.1 KH$_2$PO$_4$, optimum pH) and skim milk medium (M2) (%

w/v: 8 skim milk, 0.3 peptone, 0.2 yeast extract, 0.2 glucose, optimum concentration of NaCl, 0.01 CaCl$_2$·2H$_2$O, 0.02 MgSO$_4$·7H$_2$O, 0.672 KH$_2$PO$_4$, 3.863 NaHPO$_4$, optimum pH). A total of 50 mL of strain OA30's preculture were prepared as in Section 2.4, and were used to inoculate 250 mL of medium M1 or medium M2 contained in a sterile 2000 mL flask. The culture was incubated at optimum temperature in vigorous aeration conditions at 140 rpm for 64 h. At different time intervals, the absorbance and the proteolytic activity of the cultures were determined as described in Section 2.9. Culture supernatants were collected by centrifugation (22,000× *g* for 30 min at 4 °C) and used as the crude enzyme solution.

2.7. Purification of Protease

Proteins from culture supernatants were filtered through 0.45 μm, then 0.2 μm pore sizemembrane filters, and the filtrates were precipitated by adding ammonium sulfate at a final concentration of 80% (*w/v*) and the suspension was kept at 4 °C overnight under gentle stirring. The precipitated proteins were collected by centrifugation at 22,000× *g* for 20 min at 4 °C and then dissolved in 25 mL phosphate-buffered saline (PBS) buffer (50 mM). The enzyme solution was dialyzed overnight in a 14 kDa cut-off dialysis tubing cellulose membrane at 4 °C against 500 mL of the same buffer, which was replaced four times every 2 h. The resulting solution was filtered again through 0.2 μm pore size membrane filters.

Anion-exchange chromatography was performed using a HiTrap Q HP 5 mL column (GE Healthcare, Little Chalfont, UK). The column was equilibrated with 50 mM PBS (buffer A). Bounded proteins were eluted by applying a linear NaCl gradient (0–1 M) in buffer A and fractions were collected at 1 mL.

Active fractions were subsequently concentrated using a 10 kDa cut-off Amicon Pro Purification System (Millipore, Burlington, MA, USA). Tubes were first washed with 50 mM PBS.

A second purification of the fractions with enzymatic activity was done by gel filtration using a HiLoad 16/60 Superdex 200 prepgrade column (GE Healthcare) in 50 mM PBS buffer. Fractions were collected at 1 mL.

2.8. Molecular Weight Determination and Zymography

The SDS-PAGE method in a 10% polyacrylamide slab gel was carried out to analyze the molecular mass [39]. For zymogram analysis, protease was separated in a 10% SDS-polyacrylamide gel containing 0.5% (*w/v*) skim milk as a substrate. The samples were not heated prior to electrophoresis. Electrophoresis of both gels was run at 150 V for 120 min at room temperature. The zymography gel was washed with 2.5% (*v/v*) Triton X-100 for 1 h and incubated for 15 min in 50 mM Tris-HCl buffer, pH 7.5, and was then incubated at 50 °C overnight in a PBS solution (pH 7.5) containing 1% (*w/v*) casein. The gel was stained with Coomassie Brilliant Blue R-250 (0.2% *w/v*) for 1 h and then destained in distilled water/acetic acid (80:20). The protease band appeared as a clear zone surrounded by the blue color of the gel. NZY Color Protein Marker II (Lisbon, Portugal) was used as a molecular marker for both electrophoresis techniques and for the estimation of the molecular weight of protease.

2.9. Enzyme Assay

Protease activity was determined using azocasein as a substrate. The reaction was performed in 50 mM PBS solution at pH 7.5 with 50 μL of azocasein (30 mg/mL in water) and with 25 μL of the enzyme solution for a final volume of 750 μL. The reaction was incubated in the dark at 50 °C for 1 h and stopped by adding 125 μL of 20% (*w/v*) trichloroacetic acid. The blank assay was performed using 25 μL of culture medium or PBS buffer. After centrifugation at 15,000× *g* for 10 min, the absorbance of the supernatant was measured at 366 nm using a Synergy H1 Hybrid Multi-Mode Microplate Reader [40]. One unit of protease activity was defined as the amount required to produce enough acid-soluble material from azocasein to yield an absorbance of 0.01 at 366 nm, following 1 h of incubation.

Protein was quantified by the Bio-Rad protein assay kit (Hercules, CA, USA) [41] with BSA as the standard.

2.10. Effect of Temperature on Protease Activity and Stability

The optimum temperature was determined by measuring enzyme activity at 30–80 °C as described above. Enzyme stability was measured by incubating for 20, 30, 40, 50, 60, 90, 120, 180, and 240 min at optimum temperature in 50 mM PBS buffer pH 7.5.

2.11. Effect of pH on Protease Activity

The effect of pH on the enzymatic assay was determined by measuring the enzymatic activity using substrate solutions with different pH (4–11; Na_2HPO_4-citric acid: 5.0–7.0; Tris-HCl: 8.0–9.0; and glycine-NaOH: 10.0–11.0) at optimum temperature.

2.12. Effect of Metal Ions on Protease Activity

The effect of various metal ions on enzyme activity was determined by incubating the enzyme with 2.5 mM metal ions (Mg^{2+}, Li^{2+}, Fe^{3+}, Cu^{2+}, Zn^{2+}, Mn^{2+}, and Ca^{2+}). The protease activity without metal ions served as the control and was considered as 100% activity.

2.13. Effect of Solvents on Protease Activity

The effect of solvents ethanol, methanol, and acetone on protease activity was measured by incubating the enzyme with 1% (v/v) of the solvents. The protease activity without solvents served as the control, which was considered as 100% activity.

2.14. Effect of Detergents and Chemicals on Protease Activity

The effect of different concentrations of surfactants: SDS, Tween-20, Tween-80, Triton X-100, metal ion chelators ethylenediaminetetraacetic acid (EDTA), dithiothreitol (DTT), protease inhibitors phenylmethylsulfonyl fluoride (PMSF), pepstatin A, trypsin inhibitor and dimethyl sulfoxide (DMSO) were studied by incubating enzyme with 1% of chemical or 1 mM of EDTA, trypsin inhibitor, and pepstatin A. The protease activity without chemicals served as the control, which was considered as 100% activity.

2.15. Statistical Analysis

Three replicates of each sample were used for statistical analysis using STATISTICA 12 software [42]. Statistical analysis was conducted by Student's t test. A probability level of $p < 0.05$ was considered statistically significant.

3. Results and Discussion

3.1. Isolation and Characterization of Protease-Producing Strain OA30

Strain OA30 was isolated on a nutritive agar medium used for the isolation of aerobic heterotrophic bacteria, from water samples of an Algerian terrestrial hot spring with moderate temperature (54 °C), chloride–calcica water type [43], and neutral pH (7.0). Screening for extracellular protease activity on casein agar was positive. Large, clear zones appeared around the colonies indicating the production of an extracellular enzymatic activity against casein.

3.1.1. Phenotypic Characterization

Strain OA30 produced smooth, flat, spreading colonies, with yellowish color and no particular pigmentation on *Thermus* medium. Cells were rod-shaped, Gram-positive (Figure 1), and motile, with the presence of terminally-born ellipsoidal spores. The strain was aerobic, catalase, and oxidase positive. It was able to use all the tested sugars as carbon sources, but was unable to ferment them and could not use citrate. ONPG, VP, and MR reactions were negative. The strain was unable to produce indole, but it could use urea and hydrolyze, in addition to casein, gelatin, and starch, but did not degrade pectin.

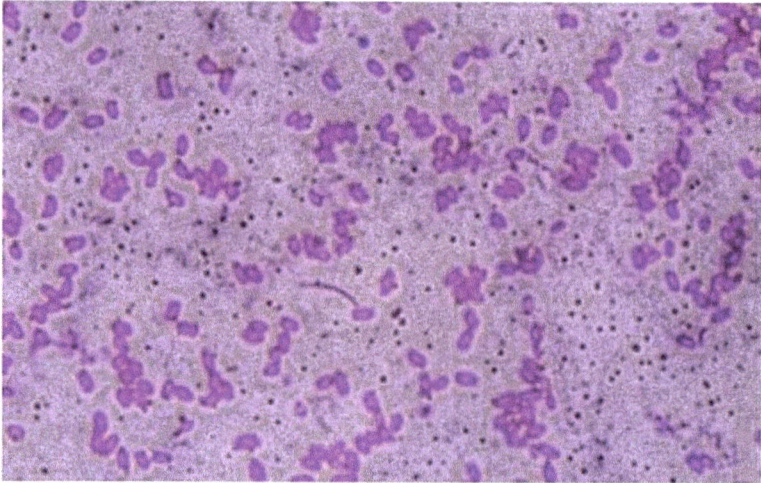

Figure 1. Cells of strain OA30 viewed by light microscopy (1000×) after Gram staining.

The growth of strain OA30 occurred at 30–70 °C, a pH range from 6.0 to 8.6, and was stopped by a 5% concentration of NaCl. Optimal growth on liquid medium was studied by measuring the absorbance of the cultures under variable physiological conditions and significantly occurred at 55 °C, pH 7.0, and at a concentration of NaCl of 1% (w/v) (Figure 2). These values were very close to those from the isolation site and were used as parameters to optimize the growth medium for protease production.

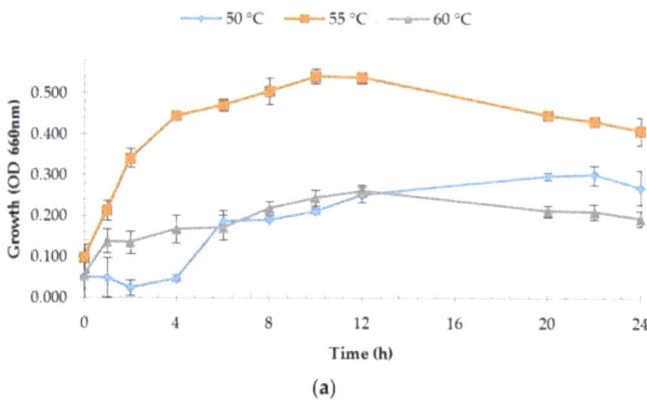

(a)

Figure 2. *Cont.*

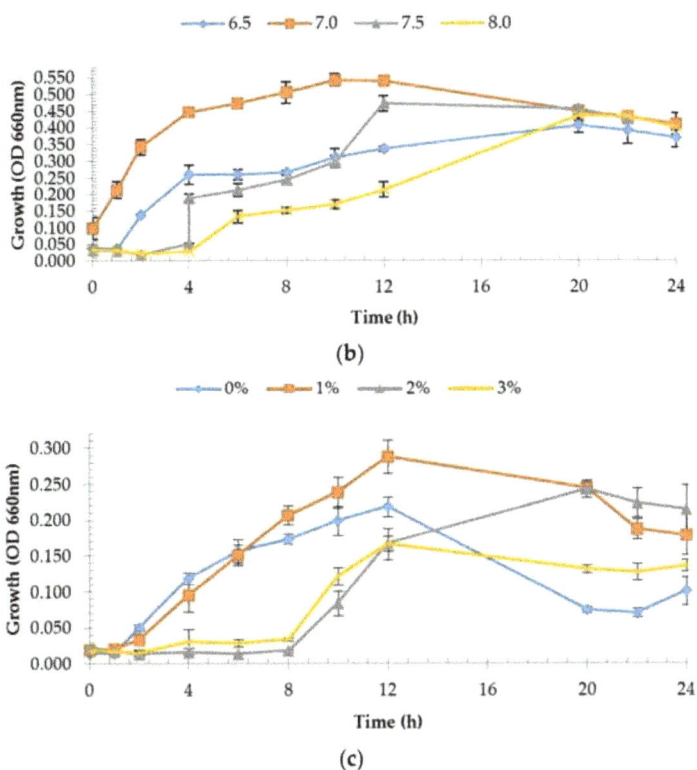

Figure 2. Growth of strain OA30 under different physiological conditions: (**a**) growth at different temperatures; (**b**) growth at different pH values; and (**c**) growth at different NaCl concentrations.

3.1.2. Genotypic Characterization

The 16S rRNA gene sequences of strain OA30 have been deposited in the NCBI database under accession number MF136824. Based on the 16S rRNA gene sequence similarity searches by the nBLAST tool against the EzBioCloud database, strain OA30 showed 92 to 99% sequence similarity to members of the genus *Brevibacillus*. A 16S rRNA gene-based phylogenetic tree of *Brevibacillus* sp. strain OA30 was constructed (Figure 3). The *Brevibacillus* sp. strain OA3016S rRNA gene sequence exhibited high identity (99%) with type strain *Brevibacillus thermoruber*, strain DSM 7064^{T} (Z26921), the closest validly published *Brevibacillus* species.

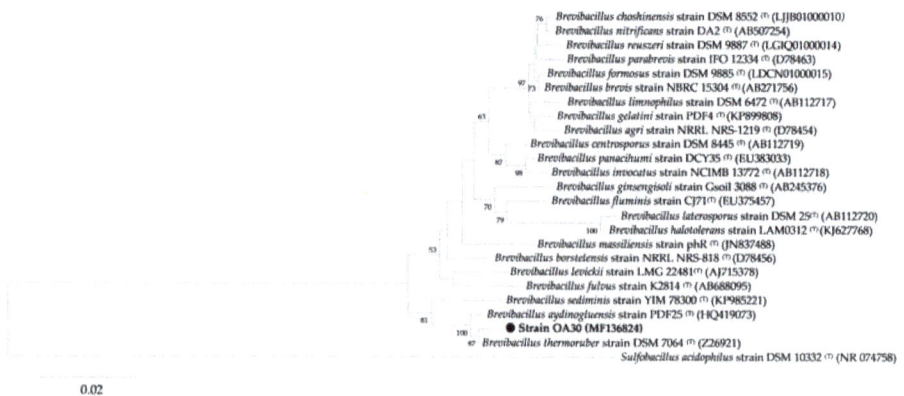

Figure 3. Phylogenetic tree based on 16S rRNA gene sequences showing the relationship between strain OA30 (1489 bp) and strains of the genus *Brevibacillus*. The strains and their corresponding Genbank accession numbers are shown following the organism name and indicated in parentheses. The phylogenetic tree was made using the neighbor-joining method with maximum composite likelihood model implemented in MEGA 7. The tree includes the 16S rRNA gene sequences of *Sulfobacillus acidophilus* DSM 10332^T as an outgroup. Bootstrap consensus trees were inferred from 1000 replicates, and only bootstrap values >50% are indicated. The scale bar represents 0.02 nucleotide changes per position. (●) indicates the isolate assessed in the current study, *Brevibacillus* sp. strain OA30.

3.2. Protease Purification

3.2.1. Enzyme Production and Medium Composition Effect

Extracellular protease production assay was performed on two different media, M1 (casein) and M2 (skim milk) (at 55 °C, pH 7.0, NaCl 1% *w/v*). Good enzyme activity and growth rates were obtained on M1, while growth on M2 was very low with no enzyme production, which appears to be unsuited for growth of strain OA30 (Figure 4). Extracellular protease activity appeared on M1 after 36 h of cultivation with the best, significantly higher value after 48 h. Purification was consequently performed on the crude enzyme solution collected from M1 (Table 2).

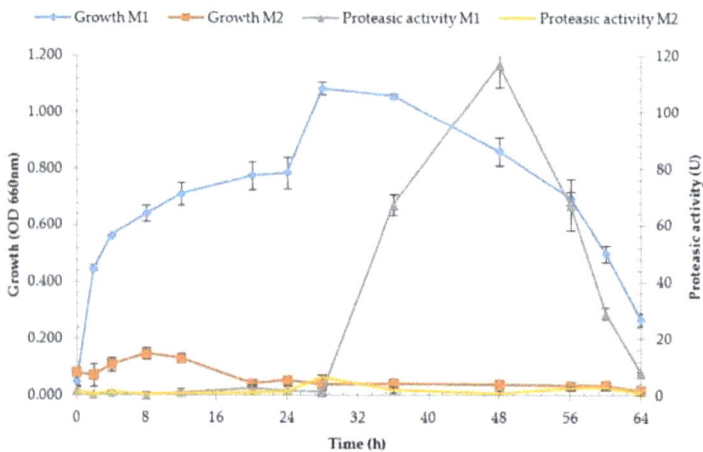

Figure 4. Protease activities and growth rates of strain OA30 on M1 and M2.

Table 2. Purification summary of protease 32-F38 from *Brevibacillus* sp. strain OA30.

Purification Step	Total Protein (mg)	Total Activity (U)	Specific Activity (U/mg)	Yield (%)	Purification (Fold)
Cell-free supernatant	131.78	147.2	1.12	100.00	1.00
80% ammonium sulfate	32.09	130.1	4.05	88.38	3.63
Dialysis	11.90	128.1	10.76	87.02	9.64
AE chromatography	2.54	30.0	11.79	20.38	10.56
Gel filtration	1.68	27.4	16.31	18.61	14.60

3.2.2. Protease Precipitation, Anion Exchange, and Gel Filtration Chromatography

Four fractions with protease activity were obtained after first purification with a HiTrap Q HP column. Fraction 32 had the highest enzyme activity (11.79 U/mg) (Figure 5a) and was selected for a second round of purification with the HiLoad 16/60 Superdex 200 prepgrade column, which gave two fractions with protease activity. Fraction 32-F38 had, significantly, the best activity (Figure 5b). This protease, named protease 32-F38, was examined for further characterization (Table 2). The presence of multiple extracellular proteases was reported for *Brevibacillus* strains isolated from similar environments. Thermophilic *Brevibacillus* species are well-known hydrolase producers [24,44].

3.3. Molecular Weight Determination and Zymoraphy

Electrophoretic profiles of fraction 32-F38 were studied. Zymography revealed protease activity for the band corresponding to an estimated molecular weight of 64.4 kDa on the corresponding SDS-PAGE gel led in the same migration conditions for fraction 32-F38 after gel filtration chromatography (Figure 6). Monomeric proteases with molecular weights between 60 and 66 kDa from *Brevibacillus* sp. have been reported in literature [45–47].

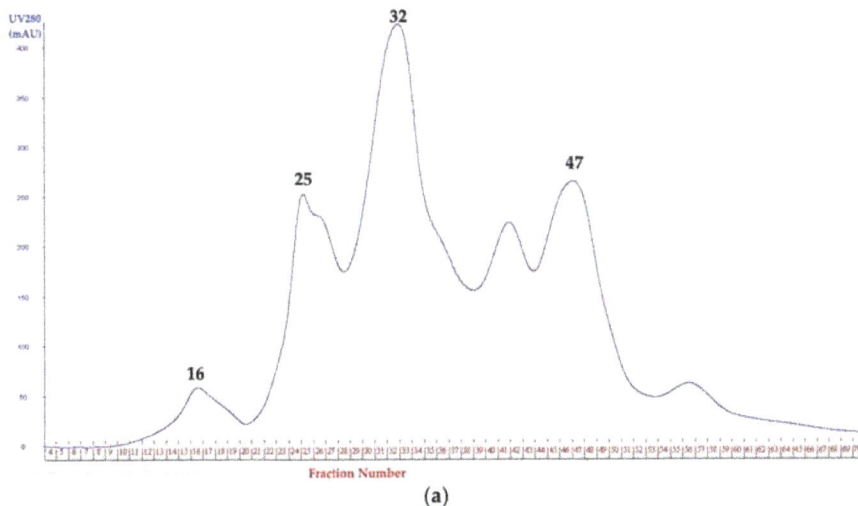

(a)

Figure 5. *Cont.*

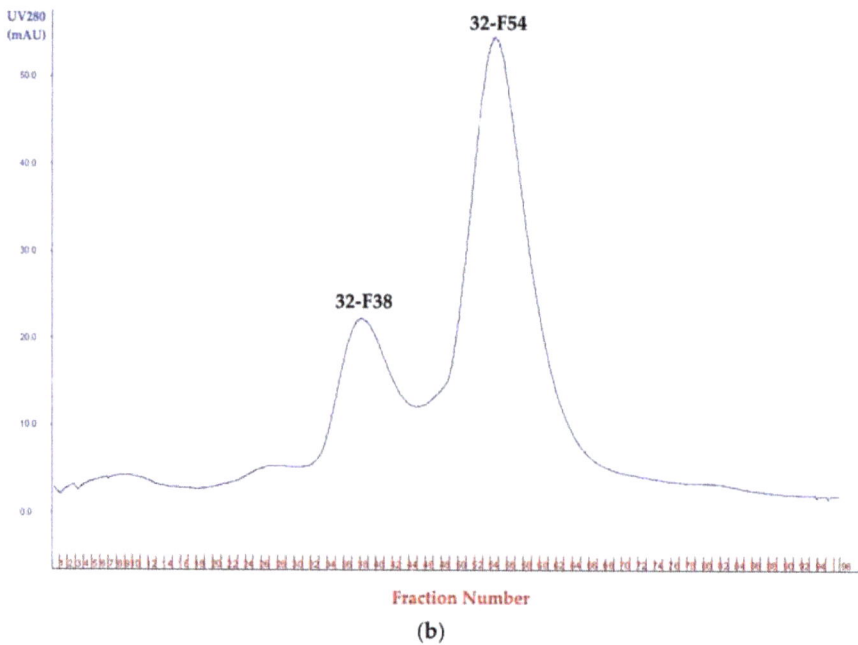

(b)

Figure 5. Purification of extracellular protease from the culture supernatant of strain OA30. (a) Anion-exchange chromatography on a HiTrap Q HP column of the dissolved precipitate from the culture supernatant. Protease activity was detected in fractions 16, 25, 32, and 47. Fraction 32 was selected for a second purification step; and (b) gel filtration chromatography on a HiLoad 16/60 Superdex 200 prepgrade column of fraction 32 containing the highest protease activity from anion-exchange chromatography. Protease activity was detected in peaks 32-F38 and 32-F54 and protease 32-F38 was selected for further investigation.

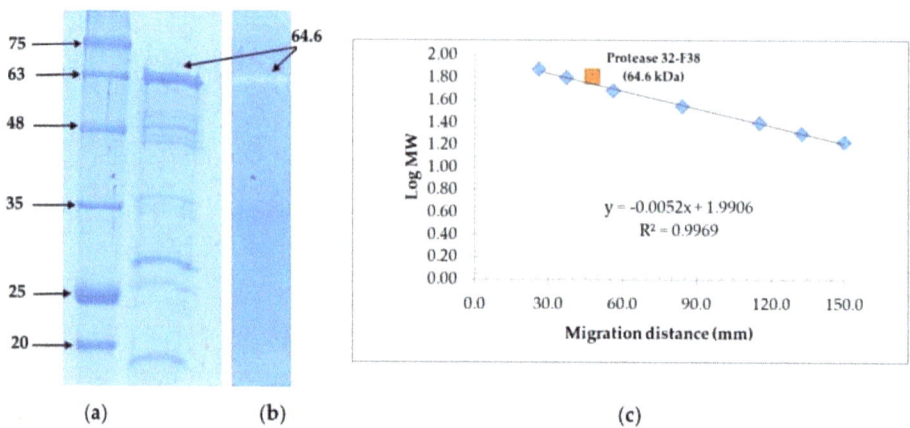

(a)   (b)   (c)

Figure 6. Electrophoresis analysis and purification of protease 32-F38 from strain OA30. (a) SDS-PAGE of the purified protease. Lane 1, protein markers (kDa). Lane 2, partially-purified protease 32-F38 obtained after gel filtration; (b) Zymogram activity staining of the purified protease; (c) Estimation of the molecular mass of protease 32-F38.

3.4. Biochemical Characterization

3.4.1. Effect of Temperature and Thermostability of the Enzyme

Relative activity of protease 32-F38 at different temperatures is shown in Figure 7. Protease activity was significantly higher at 50 °C, so this temperature was considered as the optimum temperature for the enzyme and was similar to the optimum growth temperature of strain OA30. Protease 32-F38 remained at least 80% active in the range between 40 °C and 55 °C. Thermostability tests at 50 °C revealed that the activity was the highest after 120 min of heating and relative activity lost only 16% of its value after 240 min (Figure 8). It is not uncommon that relative enzymatic activity increases after short-term heating in the case of thermoactive enzymes, as it happens with protease 32-F38 (Figure 8), and examples of other proteases can be cited [48,49].

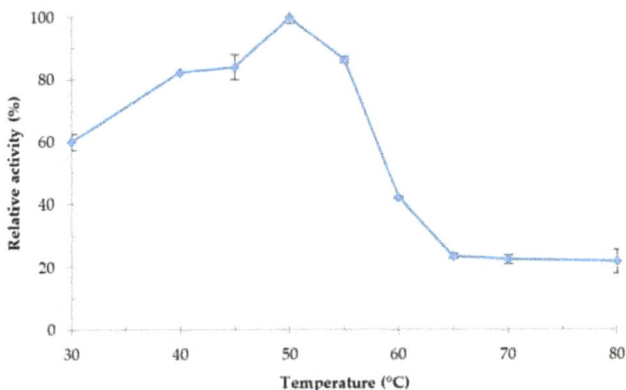

Figure 7. Optimum temperature of protease 32-F38 activity. Relative activity is expressed as a percentage of the maximum.

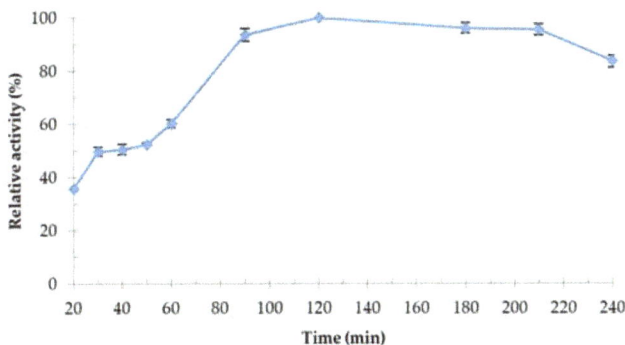

Figure 8. Thermostability of protease 32-F38 at optimum temperature (50 °C). Relative activity is expressed as a percentage of the maximum (activity after 120 min).

3.4.2. Effect of pH

The effect of pH buffers is illustrated in Figure 9. Protease 32-F38 was found to be an acid protease; optimum pH was 6.0 in Na_2HPO_4-citric acid buffer with a relative activity 55% and 58% higher than at pH 7.0 and pH 8.0 in Tris-HCl buffer, respectively. Weak activities were still observed at pH 5.0 and 11.0.

The majority of extracellular proteases reported in literature from *Brevibacillus* members were alkaline proteases with optimal pH around 8.0 [46,47,50,51], and acid proteases were rarely reported [45].

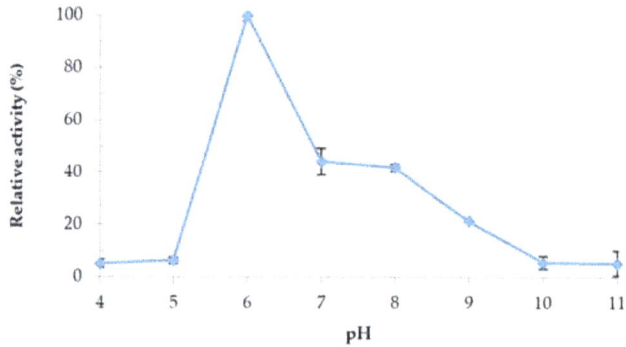

Figure 9. Effect of pH on protease 32-F38 activity. Relative activity is expressed as a percentage of the maximum.

3.4.3. Effect of Various Chemicals on Protease 32-F38 Activity

Relative activities of protease 32-F38 in the presence of different chemical agents are shown in Table 3. Ions of Mg^{2+} and Mn^{+2} showed a significant enhancing effect on the enzymatic activity with the best effect with magnesium ions, which was reported for other proteases from *Brevibacillus* spp. [50,51], and might have a protective effect against thermal denaturation of the enzyme [52]. Li^{2+} had no significant effect on protease 32-F38 while a slight inhibition was noted with calcium ions. Protease activity was completely inhibited by Zn^{2+}, Cu^{2+}, and Fe^{3+}. The same inhibitory effect of heavy metal ions was observed on a thermostable protease of several *Brevibacillus* species and is probably related to a reaction with the protein thiol groups (converting them to mercaptides), as well as with histidine and tryptophan residues. Metal ions are potent inhibitors of protein folding [46,47,51]. Protease 32-F38 activity was not affected by a 1% concentration of methanol, ethanol, or acetone. The addition of 1% SDS caused a four-fold increase in the enzyme activity, and SDS is known to denature protein substrates, such as casein, resulting in increased digestion rates. The remarkable resistance of protease 32-F38 toward SDS denaturation was also observed for other thermoactive proteases, like proteinase K and thermopsin, and might be related to a modification of the protein structure due to a high number of tyrosine residues [53–55]. Enzyme activity was significantly enhanced by 1% of Tween-80 and Triton X-100 while the presence of Tween-20 was strongly inhibiting the enzyme activity. The presence of metalloprotease inhibitors EDTA and DTT inhibited the protease activity completely and 1% of DMSO reduced it by 12%. On the other hand, the inhibitor of aspartyl proteases, pepstain A, and the inhibitors of serine proteases, trypsin inhibitor, and PMSF, increased the enzyme activity significantly. These results indicate that protease 32-F38 is most probably a metalloprotease [56,57].

Table 3. Effects of various metal ions, solvents, detergents, and other chemicals on protease from fraction 32-F38 stability.

Reagent	Concentration	Relative Activity [1] (%)
Mg^{2+}	2.5 mM	143.24 ± 2.13
Li^{2+}	2.5 mM	100.29 ± 3.25
Fe^{3+}	2.5 mM	0.00 ± 0.00
Cu^{2+}	2.5 mM	7.35 ± 0.00
Zn^{2+}	2.5 mM	0.00 ± 0.00

Table 3. *Cont.*

Reagent	Concentration	Relative Activity [1] (%)
Mn^{2+}	2.5 mM	113.24 ± 1.85
Ca^{2+}	2.5 mM	90.89 ± 3.45
Ethanol	1%	96.46 ± 2.06
Methanol	1%	106.19 ± 2.47
Acetone	1%	103.54 ± 3.06
SDS	1%	438.35 ± 3.56
Tween-20	1%	27.73 ± 2.95
Tween-80	1%	144.54 ± 4.25
Triton X-100	1%	144.25 ± 3.12
EDTA	1.0 mM	0.88 ± 0.26
DTT	1%	0.36 ± 0.02
PMSF	1%	193.59 ± 3.15
Pepstatin A	1.0 mM	111.39 ± 3.23
Trypsin inhibitor	1.0 mM	186.48 ± 2.23
DMSO	1%	87.90 ± 3.56

[1] The activity is expressed as a percentage of the activity level in the absence of reagent.

4. Conclusions

Geothermal sites are an important source of valuable molecules and only a few studies were interested in the biotechnological potential of thermophilic microorganisms isolated from Algerian hot springs. *Brevibacillus* sp. strain OA30 studied in this work showed multiple extracellular protease activities that should be investigated more closely with biochemical and genomic methods.

The characterization of the purified protease 32-F38 revealed interesting abilities of the enzyme in acid conditions and in the presence of solvents and detergents with a high stability at 50 °C. Protease 32-F38 seems to be related to other metalloproteases, such as thermolysin (EC 3.4.24.27) and bacillolysin (EC 3.4.24.28) produced by several strains of the aerobic thermophilic genera *Bacillus*, *Geobacillus* and *Brevibacillus*. Thermolysin-like proteases have potential applications in the degradation of gelatin, keratin, and other raw materials, like wheat bran and fish scales in biotechnological applications [19]. They can also act as peptide and ester synthetases and can be used for peptide synthesis and the production of a precursor of the artificial sweetener aspartame [58], and have been used for the hydrolysis of plant cell walls in order to assist in aqueous extraction processes [59]. Proteases with similar characteristics and molecular weight also showed anti-biofilm activity against pathogenic bacteria such as *Listeria monocytogenes*, *Escherichia coli*, and *Salmonella typhi* [49] and are potential candidates for the production of bioactive peptides from casein [60]. Our initial promising results contribute towards the application of protease 32-F38 in industrial and analytical processes.

Acknowledgments: We wish to acknowledge the following organizations for providing financial support for this study: the Algerian Ministry of Higher Education and Scientific Research, and the Genomics Research Institute (University of Pretoria, South Africa). The laboratory of the group EXPRELA (Universidade da Coruña, Spain) during 2017 was funded by European Union Seventh Framework Programme (FP7/2007-2013) under grant agreement number 324439 and by the Xunta de Galicia (Consolidación Grupos Referencia Competitiva contract number ED431C2016-012), co-financed by FEDER (EEC). We want to extend special acknowledgement to Thulani P. Makhalanyane from the Centre for Microbial Ecology and Genomics (CMEG, Department of Genetics, University of Pretoria), who provided assistance in the genotypic characterization of the strain.

Author Contributions: M.A.G. isolated the strain, designed and performed this work during a research stay at the laboratory of the group EXPRELA, and led the drafting of the manuscript. A.R.-D. assisted with protease production and purification and helped with the draft. J.-J.E.-R. helped with the enzyme assay and assisted with the manuscript. T.E.M.K. helped with isolation and the manuscript draft. M.-I.G.-S. and K.K. conceived the study and provided support in drafting the manuscript. All authors read and approved the final version of the manuscript.

Conflicts of Interest: The authors declare no conflict of interest.

References

1. Contesini, F.J.; Melo, R.R.D.; Sato, H.H. An overview of *Bacillus* proteases: From production to application. *Crit. Rev. Biotechnol.* **2017**, 1–14. [CrossRef] [PubMed]
2. Rawlings, N.D.; Barrett, A.J.; Thomas, P.D.; Huang, X.; Bateman, A.; Finn, R.D. The MEROPS database of proteolytic enzymes, their substrates and inhibitors in 2017 and a comparison with peptidases in the PANTHER database. *Nucleic Acids Res.* **2018**, *46*, D624–D632. [CrossRef] [PubMed]
3. Souza, P.M.D.; Bittencourt, M.L.D.A.; Caprara, C.C.; Freitas, M.D.; Almeida, R.P.C.D.; Silveira, D.; Fonseca, Y.M.; Ferreira Filho, E.X.; Pessoa Junior, A.; Magalhães, P.O. A biotechnology perspective of fungal proteases. *Braz. J. Microbiol.* **2015**, *46*, 337–346. [CrossRef] [PubMed]
4. Chitte, R.; Chaphalkar, S. The World of Proteases Across Microbes, Insects, and Medicinal Trees. In *Proteases in Physiology and Pathology*; Springer: New York, NY, USA, 2017; pp. 517–526.
5. Tavano, O.L.; Berenguer-Murcia, A.; Secundo, F.; Fernandez-Lafuente, R. Biotechnological Applications of Proteases in Food Technology. *Compr. Rev. Food Sci. Food Saf.* **2018**. [CrossRef]
6. Kumar, K.; Arumugam, N.; Permaul, K.; Singh, S. Thermostable enzymes and their industrial applications. *Microb. Biotechnol. Interdiscip. Approach* **2016**, 115–162. [CrossRef]
7. Liu, X.; Kokare, C. Microbial Enzymes of Use in Industry. In *Biotechnology of Microbial Enzymes*; Elsevier: New York, NY, USA, 2017; pp. 267–298. [CrossRef]
8. Feijoo-Siota, L.; Blasco, L.; Rodriguez-Rama, J.; Barros-Velazquez, J.; Miguel, T.; Sanchez-Perez, A.; Villa, T. Recent Patents on Microbial Proteases for the Dairy Industry. *Recent Adv. DNA Gene Seq.* **2014**, *8*, 44–55. [CrossRef] [PubMed]
9. Sawant, R.; Nagendran, S. Protease: An enzyme with multiple industrial applications. *World J. Pharm. Sci.* **2014**, *3*, 568–579.
10. Haki, G.; Rakshit, S. Developments in industrially important thermostable enzymes: A review. *Bioresour. Technol.* **2003**, *89*, 17–34. [CrossRef]
11. Mishra, S.S.; Ray, R.C.; Rosell, C.M.; Panda, D. Microbial Enzymes in Food Applications. *Microb. Enzyme Technol. Food Appl.* **2017**, 1. [CrossRef]
12. Hussein, A.H.; Lisowska, B.K.; Leak, D.J. The genus *Geobacillus* and their biotechnological potential. In *Advances in Applied Microbiology*; Elsevier: New York, NY, USA, 2015; Volume 92, pp. 1–48.
13. Elleuche, S.; Schäfers, C.; Blank, S.; Schröder, C.; Antranikian, G. Exploration of extremophiles for high temperature biotechnological processes. *Curr. Opin. Microbiol.* **2015**, *25*, 113–119. [CrossRef] [PubMed]
14. Vos, P.D.; Ludwig, W.; Schleifer, K.-H.; Whitman, W.B. "Paenibacillaceae" fam. nov. In *Bergey's Manual of Systematics of Archaea and Bacteria*; John Wiley & Sons, Ltd.: Hoboken, NJ, USA, 2015.
15. Manachini, P.; Fortina, M.; Parini, C.; Craveri, R. *Bacillus thermoruber* sp. nov., nom. rev., a red-pigmented thermophilic bacterium. *Int. J. Syst. Evol. Microbiol.* **1985**, *35*, 493–496. [CrossRef]
16. Allan, R.; Lebbe, L.; Heyrman, J.; De Vos, P.; Buchanan, C.; Logan, N.A. *Brevibacillus levickii* sp. nov. and *Aneurinibacillus terranovensis* sp. nov., two novel thermoacidophiles isolated from geothermal soils of northern Victoria Land, Antarctica. *Int. J. Syst. Evol. Microbiol.* **2005**, *55*, 1039–1050. [CrossRef] [PubMed]
17. Sanders, M.; Morelli, L.; Tompkins, T. Sporeformers as human probiotics: *Bacillus*, *Sporolactobacillus*, and *Brevibacillus*. *Compr. Rev. Food Sci. Food Saf.* **2003**, *2*, 101–110. [CrossRef]
18. Singh, P.K.; Sharma, V.; Patil, P.B.; Korpole, S. Identification, purification and characterization of laterosporulin, a novel bacteriocin produced by *Brevibacillus* sp. strain GI-9. *PLoS ONE* **2012**, *7*, e31498. [CrossRef] [PubMed]
19. Panda, A.K.; Bisht, S.S.; DeMondal, S.; Kumar, N.S.; Gurusubramanian, G.; Panigrahi, A.K. *Brevibacillus* as a biological tool: A short review. *Antonie Leeuwenhoek* **2014**, *105*, 623–639. [CrossRef] [PubMed]
20. Yildiz, S.Y.; Radchenkova, N.; Arga, K.Y.; Kambourova, M.; Oner, E.T. Genomic analysis of *Brevibacillus thermoruber* 423 reveals its biotechnological and industrial potential. *Appl. Microbiol. Biotechnol.* **2015**, *99*, 2277–2289. [CrossRef] [PubMed]
21. Atlas, R.M. *Handbook of Media for Environmental Microbiology*; CRC Press: Boca Raton, FL, USA, 2005.
22. Atlas, R.M. *Handbook of Microbiological Media for the Examination of Food*; CRC Press: Boca Raton, FL, USA, 2006.
23. Logan, N.A.; Berge, O.; Bishop, A.H.; Busse, H.-J.; De Vos, P.; Fritze, D.; Heyndrickx, M.; Kämpfer, P.; Rabinovitch, L.; Salkinoja-Salonen, M.S. Proposed minimal standards for describing new taxa of aerobic, endospore-forming bacteria. *Int. J. Syst. Evol. Microbiol.* **2009**, *59*, 2114–2121. [CrossRef] [PubMed]

24. Logan, N.A.; Vos, P.D. Brevibacillus. In *Bergey's Manual of Systematics of Archaea and Bacteria*; John Wiley & Sons, Ltd.: Hoboken, NJ, USA, 2015.

25. Bartholomew, J.W.; Mittwer, T. A simplified bacterial spore stain. *Stain Technol.* **1950**, *25*, 153–156. [CrossRef] [PubMed]

26. Frazier, W.C. A method for the detection of changes in gelatin due to bacteria. *J. Infect. Dis.* **1926**. [CrossRef]

27. Soares, M.M.; Silva, R.d.; Gomes, E. Screening of bacterial strains for pectinolytic activity: Characterization of the polygalacturonase produced by *Bacillus* sp. *Rev. Microbiol.* **1999**, *30*, 299–303. [CrossRef]

28. Gordon, R.E.; Haynes, W.; Pang, C.H.-N.; Smith, N. The genus *Bacillus*. In *US Department of Agriculture Handbook*; Agricultural Research Service: Washington, DC, USA, 1973; pp. 109–126.

29. Harley, J.P. *Laboratory Exercises in Microbiology*; McGraw-Hill Science, Engineering & Mathematics: New York, NY, USA, 2004.

30. Miller, D.; Bryant, J.; Madsen, E.; Ghiorse, W. Evaluation and optimization of DNA extraction and purification procedures for soil and sediment samples. *Appl. Environ. Microbiol.* **1999**, *65*, 4715–4724. [PubMed]

31. Farrelly, V.; Rainey, F.A.; Stackebrandt, E. Effect of genome size and rrn gene copy number on PCR amplification of 16S rRNA genes from a mixture of bacterial species. *Appl. Environ. Microbiol.* **1995**, *61*, 2798–2801. [PubMed]

32. Reysenbach, A.; Pace, N. Reliable amplification of hyperthermophilic archaeal 16S rRNA genes by the polymerase chain reaction. In *Archaea: A Laboratory Manual*; Cold Spring Harbor Laboratory Press: Cold Spring Harbor, NY, USA, 1995; pp. 101–107.

33. Yoon, S.-H.; Ha, S.-M.; Kwon, S.; Lim, J.; Kim, Y.; Seo, H.; Chun, J. Introducing EzBioCloud: A taxonomically united database of 16S rRNA and whole genome assemblies. *Int. J. Syst. Evol. Microbiol.* **2016**. [CrossRef]

34. Thompson, J.D.; Higgins, D.G.; Gibson, T.J. CLUSTAL W: Improving the sensitivity of progressive multiple sequence alignment through sequence weighting, position-specific gap penalties and weight matrix choice. *Nucleic Acids Res.* **1994**, *22*, 4673–4680.

35. Saitou, N.; Nei, M. The neighbor-joining method: A new method for reconstructing phylogenetic trees. *Mol. Biol. Evol.* **1987**, *4*, 406–425. [PubMed]

36. Tamura, K.; Nei, M.; Kumar, S. Prospects for inferring very large phylogenies by using the neighbor-joining method. *Proc. Natl. Acad. Sci. USA* **2004**, *101*, 11030–11035. [CrossRef] [PubMed]

37. Felsenstein, J. Confidence limits on phylogenies: An approach using the bootstrap. *Evolution* **1985**, *39*, 783–791. [CrossRef] [PubMed]

38. Kumar, S.; Stecher, G.; Tamura, K. MEGA7: Molecular Evolutionary Genetics Analysis version 7.0 for bigger datasets. *Mol. Biol. Evol.* **2016**. [CrossRef] [PubMed]

39. Laemmli, U.K. Cleavage of Structural Proteins during the Assembly of the Head of Bacteriophage T4. *Nature* **1970**, *227*, 680–685. [CrossRef] [PubMed]

40. Guendouze, A.; Plener, L.; Bzdrenga, J.; Jacquet, P.; Rémy, B.; Elias, M.; Lavigne, J.-P.; Daudé, D.; Chabrière, E. Effect of Quorum Quenching Lactonase in Clinical Isolates of *Pseudomonas aeruginosa* and Comparison with Quorum Sensing Inhibitors. *Front. Microbiol.* **2017**, *8*, 227. [CrossRef] [PubMed]

41. Bradford, M.M. A rapid and sensitive method for the quantitation of microgram quantities of protein utilizing the principle of protein-dye binding. *Anal. Biochem.* **1976**, *72*, 248–254. [CrossRef]

42. StatSoft Inc. *Statistica*, version 12; Data Analysis Software System; StatSoft Inc.: Tulsa, OK, USA, 2013.

43. Bahri, F.; Saibi, H. Characterization, classification, and determination of drinkability of some Algerian thermal waters. *Arab. J. Geosci.* **2011**, *4*, 207–219. [CrossRef]

44. Sinsuwan, S.; Raksakulthai, N.; Rodtong, S.; Yongsawatdigul, J. Production and characterization of proteinases from *Brevibacillus* sp. isolated from fish sauce fermentation. *J. Food Sci.* **2006**. [CrossRef]

45. Seswita Zilda, D.; Harmayani, E.N.I.; Widada, J.; Asmara, W.; Irianto, E.K.O.; Patantis, G.; Nuri Fawzya, Y. Optimization of Culture Conditions to Produce Thermostable Keratinolytic Protease of *Brevibacillus thermoruber* LII, Isolated from the Padang Cermin Hot Spring, Lampung, Indonesia. *Microbiol. Indones.* **2012**, *6*, 194–200. [CrossRef]

46. Wang, S.; Lin, X.; Huang, X.; Zheng, L.; Zilda, D.S. Screening and characterization of the alkaline protease isolated from PLI-1, a strain of *Brevibacillus* sp. collected from Indonesias hot springs. *J. Ocean. Univ. China* **2012**, *11*, 213–218. [CrossRef]

47. Zilda, D.Z.; Harmayani, E.; Widada, J.; Asmara, W.; Irianto, H.E.; Patantis, G.; Fawzya, Y.N. Purification and characterization of the newly thermostable protease produced by *Brevibacillus thermoruber* LII isolated From Padang Cermin Hotspring, Indonesia. *Squalen Bull. Mar. Fish. Postharvest Biotechnol.* **2014**, *9*, 1–10. [CrossRef]

48. Kanekar, P.; Nilegaonkar, S.; Sarnaik, S.; Kelkar, A. Optimization of protease activity of alkaliphilic bacteria isolated from an alkaline lake in India. *Bioresour. Technol.* **2002**, *85*, 87–93. [CrossRef]

49. Esakkiraj, P.; Meleppat, B.; Lakra, A.K.; Ayyanna, R.; Arul, V. Cloning, expression, characterization and application of protease produced by *Bacillus cereus* PMW8. *RSC Adv.* **2016**, *6*, 38611–38616. [CrossRef]

50. Jaouadi, N.Z.; Rekik, H.; Badis, A.; Trabelsi, S.; Belhoul, M.; Yahiaoui, A.B.; Aicha, H.B.; Toumi, A.; Bejar, S.; Jaouadi, B. Biochemical and molecular characterization of a serine keratinase from *Brevibacillus brevis* US575 with promising keratin-biodegradation and hide-dehairing activities. *PLoS ONE* **2013**, *8*, e76722. [CrossRef] [PubMed]

51. Nascimento, W.C.A.D.; Martins, M.L.L. Production and properties of an extracellular protease from thermophilic *Bacillus* sp. *Braz. J. Microbiol.* **2004**, *35*, 91–96. [CrossRef]

52. Kumar, C.G.; Takagi, H. Microbial alkaline proteases: From a bioindustrial viewpoint. *Biotechnol. Adv.* **1999**, *17*, 561–594. [CrossRef]

53. Liu, Y.; Guo, R. Interaction between casein and sodium dodecyl sulfate. *J. Colloid Interface Sci.* **2007**, *315*, 685–692. [CrossRef] [PubMed]

54. Lin, X.; Fusek, M.; Tang, J. Thermopsin, a thermostable acid protease from *Sulfolobus acidocaldarius*. In *Structure and Function of the Aspartic Proteinases*; Springer: New York, NY, USA, 1991; pp. 255–257.

55. Bajorath, J.; Hinrichs, W.; Saenger, W. The enzymatic activity of proteinase K is controlled by calcium. *FEBS J.* **1988**, *176*, 441–447. [CrossRef]

56. Salleh, A.B.; Rahman, N.Z.R.A.; Basri, M. *New Lipases and Proteases*; Nova Publishers: Hauppauge, NY, USA, 2006.

57. Rawlings, N.D. Protease families, evolution and mechanism of action. In *Proteases: Structure and Function*; Springer: New York, NY, USA, 2013; pp. 1–36.

58. Van den Burg, B.; Eijsink, V. Thermolysin and related *Bacillus* metallopeptidases. In *Handbook of Proteolytic Enzymes*, 3rd ed.; Elsevier: New York, NY, USA, 2013; pp. 540–553.

59. Casas, M.P.; González, H.D. Enzyme-Assisted Aqueous Extraction Processes. In *Water Extraction of Bioactive Compounds*; Elsevier: New York, NY, USA, 2018; pp. 333–368.

60. Guinane, C.M.; Kent, R.M.; Norberg, S.; O'Connor, P.M.; Cotter, P.D.; Hill, C.; Fitzgerald, G.F.; Stanton, C.; Ross, R.P. Generation of the antimicrobial peptide caseicin A from casein by hydrolysis with thermolysin enzymes. *Int. Dairy J.* **2015**, *49*, 1–7. [CrossRef]

![microorganisms logo] *microorganisms*

MDPI

Review

Transfer RNA Modification Enzymes from Thermophiles and Their Modified Nucleosides in tRNA

Hiroyuki Hori *[ORCID], Takuya Kawamura, Takako Awai, Anna Ochi, Ryota Yamagami, Chie Tomikawa and Akira Hirata

Department of Materials Science and Biotechnology, Graduate School of Science and Engineering, Ehime University, Bunkyo 3, Matsuyama, Ehime 790-8577, Japan; t.kwmr.0115@gmail.com (T.K.); takak0v0@gmail.com (T.A.); annaoti009@yahoo.co.jp (A.O.); w844038k@yahoo.co.jp (R.Y.); tomikawa.chie.mm@ehime-u.ac.jp (C.T.); hirata.akira.mg@ehime-u.ac.jp (A.H.)
* Correspondence: hori.hiroyuki.my@ehime-u.ac.jp; Tel.: +81-89-927-8548

Received: 12 September 2018; Accepted: 17 October 2018; Published: 20 October 2018

Abstract: To date, numerous modified nucleosides in tRNA as well as tRNA modification enzymes have been identified not only in thermophiles but also in mesophiles. Because most modified nucleosides in tRNA from thermophiles are common to those in tRNA from mesophiles, they are considered to work essentially in steps of protein synthesis at high temperatures. At high temperatures, the structure of unmodified tRNA will be disrupted. Therefore, thermophiles must possess strategies to stabilize tRNA structures. To this end, several thermophile-specific modified nucleosides in tRNA have been identified. Other factors such as RNA-binding proteins and polyamines contribute to the stability of tRNA at high temperatures. *Thermus thermophilus*, which is an extreme-thermophilic eubacterium, can adapt its protein synthesis system in response to temperature changes via the network of modified nucleosides in tRNA and tRNA modification enzymes. Notably, tRNA modification enzymes from thermophiles are very stable. Therefore, they have been utilized for biochemical and structural studies. In the future, thermostable tRNA modification enzymes may be useful as biotechnology tools and may be utilized for medical science.

Keywords: archaea; methylation; pseudouridine; RNA modification; tRNA methyltransferase; tRNA modification

1. Introduction

Transfer RNA is an adaptor molecule required for the conversion of genetic information encoded by nucleic acids into amino acid sequences of proteins [1,2]. Figure 1A shows typically conserved nucleosides in a tRNA molecule, which is represented as a cloverleaf structure (herein, the nucleotide positions in tRNA are numbered, according to Sprinzl et al. [3]). These conserved nucleotides are important for tRNA folding and for stabilization of the L-shaped tRNA structure (Figure 1B) [4–6]. In addition to the standard nucleosides, numerous modified nucleosides in tRNA (for structures, see the MODOMICS and tRNAmodviz databases: http://modomics.genesilico.pl/; http://genesilico.pl/trnamodviz [7]) have been discovered in both thermophilic and mesophilic tRNAs [7,8] (see Supplementary Table S1 for abbreviations of modified nucleosides).

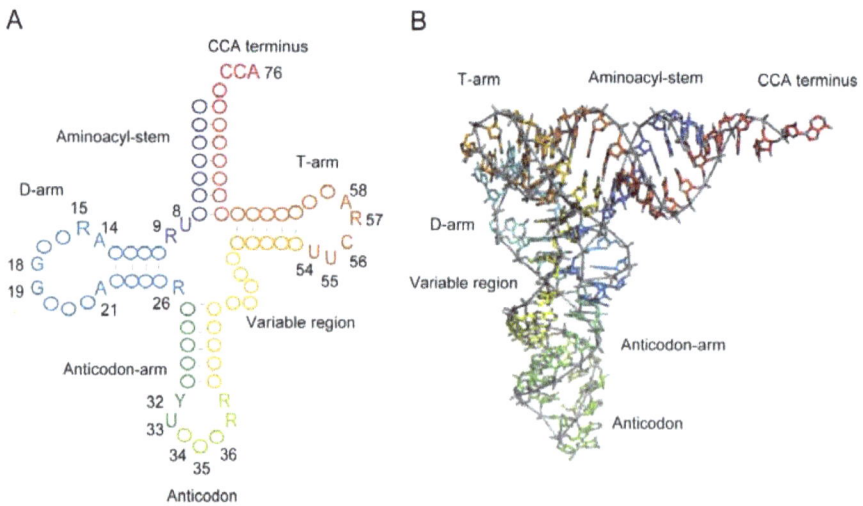

Figure 1. The structure of tRNA. (**A**) Representation of secondary structure of tRNA in a cloverleaf structure. This figure shows tRNA with a short variable region. Conserved nucleosides are shown with position numbers. Abbreviations: R, purine. Y, pyrimidine. (**B**) The L-shaped structure of *Saccharomyces cerevisiae* tRNA^{Phe}. The colors of nucleosides correspond to those in (**A**).

A comprehensive review of the modified nucleosides in tRNA from thermophiles and their positions, distribution, predicted (or confirmed) tRNA modification enzymes and structural effects (Table 1) [9–264], which suggests that the majority of modified nucleosides in tRNA from thermophiles are common to those in tRNA from mesophiles. The functions of modified nucleosides in tRNAs have been gradually elucidated by biochemical and structural studies, physicochemical measurements, and analyses of gene disruption strains. The modified nucleosides primarily function in protein synthesis (e.g., stabilization of tRNA structure [88,265–267], correct folding of tRNA [88,265–267], reinforcement, restriction, and/or alteration of codon-anticodon interaction [108,109,114–117,120,124,268–270], recognition by aminoacyl-tRNA synthetases [109,116,117,271], recognition by translation factors [272], and prevention of the frameshift error [122,123,157,158] among others). In short, living organisms cannot synthesize proteins correctly or efficiently without modifications in tRNA.

For some organisms, modifications in tRNA have not been confirmed, but the tRNA modification enzymes have been studied. For example, although no tRNA sequences from *Thermotoga maritima* have been reported, the properties of several tRNA modification enzymes of this hyper-thermophilic eubacterium have been documented and, thus, the nucleoside modifications are predicted. Although many of the functions and biosynthesis pathways of modified nucleosides in tRNA from thermophiles have not yet been investigated, most of them are considered to be basically common to those from mesophiles. However, thermophiles live in extreme environments (e.g., high temperature, anaerobic conditions, extreme pH, and high pressure). Therefore, it is possible that tRNA modifications observed in thermophiles may have novel functions. Furthermore, in some cases, the biosynthesis pathways of some modifications may differ between thermophiles and mesophiles. Moreover, in eukaryotes, tRNA modifications are related to higher biological processes such as cellular transport of tRNA [273–278], RNA quality control [274,279–281], infection [282–286], and the immune response [287–290]. As yet, modified nucleosides in tRNA from thermophilic eukaryotes have not been investigated, but it is possible that a relationship between modified nucleosides in tRNA and these biological phenomena may also be discovered in thermophilic eukaryotes.

In this review, we focus on the modified nucleosides and tRNA modification enzymes from thermophiles including the difficulties in sequencing the rigid and stable tRNAs from thermophiles.

Since the tRNA modifications in moderate thermophiles are essentially similar to those in mesophiles, we describe them separately from extreme-thermophiles and hyper-thermophiles. We focus on the strategies for tRNA stabilization of extreme hyperthermophiles. Furthemrore, we describe the potential effects of these modifications during oxidative and other environmental stresses at high temperatures. Lastly, we describe biotechnological and therapeutic uses for tRNA modification enzymes. To avoid overlap with previous publications, we intentionally refer to representative articles and reviews of modified nucleosides in tRNA and tRNA modification enzymes from mesophiles (main text and Table 1) to aid understanding by the readers. For example, tRNA modifications in archaea including mesophiles have been extensively covered [48,87,291–294] and pseudouridine modifications and methylated nucleosides in tRNA are reviewed elsewhere [87,203,295,296]. Furthermore, the stability of nucleic acids at high temperatures has been reviewed [297]. Other useful publications are pointed out in the appropriate sections throughout the review.

Table 1. Modified nucleosides in tRNA from thermophiles.

Modified Nucleoside and Position	Distrib.	Modification Enzyme	Predicted Functions and Additional Information	References
Am_6	A	Unknown	Stabilization of aminoacyl-stem Enzymatic activity for Am_6 formation has been detected in the cell extract of *Pyrococcus furiosus*	[9]
m^2G_6	B/A	TrmN/Trm14	Stabilization of aminoacyl-stem	[10–15]
U_8	A	CDAT8	Increasing G–C content in tRNA genes In *Methanopyrus kandleri*, U_8 in several tRNA is produced from C_8 by the deamination [16] In *Methanopyrus kandleri*, numerous nucleosides in RNA may be 2-O-methylated (see main text) [17]	[16,17]
s^4U_8	B/A	ThiI + IscS/ThiI	UV resistance in *E. coli* and *Salmonella typhimurium* (see main text) Stabilization of D-arm structure in *E. coli* (see main text)	[10,11,18–35]
s^4U_8 and s^4U_9	A	ThiI + α?	UV resistance Stabilization of D-arm structure (see main text) Sulfur-containing modifications in tRNA are reviewed in Reference [35].	[36]
m^1A_9	A	Archaeal Trm10	Stabilization of the D-arm structure Prevention of formation of a Watson–Crick base pair Correct folding of the D-arm region	[37,38]
m^1G_9 and m^1A_9	A	archaeal Trm10	Stabilization of D-arm structure Prevention of formation of a Watson–Crick base pair Correct folding of D-arm region *Thermococcus kodakarensis* Trm10 forms m^1G_9 and m^1A_9, whereas *Sulfolobus acidocaldarius* Trm10 forms only m^1A_9	[37,39]
(m^2G_{10} and) $m^2_2G_{10}$	A	archaeal Trm11 (Trm-G_{10}; Trm-$m^2_2G_{10}$ enzyme)	Prevention of formation of a Watson–Crick base pair Correct folding of tRNA in *Pyrococcus furiosus* Correct folding of the D-arm region	[40–43]
Ψ_{13}	B/A	TruD/TruD or archaeal Pus7	Stabilization of D-stem structure Archaeal Pus7 generally catalyzes formation of $\Psi 35$ in tRNA^{Tyr}, but *Sulfolobus solfataricus* Pus7 has weak $\Psi 13$ formation activity [46]	[23,44–46]
G^+_{13}	A	ArcTGT + ArcS?	Stabilization of the D-arm structure *Thermoplasma acidophilum* tRNA^{Leu} exceptionally possesses a G^+_{13} modification and *T. acidophilum* ArcTGT acts on positions 13 and 15 in this tRNA [47]	[36,47]

Table 1. *Cont.*

Modified Nucleoside and Position	Distrib.	Modification Enzyme	Predicted Functions and Additional Information	References
G^+_{15}	A	ArcTGT + ArcS or QueF-like protein	Stabilization of interaction between the D-arm and the variable region. Several archaea possess a split-type ArcTGT [60,61]. Several species in Crenarchaeota possess a QueF-like protein instead of ArcS [60,62,63]. G^+ is not found in nucleosides from a *Stetteria hydrogenophila* tRNA mixture [36].	[25,36,47–63]
D_{17}	B	Dus family protein?	Maintenance of D-loop flexibility. D_{17} and D_{20} modifications have been reported in *Geobacillus stearothermophilus* tRNA. However, D_{17} and D_{20} are formed by DusB and DusA, respectively, in *Escherichia coli* [65,66] and the *G. stearothermophilus* genome possesses only one *dus*-like gene. This is also observed in *Bacillus subtilis*, which is a mesophilic eubacterium.	[18,19,64–66]
Gm_{18}	B	TrmH	Stabilization of the D-arm and the T-arm interaction. TrmH from thermophiles possess broad substrate tRNA specificities as compared with TrmH from *E. coli*. The substrate tRNA specificities of TrmH enzymes differ among thermophiles. TrmH from *Thermus thermophilus* can methylate all tRNA species.	[10,11,20,21,23,24,30,67–81]
D_{20}	B	Dus family protein	Stabilization of local structure of D-loop in *E. coli*? In *A. aeolicus*, the nucleosides at positions 20 and 20a in tRNA^{Cys} are D_{20} and U_{20a}, respectively. Therefore, Dus from *A. aeolicus* may act only on U_{20} in tRNA.	[24,33,65,66,82]
D_{20} and D_{20a}	B	DusA	Stabilization of local structure of the D-loop. The melting temperature of a tRNA mixture from the *E. coli dusA* gene disruptant strain is lower than that from the wild-type strain [33]. Therefore, D_{20} and D_{20a} modifications may contribute to stabilize local structure of the D-loop. *Thermus thermophilus* Dus was recently confirmed as a member of the DusA family [65,66,84,85].	[21–23,33,65–67,83–85]
m^1A_{22}	B	TrmK	Prevention of formation of a Watson–Crick base pair	[18,20,86]
Ψ_{22}	A	Unknown	The Ψ_{13}-Ψ_{22} base pair may stabilize D-arm structure [88]	[87,88]

Table 1. *Cont.*

Modified Nucleoside and Position	Distrib.	Modification Enzyme	Predicted Functions and Additional Information	References
m^2G_{26} and $m^2_2G_{26}$	A	Trm1	Stabilization of three-dimensional core structure Correct folding of tRNA Recently, it has been reported that m^2G_{26} modification is required for correct folding of precursor tRNA^{Ser} from *Schizosaccharomyces pombe* [94]. Therefore, a similar phenomenon may take place in thermophiles.	[9,25,44,89–94]
m^2G_{26}, $m^2_2G_{26}$, m^2G_{27} and $m^2_2G_{27}$	B	Trm1	Stabilization of three-dimensional core structure in *A. aeolicus*. In the case of m^2G_{26} and $m^2_2G_{27}$, stabilization of aminoacyl-stem	[24,95]
$m^2_2Gm_{26}$	A	Trm1 + unknown MT	Stabilization of three-dimensional core structure The presence of m^2_2Gm has been confirmed in nucleosides of a tRNA mixture from several thermophilic archaea [56,97–100]. Although the nucleoside at position 26 in *S. acidocaldarius* tRNA$^{Met}_i$ was originally reported as an unidentified G modification [44], it was recently described as m^2_2Gm [96]. The MT for 2'-*O*-methylation is unknown.	[44,96]
Cm_{32}	A	archaeal TrmJ	Stabilization of anticodon-loop	[96]
Cm_{32} and Nm_{32}	B	TrmJ	Stabilization of anticodon-loop TrmJ from *E. coli* does not recognize the base at position 32 [9,102]. Um_{32} and Am_{32} have not been reported in tRNAs from thermophilic eubacteria.	[9,101,102]
I_{34}	B	TadA	Alteration of codon–anticodon interaction A-to-I editing in tRNA is reviewed in Reference [107]	[103–107]
k^2C_{34}	B	TilS	Alteration of codon–anticodon interaction (*E. coli* and *B. subtilis*) Change of recognition by aminoacyl-tRNA synthetase (*E. coli* and *B. subtilis*) Decoding of AUA codons by k^2C_{34} and agm^2C_{34} modifications is reviewed in References [114,115].	[108–113]
agm^2C_{34}	A	TiaS	Alteration of codon–anticodon interaction (*Arhaeoglobus fulgidus* and *Haloarcula marismourtui*) Change of recognition by aminoacyl-tRNA synthetase (*Arhaeoglobus fulgidus* and *Haloarcula marismourtui*) Decoding of AUA codons by k^2C_{34} and agm^2C_{34} modifications is reviewed in References [114,115].	[114–120]

Table 1. *Cont.*

Modified Nucleoside and Position	Distrib.	Modification Enzyme	Predicted Functions and Additional Information	References
xm^5U_{34} derivatives	B/A	MnmE + MnmG + MnmC (for mnm^5U_{34} in eubacteria)/Elp3? + α (for cm^5U_{34} in archaea) IscS + TusA + TusBCD + TusE + mmmA (for 2-thiolation in *E. coli*) or YrvO + mmmA (for 2-thiolation in *B. subtilis*) SAMP2 + UbaA + NcsA (for 2-thiolation in *M. maripuludis*)	Reinforcement of codon–anticodon interaction (*E. coli* and other mesophiles) Restriction of wobble base pairing (*E. coli* and other mesophiles) Prevention of frameshift errors (*E. coli* and other mesophiles) Biosynthesis pathways of xm^5U_{34} derivatives are not completely clarified. Although the information on xm^5U_{34} derivatives in tRNA from thermophiles is limited, the biosynthesis pathways may be common with those from mesophiles. For the functions and biosynthesis pathways for xm^5U_{34} derivatives, see References [121–132, 136–139,142]. For the thiolation of xm^5s^2U34 derivatives, see References [35,133–135]. *Aquifex aeolicus* exceptionally possesses a DUF752 protein, which is an MT for the xm^5U_{34} modifications without an oxidase domain [136]. A mnm^5U nucleoside has been found in modified nucleosides from unfractionated tRNA in several methane archaea [56]. *Thermoplasma acidophilum* $tRNA^{Leu}$ possesses ncm^5U_{34} [36]. Some thermophiles in Euryarchaea may have a cnm^5U_{34} modification in tRNA [137]. The cm^5U_{34} formation activity of Elp3 from *Methanocaldococcus infernus* has been reported [142]. Several related proteins for synthesis of xm^5U_{34} derivatives from thermophiles have been used for structural studies [136,138–141].	[34–36,56,121–142]
Cm_{34} and $cmnm^5Um_{34}$	B	TrmL	Reinforcement of codon–anticodon interaction (*E. coli*)	[18,143,144]
Gm_{34}	B	Unknown	Reinforcement of codon–anticodon interaction (*G. stearothermophilus*)	[19]
Q_{34} derivatives	B	Tgt + QueA + QueG	Reinforcement of codon–anticodon interaction (*E. coli*) Prevention of frame-shift error (*E. coli*) Biosynthesis pathways and functions of Q derivatives are reviewed in References [152,153]. A crystal structure of QueA from *T. maritima* has been reported [151].	[20,122,145–153]
Cm_{34} and Um_{39} (or Cm_{39})	A	L7Ae + Nop5 + archaeal fibrillarin + Box C/D guide RNA (intron)	Reinforcement of codon–anticodon interaction Reinforcement of anticodon-arm In several archaea, an intron in precursor $tRNA^{Trp}$ functions as a Box C/D guide RNA.	[9,154,155]
Ψ_{35}	A	aPus7 and H/ACA guide RNA system	Reinforcement of codon–anticodon interaction	[46]

Table 1. *Cont.*

Modified Nucleoside and Position	Distrib.	Modification Enzyme	Predicted Functions and Additional Information	References
m^1G_{37}	B/A	TrmD/Trm5	Prevention of frame-shift error (*E. coli* and other mesophiles) Recognition by aminoacyl-tRNA synthetase (*Saccharomyces cerevisiae*)	[36,150–171]
wyosine$_{37}$ derivatives	A	Trm5 + Taw1 + Taw2 + Taw3	Reinforcement of codon–anticodon interaction Prevention of the frame-shift error In several archaea, m^1G_{37} in tRNA^{Phe} is further modified to wyosine derivatives. For the biogenesis pathway of wyosine derivatives, see References [181–183].	[172–183]
t^6A_{37} derivatives	B/A	TsaB, TsaC (TsaC2), TsaD and TsaE/KEOPS complex: Kae1, Bud32, Cgi121 and Pcc1 + Sua5	Reinforcement of codon–anticodon interaction Prevention of frame-shift error Recognition by aminoacyl-tRNA synthetases The biogenesis pathway for t^6A derivatives is reviewed in Reference [191]	[68,184–191]
i^6A_{37} derivatives	B	MiaA + MiaB	Prevention of frame-shift error Reinforcement of codon–anticodon interaction Recognition by aminoacyl-tRNA synthetases i^6A derivatives are reviewed in Reference [197]	[10,11,18–20,24,192–197]
m^6A_{37}	B	YfiC (TrmG2?)		[64,198]
Ψ_{38}, Ψ_{39} and Ψ_{40}	B/A	TruA/Pus3	Prevention of frame-shift error (*E. coli*) Reinforcement of anticodon-arm	[10,11,18–20,23,87,199–203]
m^7G_{46}	B	TrmB	Stabilization of three-dimensional core In the case of *T. thermophilus*, m^7G_{46} modification functions a key factor in a network between modified nucleosides in tRNA and tRNA modification enzymes (see main text) [11]	[10,11,19,67,204–208]
m^5C_{48} and m^5C_{49}	A	archaeal Trm4	Stabilization of three-dimensional core	[9,209,210]
m^7G_{49}	A	Unknown		[36]
m^5C_{51}	A	Unknown	Stabilization of T-arm structure	[209]
m^5C_{52}	A	Unknown	Stabilization of T-arm structure	[209]
Ψ_{54} and Ψ_{55}	A	Pus10	Stabilization of D-arm and T-arm interaction	[211–214]
$m^1\Psi_{54}$	A	Pus10 + TrmY	Stabilization of D-arm and T-arm interaction	[215–217]

Table 1. *Cont.*

Modified Nucleoside and Position	Distrib.	Modification Enzyme	Predicted Functions and Additional Information	References
$m^5U_{54} + m^5s^2U_{54}$	B/A	TrmFO + TtuA + TtuB + TtuC + TtuD + IscS/TrmA + TtuA? + TtuB? + α	Stabilization of D-arm and T-arm interaction (see main text) 2-Thiolation of $m^5s^2U_{54}$ in tRNA is reviewed in Reference [239]	[10,11,21–24,67,97,98,134, 218–239]
Um_{54}	A	Unknown	Stabilization of D-arm and T-arm interaction	[44]
Ψ_{55}	B/A	TruB/Pus10 or archaeal Cbf5 + α	Stabilization of D-arm and T-arm interaction In the case of *T. thermophilus*, Ψ_{55} is required for low-temperature adaptation (see main text) [248].	[10,11,18–20,23,25,36,44,64, 67,211–214,240–248]
Cm_{56}	A	Trm56	Stabilization of D-arm and T-arm interaction	[9,25,36,44,48,89,249–251]
m^2G_{57}	A	Unknown		[44,252]
m^1I_{57}	A	archaeal TrmI + unknown deaminase	Stabilization of T-arm structure	[44,253,254]
m^1A_{57} and m^1A_{58}	A	archaeal TrmI	Stabilization of T-arm structure	[44,255–258]
m^1A_{58}	B	TrmI	Stabilization of T-arm structure	[11,23,67,204,259–264]

This table shows the nucleosides that are modified in tRNA from thermophiles. Most modifications are common to those in tRNA from mesophiles. Several modifications include derivatives and they are summarized as the derivatives (e.g., xm^5U_{34} derivatives). In some cases, only modification enzymes from thermophiles have been reported. For example, although Q derivatives have not been confirmed in tRNA from *T. maritima*, the structure of QueA from *T. maritima* has been reported. In these cases, the modifications are listed here. The references for tRNA modifications and tRNA modification enzymes are mainly those for thermophiles. While there are many references for mesophiles, only representative references are cited. Where available, reviews of a modification and related proteins have been cited. Since modified nucleosides in tRNA from thermophilic eukaryotes have not been reported, modified nucleosides in eukaryotic tRNA have not been included here. The following modified nucleosides have been found in unfractionated tRNA from thermophiles. However, their positions and modified tRNA species are unknown: ac^6A, hn^6A, ms^2hn^6A, methyl-hn^6A, $m^{2.7}Gm$, s^2Um, and ac^4Cm [56,97–100]. Abbreviations are as follows: A, archaea, B, eubacteria, and MT, methyltransferase. The "?" mark indicates the potential function speculated from the structure of the modified nucleosides.

2. Sequencing of tRNA from Thermophiles

The sequence of tRNA provides the most basic information of tRNA. However, as shown in Figure 2, which displays nucleotide sequences of tRNAs from thermophilic eubacteria [10,11,18–24,64,67] and archaea [25,36,44,252], the sequences of only 14 tRNA species have been reported from thermophiles. In addition, in the case of *Aquifex aeolicus* tRNA^{Cys}, the sequence has been only partially determined [24].

In general, sequencing of tRNA from thermophiles is difficult for the following reasons. First, purification of specific tRNA from thermophiles is not easy. Currently, tRNA is purified by the solid DNA probe method [298–300]. In this method, the solid-phase complementary DNA probe is placed in a column and hybridized with the target tRNA and then the target tRNA is eluted from the column. Since the structures of tRNA from thermophiles are very rigid, denaturing the tRNA to allow hybridization is difficult. This problem has been solved by incorporating tetraalkyl-ammonium salt in the hybridization buffer [301]. This salt destabilizes the secondary and tertiary structures of tRNA and promotes formation of the RNA-DNA hetero-duplex. This alteration enabled us to purify *A. aeolicus* tRNA^{Cys} [24], *Thermus thermophilus* tRNA^{Phe} [11], tRNA$^{Met}_{f}$1 [248] and tRNA^{Thr} [263,302], *Thermoplasma acidophilum* initiator tRNA^{Met} [89], elongator tRNA^{Met} [89], and tRNA^{Leu} [36]. Even with the use of tetraalkyl-ammonium salt, however, the solid DNA probe method is not versatile. For example, because the difference between *T. thermophilus* tRNA$^{Met}_{f}$1 and tRNA$^{Met}_{f}$2 is only one G-C base pair in the T-stem (Figure 2H) [21], purification of tRNA$^{Met}_{f}$1 required its separation from tRNA$^{Met}_{f}$2 by BD-cellulose column chromatography before the solid DNA probe method could be applied [248].

Second, since the structure of tRNA from thermophiles is rigid, limited cleavage by formamide [303,304] is difficult. Therefore, it is difficult to apply the classical technique used for RNA sequencing to tRNA from thermophiles. Liquid-chromatography/mass-spectrometry (LC/MS) has been found to be the most reliable method to overcome this problem [305,306]. In general, LC/MS requires prior cleavage of tRNA by RNases. However, because the G-C content in the stem regions of tRNA from thermophiles is very high (Figure 2), RNA fragments with the same sequences are often generated by RNase cleavage. Therefore, use of multiple RNases and/or preparation of gene disruptant strains are required to overcome this problem.

Furthermore, given that it is not possible to distinguish uridine and pseudouridine by MS, cyanoethylation of tRNA is generally required to detect this nucleoside [307]. In the sequencing of *T. acidophilum* tRNA^{Leu} [36], we used a combination of the cyanoethylation and classical formamide method for detection of Ψ_{54} because the efficiency of cyanoethylation of Ψ_{54} was low. Thus, specific techniques are required even if an LC/MS system is available.

Third, to determine the modified nucleoside precisely, preparation of a standard compound is often required. For example, it was necessary to prepare the standard ncm^{5}U nucleoside from the *Saccharomyces cerevisiae trm9* gene disruptant strain [308] to determine the anticodon modification of *T. acidophilum* tRNA^{Leu} [36]. In some cases, synthesis of a standard compound by organic chemistry may be required. Lastly, preparing cultures of thermophiles is not so easy for general biochemical researchers (e.g., under anaerobic conditions at high temperatures).

To overcome these problems, the cooperation of researchers in different fields is required. At present, the solid DNA probe method with tetraalkyl-ammonium coupled with LC/MS is the main method for sequencing tRNA from thermophiles. Therefore, it is anticipated that a large numbers of sequences of tRNA from thermophiles will be reported by using this approach in the future.

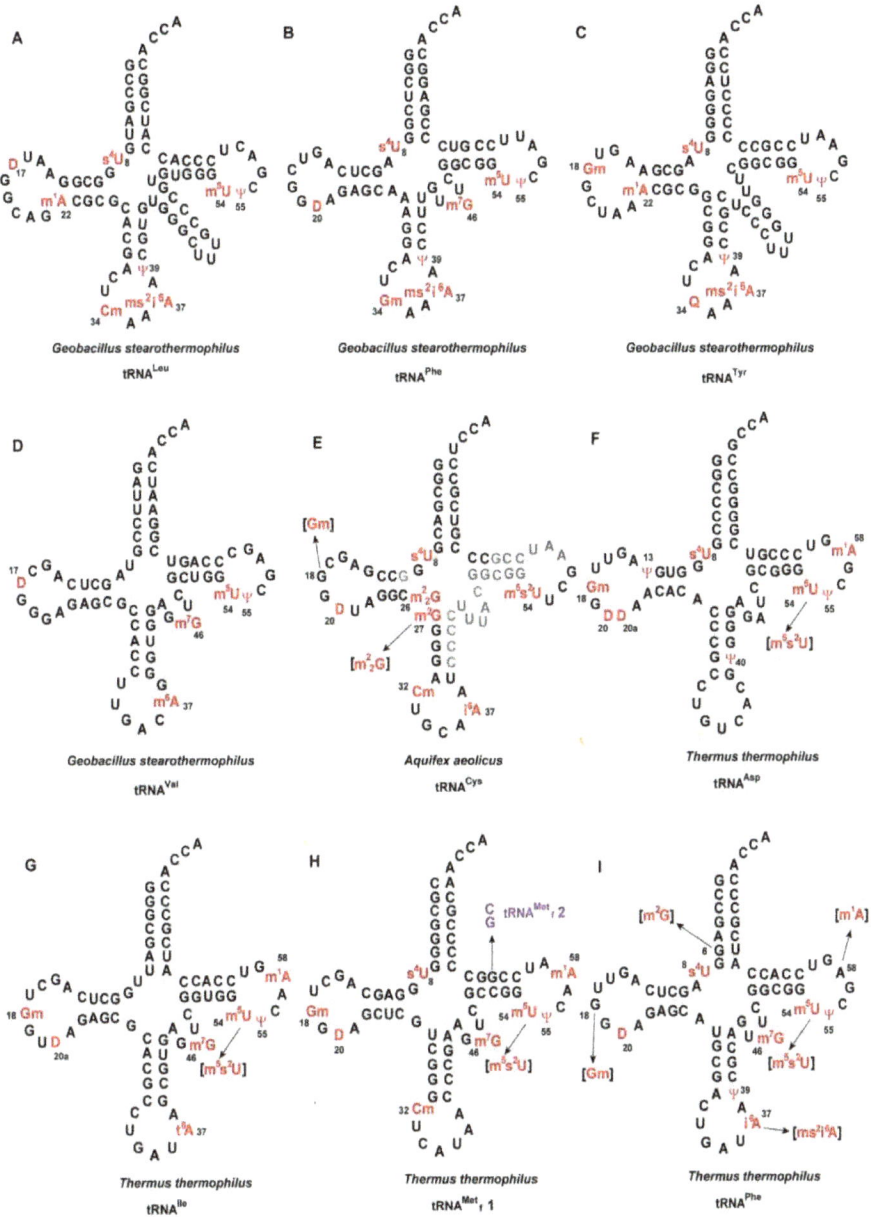

Figure 2. *Cont.*

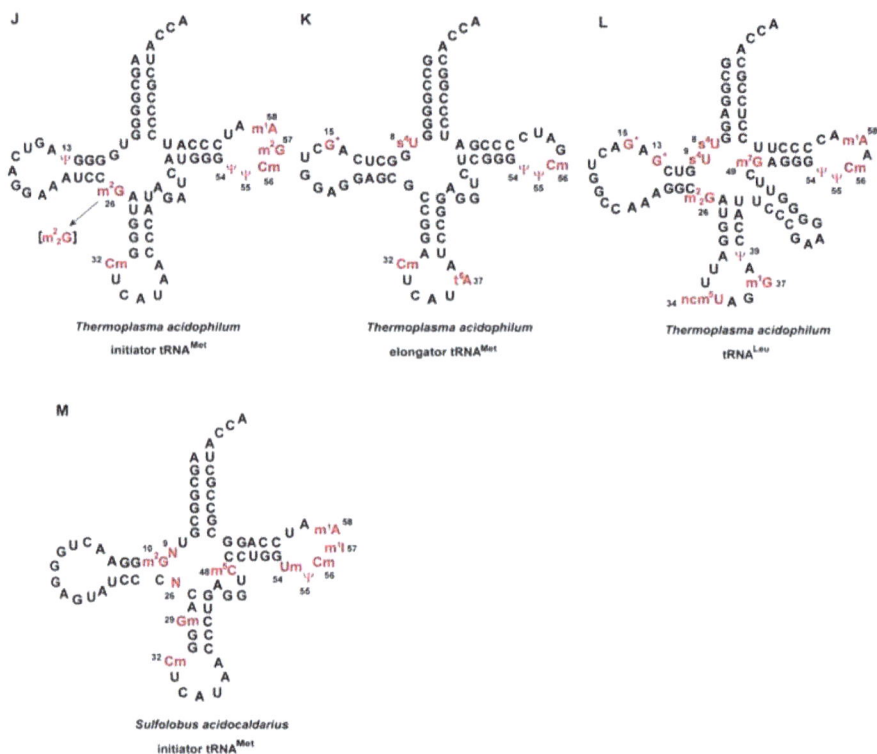

Figure 2. Sequences of tRNA from thermophiles. The modified nucleosides are indicated in red with their positions. Parentheses indicate that a portion of the modified nucleoside is further modified to its derivative. Abbreviations of modified nucleosides are given in Supplementary Table S1. (**A**) *Geobacillus stearothermophilus* tRNA^{Leu}. (**B**) *G. stearothermophilus* tRNA^{Phe}. (**C**) *G. stearothermophilus* tRNA^{Tyr}. (**D**) *G. stearothermophilus* tRNA^{Val}. (**E**) *Aquifex aeolicus* tRNA^{Cys}. (**F**) *Thermus thermophilus* tRNA^{Asp}. (**G**) *T. thermophilus* tRNA^{Ile}. (**H**) *T. thermophilus* tRNA^{Met}f1. (**I**) *T. thermophilus* tRNA^{Phe}. (**J**) *Thermoplasma acidophilum* initiator tRNA^{Met}. (**K**) *T. acidophilum* elongator tRNA^{Met}. (**L**) *T. acidophilum* tRNA^{Leu}. (**M**) *Sulfolobus acidocaldarius* initiator tRNA^{Met} In *A. aeolicus* tRNA^{Cys} (E) the nucleotides shown in gray could not be determined and cyanoethylated tRNA^{Cys} was not analyzed. Therefore, this tRNA may possess additional modifications (e.g., Ψ_{39}, Ψ_{55} and m^{1}A58). *Thermus thermophilus* possesses two tRNA^{Met}f species. The difference of tRNA^{Met}f2 is single G-C base pair, which is indicated in purple in (H). In *S. acidocaldarius* initiator tRNA^{Met} (M), the nucleosides at positions 9 and 26 may be m^{1}A$_9$ and m$^2_2$Gm$_{26}$, respectively.

3. Modified Nucleosides in tRNA from Moderate Thermophiles Are Common to Those from Mesophiles

Seven tRNA sequences from moderate thermophiles (*Geobacillus stearothermophilus* and *T. acidophilum*), which live at below 75 °C, have been reported (Figure 2). Furthermore, the modified nucleosides in unfractionated tRNA from moderate thermophiles (*Methanobacterium thermoaggregans*, *Methanobacterium thermoautotrophicum*, and *Methanococcus thermolithotrophicus*) were analyzed [97,99]. These studies have shown that the modified nucleosides in tRNA from moderate thermophiles are typically common to those in tRNA from mesophiles. In summarizing the information on tRNA modifications and tRNA modification enzymes by thermophilic species [8–330] (Table 2), we have separately considered moderate thermophiles, extreme-thermophiles, and hyper-thermophiles. However, there are some differences between moderate thermophiles and mesophiles. For example, the degree of 2′-*O*-methylation in tRNA from *G. stearothermophilus* is increased at high temperatures [309]. Furthermore, several modifications (Gm_{18}, D modifications, and Gm_{34}) in tRNA from *G. stearothermophilus* cannot be explained by the enzymatic activities of the already-known tRNA modification enzymes, which is described in Table 2. Moreover, *T. acidophilum* possesses several distinct tRNA modifications such as G^+_{13} and m^7G_{49} (Table 2) [36]. Although these differences are present, thermophile-specific modified nucleosides have not been found in tRNA from moderate thermophiles, which suggests that living organisms can survive at 75 °C via the tRNA modifications in mesophiles.

Table 2. Thermophiles: their tRNA modifications and tRNA modification enzymes.

Species	Predicted Enzyme	Distinct tRNA Modifications and General Information	References
Moderate Thermophiles			
Eubacteria			
Geobacillus stearothermophilus (*Bacillus stearothermophilus*) 30–75 °C		Sequences of tRNA^{Leu} [18], tRNA^{Phe} [19], tRNA^{Tyr} [20], and tRNA^{Val2} [64] have been reported (Figure 2). The majority of modifications in tRNA are similar to those in *B. subtilis*. With increasing culture temperature, the extent of 2′-O-methylation in the tRNA mixture increases [309].	
	Gm$_{18}$ (TrmH?)	Although *trmH* is not encoded in the *B. subtilis* genome, a *trmH*-like gene is encoded in the *G. stearothermophilus* genome. Gm$_{18}$ has been found in tRNA^{Tyr} but not in tRNA^{Leu}. This modification pattern suggests that the substrate tRNA specificity of *G. stearothermophilus* TrmH may be different from that of other known TrmH enzymes.	[20]
	D$_{17}$, D$_{20}$ and D$_{20a}$ (Dus family protein?)	In *G. stearothermophilus* tRNA, D$_{17}$, D$_{20}$, and D$_{20a}$ modifications have been reported. In *E. coli*, three Dus family proteins known as DusA, DusB, and DusC, produce D$_{20}$ and D$_{20a}$, D$_{17}$, and D$_{16}$, respectively [65,66]. In the *G. stearothermophilus* genome, however, only one gene is annotated as a *dus*-like gene. Therefore, D modifications in *G. stearothermophilus* cannot be explained by the tRNA substrate specificity of the known Dus proteins.	[17,19,64]
	m^{1}A$_{22}$ (TrmK?)	The m^{1}A$_{22}$ modification has been found in tRNA^{Tyr} and tRNA^{Ser} from *B. subtilis* and *Mycoplasma capricolum* [310,311]. *G. stearothermophilus* tRNA^{Leu} and tRNA^{Tyr} possess m^{1}A$_{22}$ [18,20]. The presence of a *trmK*-like gene in the genome of *G. stearothermophilus* has been reported [86].	[20,86]
	Gm$_{34}$ (unknown MT)	*G. stearothermophilus* tRNA^{Phe} possesses Gm$_{34}$ (Figure 2B) [19]. In contrast, the nucleoside at position 34 in *E. coli* tRNA^{Phe} is unmodified G. Given that *E. coli* TrmL acts only on tRNA^{Leu} isoacceptors [143], the 2′-O-methylation of G$_{34}$ in tRNA^{Phe} from *G. stearothermophilus* is cannot be simply explained by the activity of known TrmL.	[19]
	m^{6}A$_{37}$ (YfiC; TrmG?)		[198]
Archaea			
Methanobacterium thermoaggregans Optimum growth temperature 60 °C		Sequences of tRNA^{Asn} and tRNA^{Gly} have been reported [8].	
Methanobacterium thermoautotrophicum 45–75 °C		The modified nucleosides in unfractionated tRNA are essentially common to those in tRNA from mesophilic methane archaea [97].	

Table 2. Cont.

Species	Predicted Enzyme	Distinct tRNA Modifications and General Information	References
Methanococcus thermolithotrophicus 17–62 °C		The modified nucleosides in unfractionated tRNA are essentially common to those in tRNA from mesophilic methane archaea [99].	
Thermoplasma acidophilum Optimum growth temperature 55–60 °C		Sequences of tRNA^{Met}i [44,252], tRNA^{Met}m [25], and tRNA^{Leu} [36] have been reported. Several recombinant tRNA modification enzymes have been used for biochemical studies.	
	s^{4}U$_8$ and s^{4}U$_9$ (ThiI? + α)	The s^{4}U$_9$ modification has been found in tRNA^{Leu} [36]. The sulfur donor for s^{4}U formation is unknown [35].	[36]
	G$^+$$_{13}$ and G$^+$$_{15}$ (ArcTGT + ArcS?)	The G$^+$$_{13}$ modification has been found only in tRNA^{Leu} from *T. acidophilum* [36]. ArcTGT from *T. acidophilum* acts on both G13 and G15 in tRNA^{Leu} [47].	[36,47]
	m$^2_2$G$_{26}$ (Trm1)		[89]
	ncm^{5}U$_{34}$ (Elp3?)		[36]
	m^{1}G$_{37}$ (Trm5)		[89]
	m^{7}G$_{49}$ (unknown MT)		[36]
	Cm$_{56}$ (Trm56)	The presence of unusual *Trm56*-like gene in the *T. acidophilum* genome has been reported in a bioinformatics study [250]. The Trm56 enzymatic activity has been confirmed via the recombinant protein [89]. *T. acidophilum* Trm56 exceptionally possesses a long C-terminal region in the SPOUT tRNA MT [312].	[89,250,312]
Extreme-thermophiles and Hyper-thermophiles **Eubacteria**			
Aquifex aeolicus Optimum growth temperature 85–94 °C	Gm$_{18}$ (TrmH)	The partial sequence of tRNA^{Cys} has been reported [24] (Figure 2E). Several tRNA MT activities have been detected in the *A. aeolicus* cell extract using an *E. coli* tRNA mixture [24]. The tRNA modification enzymes listed below were characterized via recombinant proteins.	[74,77]
	D$_{20}$ (Dus)	D$_{20}$ exists in tRNA^{Cys}. However, the nucleoside at position 20a is unmodified U [24]. Therefore, *A. aeolicus* Dus may act only on U$_{20}$.	[24,82]
	m^{2}G$_{26}$, m$^2_2$G$_{26}$, m^{2}G$_{27}$ and m$^2_2$G$_{27}$ (Trm1)	*Aquifex aeolicus* exceptionally possesses Trm1 in eubacteria [24].	[24,95]
	I$_{34}$ (TadA)		[104,105]

Table 2. *Cont.*

Species	Predicted Enzyme	Distinct tRNA Modifications and General Information	References
	mnm^5U_{34} (MnmC2)	MnmC catalyzes the final methylation step of mnm^5U synthesis. *Aquifex aeolicus* MnmC2 comprises only an MT domain.	[136]
	(MnmD; previously called GidA)		[140,141]
	k^2C_{34} (TilS)		[111–113]
	m^1G_{37} (TrmD)	The dimer structure of *A. aeolicus* TrmD is stabilized by inter-subunit disulfide bonds [165].	[164,162,165]
	m^7G_{46} (TrmB)	TrmB proteins from thermophiles (*A. aeolicus* and *T. thermophilus*) possess a long C-terminal region.	[206–208]
	m^5U_{54} and $m^5s^2U_{54}$ (TrmFO)	The presence of *trmFO* gene in *A. aeolicus* genome was initially described in Reference [221].	[24,221]
	m^1A_{58} (TrmI)		[257,262]
Thermotoga maritima 80–90 °C	hn^6A (?)	Sequences of tRNA from *T. maritima* have not been reported. Recombinant proteins have been used for biochemical and structural studies. hn^6A was first identified in modified nucleosides from unfractionated tRNA from *T. maritima* [313]. Because hn^6A was subsequently found in modified nucleosides from psychrophilic archaea [56], it is not a thermophile-specific modification. *Thermotoga maritima* and *Thermodesulfobacterium commune* exceptionally possess hn^6A in eubacteria. The modification position in tRNA, modified tRNA species, and biosynthesis pathway of hn^6A are unknown.	[56,313]
	s^4U_8 (ThiI + IscS)		[31,32]
	oQ_{34} (QueA)		[151]
	mnm^5U_{34} (TrmE)		[138,139]
	t^6A_{37} (TsaB, TsaC/TsaC2, TsaD and TsaE)		[190]
	$ms^2t^6A_{37}$ (MiaB)		[194–196]
	m^1G_{37} (TrmD)		[171]
	m^5U_{54} and $m^5s^2U_{54}$ (TrmFO and TtuA)	The m^5s^2U nucleoside has been found in unfractionated tRNA from *T. maritima* [7].	[97,184,221,222]
	Ψ_{55} (TruD)		[244–247]

Table 2. *Cont.*

Species	Predicted Enzyme	Distinct tRNA Modifications and General Information	References
Thermodesulfobacterium commune Optimum growth temperature 70 °C	m^{1}A$_{58}$ (TrmI)	hn^{6}A and ms^{2}hn^{6}A have been found in modified nucleosides from unfractionated tRNA from *T. commune*. The ms^{2}hn^{6}A modification may be derived from hn^{6}A. So far, *T. commune* is the only eubacterium that possesses ms^{2}hn^{6}A in tRNA. The modification position in tRNA, modified tRNA species, and biogenesis pathway of hn^{6}A and ms^{2}hn^{6}A are unknown.	[257]
	hn^{6}A and ms^{2}hn^{6}A (?)		[313]
Thermus flavus Optimum growth temperature 70 °C		Partial purification of tRNA m^{1}A58 MT has been reported: the activity of tRNA m^{7}G$_{46}$ MT has also been described [204].	
Thermus thermophiles 50-83 °C		Sequences of tRNA^{Met}f1 [21], tRNA^{Met}f2 [21], tRNA^{Ile}1 [67], tRNA^{Asp} [23], and tRNA^{Phe} [10,11] have been reported (Figure 2). Partial sequences of tRNA$^{Ser}_{GGA}$ [259], tRNA$^{Pro}_{CGG}$ [314], and tRNA$^{Pro}_{CGA}$ [314] have been determined. The modification extent of Gm$_{18}$, m^{5}s^{2}U$_{54}$ and m^{1}A$_{58}$ changes with the culture temperature. At high temperatures (>75 °C), m^{7}G$_{46}$ [11], m^{5}s^{2}U$_{54}$ [230], and m^{1}A$_{58}$ [260] modifications are essential for survival. At low temperatures (<55 °C), Ψ$_{55}$ is essential for survival [248] and m^{5}U$_{54}$ supports this effect [225] (see the main text). Recombinant proteins have been used for biochemical and structural studies.	
	m^{2}G$_6$ (TrmN)		[10,11,13–15]
	Gm$_{18}$ (TrmH)		[10,11,21,23,30,69–73,75,76,78–81]
	D$_{20}$ and D$_{20a}$ (DusA)		[10,11,23,67,83–85]
	Cm$_{34}$ and cmnm^{5}Um$_{34}$ (TrmL)		[144]
	Ψ$_{39}$ and Ψ$_{40}$ (TruA)		[10,11,18,23,202]
	m^{7}G$_{46}$ (TrmB)		[10,11,21,23]
	m^{5}U$_{54}$ and m^{5}s^{2}U$_{54}$ (TrmFO + TtuA + TtuB + TtuC + TtuD + IscS)		[10,11,17,21,23,67,218–236,239]
	Ψ$_{55}$ (TruB)		[10,11,21,23,67,248]
	m^{1}A$_{58}$ (TrmI)		[11,30,257,259–261,263,264]

Table 2. *Cont.*

Species	Predicted Enzyme	Distinct tRNA Modifications and General Information	References
Archaea			
Aerophyrum pernix 80–100 °C	Ψ_{13} and Ψ_{15} (archaeal Pus7 and H/ACA guide RNA system)	A guide RNA for Ψ formation has been predicted based on genome sequencing.	[46]
Archaeoglobus fulgidus 60–95 °C	agm^2C_{34} (TiaS)	Modified nucleosides in unfractionated tRNA from *A. fulgidus* have been reported [97].	[116,118,119,315]
Methanocaldococcus igneus (*Methanococcus igneus*; *Methanotorris igneus*) 45–91 °C		Modified nucleosides in unfractionated tRNA from *M. igneus* have been reported [56,99].	
Methanocaldococcus infernus 55–92 °C	cm^5U_{34} (Elp3)		[142]
Methanocaldococcus jannashii (*Methanococcus janaschii*) 48–94 °C	m^2G_6 (Trm14)	Although sequences of tRNA are unknown, the recombinant proteins listed below have been used for biochemical and structural studies.	[12]
	G^+_{15} (ArcTGT + ArcS)		[51,59]
	Cm_{34} and Um_{39} (L7Ae, Nop5, aFib, Box C/D guide RNA system)		[316]
	m^1G_{37} (Trm5)		[159,161,163,164,166–170]
	$imG2_{37}$ (Trm5b + Taw1)		[173,179]
	$yW\text{-}86_{37}$ (Taw2)		[174]
	m^5C_{48} and m^5C_{49} (archaeal Trm4)		[210]
	Ψ_{54} and Ψ_{55} (Pus10)		[211–214]
	$m^1\Psi_{54}$ (Pus10 + TrmY)		[215–217]
	Ψ_{55} (archaeal Cbf5)		[240]
Methanopyrus kandleri 84–110 °C (Strain 116: up to 122 °C)		Many unique modified nucleosides have been found in unfractionated tRNA [100]. tRNAs likely contain many 2′-O-methylated nucleosides derived from the C/D box guide RNA system [17].	

Table 2. *Cont.*

Species	Predicted Enzyme	Distinct tRNA Modifications and General Information	References
	ac^6A (?)	The ac^6A nucleoside has been purified from the modified nucleosides in unfractionated tRNA and its structure determined. The modification site, modified tRNA species, and biosynthesis pathway are unknown.	[100]
Methanothermus fervidus 80–97 °C	U_8 (CDAT8)	Only tRNA genes were reported in an early study [317].	[16]
Nanoarchaeum equitans 70–98 °C		A unique tRNA processing system has been found [318,319]. The processing of small RNAs in *N. equitans* is reviewed in Reference [320].	
	m^1G_{37} and $imG2_{37}$ (Trm5a)		[176]
	m^5U_{54} (TrmA-like protein)		[237]
Pyrobaculum aerophilum Optimum growth temperature 100 °C	Cm_{56} (L7Ae, Nop5, aFib, Box C/D guide RNA system)	Cm_{56} in tRNA is generally produced by Trm56. However, this modification in *P. aerophilum* is synthesized by the C/D box guide RNA system.	[249]
Pyrobaculum calidifontis 90–95 °C	G^+_{15} (ArcTGT + QueF-like protein)	Eubacterial QueF catalyzes the conversion from $preQ_0$ to $preQ_1$. In *P. caldifontis*, however, QueF-like protein catalyzes the conversion from $preQ_0$ at position 15 in tRNA to G^+_{15}.	[60,62,63]
Pyrobaculum islandicum Optimum growth temperature 100 °C		Modified nucleosides in unfractionated tRNA from *P. islandicum* have been reported [97].	
Pyrococcus abyssi Optimum growth temperature 96 °C		No tRNA sequence has been determined. However, the tRNA modification enzymes listed below have been characterized.	
	m^2G_{10} and $m^2_2G_{10}$ (archaeal Trm11, Trm-G10 enzyme, Trm-m22G10 enzyme)		[40,41]
	Ψ_{13} and Ψ_{35} (archaeal Pus7 and H/ACA guide RNA system)		[46]
	Cm_{34} and Um_{39} (L7Ae, Nop5, aFib, and C/D box guide RNA system)	Cm_{34} and Um_{39} in tRNA^{Trp} are formed by the C/D box guide RNA system in which the intron functions as a guide RNA.	[154,155]
	m^1G_{37} (Trm5b)		[180]
	m^1G_{37} and $imG2_{37}$ (Trm5a)		[176,177,179]

Table 2. *Cont.*

Species	Predicted Enzyme	Distinct tRNA Modifications and General Information	References
	imG-14$_{37}$ (Taw1)		[173,175]
	t^{6}A37 (Kae1)		[185]
	(KEOPS complex)		[184]
	(Sua5 + KEOPS complex)		[187,189]
	m^{5}C$_{48}$ and m^{5}C$_{49}$ (archaeal Trm4 + archaese)		[209]
	m^{5}U$_{54}$ (TrmA-like protein, PAB0719)		[237,238]
	Ψ$_{55}$ (Cbf5 + Nop10)		[241]
	Cm$_{56}$ (Trm56)		[249]
	m^{1}A$_{57}$ and m^{1}A$_{58}$ (archaeal Trm1)		[255–258]
Pyrococcus furiosus Optimum growth temperature 100 °C		Modified nucleosides in unfractionated tRNA from *P. furiosus* have been reported [98]. Activity of several tRNA modification enzymes has been detected in the cell extract of *P. furious* [9].	
	m^{2}G$_6$ (Trm14)		[13,15]
	m^{2}G$_{10}$ and m$^2{}_2$G$_{10}$ (archaeal Trm11, Trm-G10 enzyme, Trm-m22G10 enzyme)		[42]
	G$^+{}_{15}$ (ArcTGT)		[57]
	m^{2}G$_{26}$ and m$^2{}_2$G$_{26}$ (Trm1)		[91,92]
	t^{6}A37 (KEOPS complex)		[188]
	Ψ$_{54}$ and Ψ$_{55}$ (Pus10)		[212,214]
	Ψ$_{55}$ (Cbf5 + Nop10 + Gar1)		[242]
Pyrococcus horikoshii 80–102 °C		The crystal structure of Nop5 in the C/D box guide RNA system from *P. horikoshii* has been solved [121].	
	G$^+{}_{15}$ (ArcTGT)		[50,52–55,89]
	m^{2}G$_{26}$ and m$^2{}_2$G$_{26}$ (Trm1)		[89,93]

Table 2. *Cont.*

Species	Predicted Enzyme	Distinct tRNA Modifications and General Information	References
	yW-86$_{37}$ (Taw2)		[174]
	m^{5}s^{2}U$_{54}$ (TtuA)		[233]
	Cm$_{56}$ (Trm56)		[251]
Pyrodictium occultum Optimum growth temperature 105 °C		Modified nucleosides in unfractionated tRNA have been analyzed and many 2′-O-methylated nucleosides found [97,98]. mimG was originally found among the modified nucleosides in tRNAs from *P. occultum*, *Sulfolobus solfaticus*, and *Thermoproteus neutrophilus* [322]. Although the melting temperature of *P. occultum* tRNA$^\text{Met}$i transcript is only 80 °C and that of native tRNA$^\text{Met}$i is more than 100 °C (see main text) [323].	
Pyrolobus fumarii This archaeon can survive at 113 °C.		Modified nucleosides in unfractionated tRNA have been analyzed [324].	
Stetteria hydrogenophila Optimum growth temperature 95 °C		Modified nucleosides in unfractionated tRNA have been analyzed and methyl-hn^{6}A, ms^{2}hn^{6}A, and m^{2,7}Gm identified [56].	
Sulfolobus acidocaldarius Optimum growth temperature 75–80 °C	m^{1}A$_9$ (archaeal Trm10)	Sequence of tRNA$^\text{Met}$i has been reported [44]. The m^{1}I$_{57}$ modification was originally found in tRNAs from *S. acidocaldarius* and *Haloferax volcanii* [253]. G$^+$ was first isolated from the nucleosides in *S. acidocaldarius* tRNAs and its structure determined [49]. The structures of wyosine derivatives (imG-14 and imG2) have been determined by using the nucleosides from *S. acidocaldarius* tRNAs [325].	[37,38]
	Ψ$_{13}$ and Ψ$_{35}$ (archaeal Pus7 and H/ACA guide RNA system)		[46]
	Cm$_{32}$ (archaeal TrmJ)		[96]
Sulfolobus solfaticus 55–90 °C		mimG was originally found among the modified nucleosides in tRNAs from *P. occultum*, *S. solfaticus*, and *Thermoproteus neutrophilus* [322]. The structure of box C/D RNP from *S. solfaticus* has been reported [326].	
	agm^{2}C (TiaS)	The identification of agm^{2}C$_{34}$ in *Haloarcula marismortui* tRNA$^\text{Ile2}$ and the presence of agm^{2}C in *S. solfaticus* tRNA have been reported.	[117]
	Ψ$_{13}$ and Ψ$_{35}$ (archaeal Pus7 and H/ACA guide RNA system)	Generally, Ψ$_{35}$ in tRNA$^\text{Tyr}$ is synthesized by archaeal Pus7. However, Pus7 from *S. solfaticus* possesses weak Ψ$_{13}$ formation activity but not Ψ$_{35}$ formation activity. In *S. solfaticus* and *A. pernix*, a guide RNA for Ψ$_{35}$ formation exists.	[46]

Table 2. *Cont.*

Species	Predicted Enzyme	Distinct tRNA Modifications and General Information	References
	$imG2_{37}$ (Trm5a; SSO2439 protein)	Trm5a (SSO2439 protein) does not possess m^1G_{37} formation activity and is used only for imG2 formation.	[178]
	$mimG_{37}$ (Taw3)		[180]
Sulfolobus tokodaii This archaeon can survive at 87 °C.	Ψ_{13} and Ψ_{35} (archaeal Pus7 and H/ACA guide RNA system)		[46]
	t^6A_{37} (Sua5)		[327–329]
Thermococcus celer This archaeon can survive at 85 °C.		Although tRNA genes were analyzed in an early study [330], there is no information on tRNA modifications.	
Thermococcus kodakarensis (*Thermococcus kodakaraensis*; *Pyrococcus kodakarensis*) 65–100 °C	m^1A_9 and m^1G_9 (archaeal Trm10)		[37,39]
	m^2G_{10} and $m^2_2G_{10}$ (archaeal Trm11, Trm-G10 enzyme, Trm-m22G10 enzyme)		[43]
	G^+_{15} (ArcTGT)		[47]
	m^5U_{54} (TrmA-like protein)		[237]
Thermoproteus neutrophilus Optimum growth temperature 85 °C		Modified nucleosides in unfractionated tRNA have been analyzed [97]. mimG was originally found among the modified nucleosides in tRNAs from *P. occultum*, *S. solfataricus*, and *T. neutrophilus* [322].	

Only distinct modifications that have been investigated are listed by thermophile species. In many cases, only tRNA modification enzymes (rather than modifications) have been studied by using recombinant proteins. For example, the presence of the m^7G_{46} modification has not been confirmed in tRNA from *A. aeolicus*, but TrmB (tRNA m^7G_{46} MT) has been characterized through the recombinant protein. In this case, m^7G_{46} (TrmB) is listed in the section "*Aquifex aeolicus*". The moderate thermophiles and extreme-thermophiles along with hyper-thermophiles are separated. Transfer RNA modifications in thermophilic eukaryotes are unknown. Abbreviation: MT, methyltransferase.

4. Strategies of tRNA Stabilization by Modified Nucleosides in Extreme-Thermophiles and Hyper-Thermophiles

In general, the G-C content in the stem regions of tRNA from thermophiles is very high (Figure 2). However, the stability of tRNA from thermophiles cannot be explained only by the increase in G-C content in the stem region. For example, although the melting temperature of *T. thermophilus* tRNA^{Phe} transcript is 76 °C, that of the native tRNA^{Phe} is 84.5 °C [11]. Thus, modified nucleosides are essentially required for stabilization of tRNA at high temperatures. Modified nucleosides in tRNA from thermophiles have been studied mainly from the view point of tRNA stabilization. So far, only a few modified nucleosides specific to thermophiles have been found (Figure 3). These thermophile-specific modified nucleosides seem to stabilize the tRNA structures at high temperatures. As described below, extreme-thermophiles and hyper-thermophiles possess two strategies of tRNA stabilization by modified nucleosides. One is based on thermophile-specific modification such as m^{5}s^{2}U54 (Figure 3A) and the other is based on 2′-O-methylations at multiple positions in tRNA (Figure 3B–E). Recently, the unknown modified nucleoside at position 26 in *Sulfolobus acidocaldarius* tRNA^{Met} (Figure 2M) was described as m$^2_2$Gm [96]. On the whole, however, the modification site(s), modified tRNA species, and biosynthesis pathways of most thermophile-specific modified nucleosides are unknown. Moreover, these nucleosides may have additional functions at high temperatures beyond their structural effect.

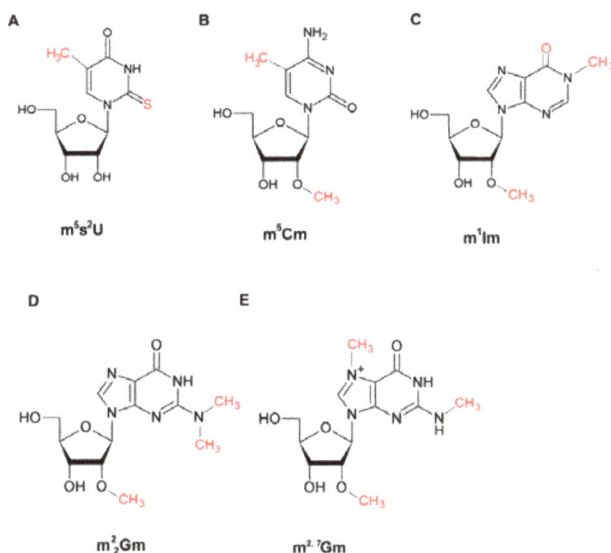

Figure 3. Thermophile-specific modified nucleosides in tRNA. Abbreviations of modified nucleosides are given in Supplementary Table S1. (**A**) m^{5}s^{2}U. (**B**) m^{5}Cm. (**C**) m^{1}Im. (**D**) m$^2_2$Gm. (**E**) m^{2,7}Gm. The modifications are indicated in red.

4.1. m^{5}s^{2}U$_{54}$ Is a Typical Thermophile-Specific Modified Nucleoside in tRNA

The m^{5}s^{2}U$_{54}$ modification was originally found in tRNA from *T. thermophilus* [331]. Subsequently, this modified nucleoside was found in tRNA from *A. aeolicus*, *T. maritima*, *Pyrococcus abyssi*, *Pyrococcus horikoshii*, and *T. kodakarensis* (Table 2) but not from mesophiles. The m^{5}s^{2}U$_{54}$ modification forms a reverse Hoogsteen base-pair with A$_{58}$ (or m^{1}A$_{58}$) in tRNA and stabilizes the tRNA structure by stacking with the G$_{51}$–C$_{61}$ base-pair [220]. Because the 2-thio-modification at position 54 increases the melting temperature of tRNA by more than 3 °C [22,218,220], the m^{5}s^{2}U$_{54}$ modification contributes to stabilization of the tRNA structure. The degree of m^{5}s^{2}U$_{54}$ modification increases with an increasing temperature [22,67,220,229]. At 80 °C, the extent of m^{5}s^{2}U$_{54}$ modification in tRNA is almost 100% [22,67,220,229]. The melting temperature of tRNA mixture is maintained above 85 °C due to the presence of $m^5s^2U_{54}$ modification [229] and *T. thermophilus* can grow at 50 to 83 °C. Thus, living organisms can survive at 80 °C due to the presence of $m^5s^2U_{54}$ modification in tRNA.

4.2. The Network Between Modified Nucleosides in tRNA and tRNA Modification Enzymes in T. thermophilus Adapts Protein Synthesis at Low and High Temperatures

Under natural conditions, the temperature of hot spring water fluctuates for several reasons including an influx of river water, snowfall, and an eruption of hot water. In accordance with these temperature changes, *T. thermophilus* can synthesize proteins efficiently at a wide range of temperatures (50 to 83 °C) by regulating the flexibility (rigidity) of its tRNA [220]. At high temperatures (above 75 °C), three modified nucleosides in tRNA, $m^5s^2U_{54}$ [230], m^1A_{58} [260], and m^7G_{46} [11] are essential for survival of *T. thermophilus*. The m^1A_{58} modification is one of the positive determinants for the two-thiolation system of $m^5s^2U_{54}$. Thus, a *T. thermophilus* disruptant strain of the *trmI* gene encoding the tRNA m^1A_{58} methyltransferase cannot grow at 80 °C [229,260]. The presence of m^7G_{46} modification in tRNA increases the speed of tRNA modification enzymes such as TrmH for Gm_{18}, TrmD for m^1G_{37}, and TrmI for m^1A_{58} [11]. The m^1A_{58} modification further increases the rate of sulfur-transfer to m^5U_{54} by the 2-thiolation system and the introduced modified nucleosides coordinately stabilize the tRNA structure. Thus, the m^7G_{46} modification produced by TrmB is a key factor in the network between modified nucleosides in tRNA and tRNA modification enzymes of *T. thermophilus* at high temperatures. In the *trmB*-gene disruptant starin, tRNA^{Phe} and tRNA^{Ile} were found to be degraded by a temperature shift from 70 °C to 80 °C and heat-shock proteins were not synthesized efficiently [11].

At low temperatures (below 55 °C), in contrast, the Ψ_{55} modification produced by TruB is essential for the survival of *T. thermophilus* [248]. The presence of Ψ_{55} stabilizes both the local structure of the T-arm and the interaction of the T-arm with the D-arm in tRNA. The local rigidity in tRNA caused by Ψ_{55} slows down the speeds of introducing modified nucleosides around Ψ_{55} (Gm_{18}, $m^5s^2U_{54}$ and m^1A_{58}), which maintains the flexibility of tRNA at low temperatures. The presence of m^5U_{54} modification by TrmFO supports this effect of Ψ_{55} [225].

It should be mentioned that D modifications are thought to bring flexibility to tRNA because D does not stack with other bases and brings about the C2'-endo form of ribose [332]. However, a *T. thermophilus* disruptant strain of the *dusA* gene encoding tRNA D_{20}/D_{20a} synthase did not show growth retardation at 50, 60, 70, or 80 °C, and abnormal modifications were not observed in tRNA from this strain [85]. Therefore, the function of D_{20} and D_{20a} modifications is unknown. Since DusA recognizes the interaction of T-arms and D-arms in tRNA [84], the stabilization of the L-shaped tRNA structure by other modified nucleosides is required for the efficient introduction of D_{20} and D_{20a} at high temperatures [85]. Thus, D_{20} and D_{20a} are relatively late modifications in *T. thermophilus* tRNA.

Although the above network is a temperature adaptation system of *T. thermophilus*, it regulates the order in which modified nucleosides are introduced into tRNA. Similar networks have been found in mesophiles [333]. In *Escherichia coli*, for example, the 2'-O-methylation at position 34 by TrmL requires an i^6A_{37} modification [334]. However, the network in *T. thermophilus* is distinct because it regulates the structure of a three-dimensional core and many modifications in tRNA are related. One of the advantages of this system is that protein synthesis is not required. The response of the system is very rapid. It is possible that thermophilic archaea possess a similar network between modified nucleosides in tRNA and tRNA modification enzymes because some of them can also grow at a wide range of temperatures.

4.3. Stabilization of tRNA Structure by 2'-O-Methylation

Because 2'-O-methylation shifts the equilibrium of ribose puckering to the C3'-endo form and enhances the hydrophobic interaction, this modification, when carried out at multiple positions, brings rigidity of tRNA. Furthermore, 2'-O-methylations prevents hydrolysis of phophodiester-bonds in tRNA at high temperatures. Therefore, 2'-O-methylations may prolong the half-lives of tRNA. Notably, there is a living organism in which tRNA is stabilized without $m^5s^2U_{54}$ modification. A hyper-thermophilic archaeon, *Pyrodictium occultum* can grow at 105 °C, and various $2'$-O-methylted nucleosides such as Ψm, m^1Im, and m^2_2Gm are present in its tRNA, but s^2U and m^5s^2U are not observed [97,98]. Notably, although the melting temperature of the *P. occultum* tRNA^{Met} transcript is 80 °C, that of the native tRNA^{Met} is more than 100 °C [323]. Thus, the melting temperature of *P. occultum* tRNA is increased by more than 20 °C through a combination of numerous $2'$-O-methylated nucleosides.

Methanopyrus kandleri can grow at more than 110 °C and tRNAs from this archaeon contain many unique modifications such as U_8 (the product of C_8 to U_8 editing) [16], ac^6A, $m^{2,7}Gm$, and methyl-hn^6A [100]. Furthermore, *M. kandleri* possesses 132 species of C/D-box guide RNAs [17], which suggests that RNAs are highly methylated by the L7Ae, Nop5, aFib, and C/D-box guide RNA system. In the case of *M. kandleri*, therefore, tRNA seems to be stabilized by unique modifications and $2'$-O-methylations.

These observations suggest that living organisms can survive at more than 100 °C by a combination of $2'$-O-methylations and other thermophile-specific tRNA modifications.

4.4. Other tRNA Stabilization Factors

RNA binding proteins, polyamines, magnesium ions, and potassium ions are all able to stabilize tRNA in thermophiles. For example, transfer RNA-binding protein 111 (Trbp111) is an RNA-binding protein that is observed only in *A. aeolicus* [335–337]. *A. aeolicus* can grow at 94 °C and modified nucleosides in tRNA of this hyperthermophilic eubacterium are not so different from those in tRNA from *T. thermophilus*, which grows at temperatures below 83 °C. Therefore, Trbp111 may provide more than 10 °C of tRNA stabilization in *A. aeolicus*. The docking model of Trbp111 and tRNA suggests that Trbp111 stabilizes the three-dimensional core of tRNA [336]. Archease is another tRNA-binding protein that can change the methylation site of *P. abyssi* Trm4 [209]. Furthermore, archease promotes the ligation of tRNA exons during tRNA splicing [338,339]. Therefore, it has the potential to stabilize the tRNA structure at high temperatures.

Many tRNA-binding proteins and RNA chaperone proteins have been identified in eukaryotic cells [340,341]. Although these types of protein are unknown in thermophilic eukaryotes, some of them may stabilize the tRNA structure (or help correct folding of tRNA) at high temperatures. Recently, it was revealed that *E. coli* TruB (tRNA Ψ_{55} synthase [243]) possesses an RNA chaperone activity [342,343]. In the case of *T. thermophilus*, although the Ψ_{55} modification is required for survival at low temperatures (below 55 °C), the *truB* gene disruptant strain shows abnormal growth at 80 °C [248]. Therefore, the RNA chaperone effect of TruB may also be expressed at high temperatures in *T. thermophilus*. Furthermore, these observations suggest that other tRNA modification enzymes have the potential to work as RNA chaperones.

In general, polyamines have the potential to interact with nucleic acids and phospholipids because they possess multiple positive charges and hydrophobic areas. There are several studies on the interaction between tRNA and polyamines [344–347]. Thermophiles produce unique polyamines including long and branched polyamines [348–351]. Therefore, polyamines probably contribute to stabilize the tRNA structure at high temperatures. Furthermore, in vitro studies have shown that thermophile-specific long and branched polyamines affect the activities of several tRNA modification enzymes [81,352]. For example, TrmH from *T. thermophilus* methylates tRNA transcript at 80 °C only in the presence of long or branched polyamines [81]. Moreover, the long and branched polyamines are required for the maintenance of several tRNAs and the 70S ribosome and are essential for the survival of *T. thermophilus* at high temperatures [353].

Lastly, magnesium ions have been shown to be a tRNA stabilization factor [6,88,354] and are very important when considering the structural effects of several modified nucleosides in tRNA [58,88,354–356]. However, the precise concentration of magnesium ions in thermophile cells is unknown. It may differ depending on the growth environments. Potassium ions also function as RNA stabilization factor [88]. Notably, the interacellular concentration of some hyperthermophilic archaea

(*M. fervidus* and *P. furiosus*) is much higher (700–900 mM) than that of mesophilic archaea [357]. In the case of *Methanothermus sociabilis*, the interacellular potassium concentration reaches 1060 mM [357]. These high concentrations of potassium ions may have effects on the stability of tRNA and the activities of tRNA modification enzymes.

5. tRNA Modifications and Environmental Stresses at High Temperatures

Recent studies have revealed that the modifications in tRNA are stress-resistance and/or stress-response factors [102,358–361]. Furthermore, a high temperature itself can be a stress factor for living organisms because some modified nucleosides (D and m^7G) are liable at high temepratures [297].

5.1. Oxidative Stress

Many thermophiles can grow under aerobic conditions. For example, *Aerophyrum pernix* can grow at 100 °C under aerobic conditions. Under such conditions at high temperatures, living organisms seem to be exposed to heavy oxidative stress, which is a typical environmental stress. The amount of antioxidant enzymes such as superoxide-dismutase, catalase, and peroxidase in *Thermus filiformis*, which is an extreme-thermophilic eubacterium, increases at high temperatures [362].

Among tRNA modification enzymes, both Fe-S cluster proteins [34,130,134,142,150,173,196,236,363] for sulfur-transfer, reduction of base and/or radical S-adenosyl-l-methionine (SAM) reaction, and enzymes with catalytic cysteine residues [141,210,364–366], seem to be easily changed under oxidative stress. In some cases, the substrate (e.g., electron donors and folate derivatives [126,221,227,367]) may be unstable under aerobic conditions at high temperatures. Similarly, several modified nucleosides such as D and s^4U may be labile under oxidative stress at high temperatures. Therefore, aerobic thermophiles need to protect their cellular components from oxidative stress and their tRNA modifications may respond to such stress as in mesophiles. Overall, however, the relationship between oxidative stress and tRNA modifications in thermopiles is unclear. In addition, tRNA modification systems in some thermophiles may utilize aerobic conditions at high temperatures. For example, *A. aeolicus* grows under microaerophilic conditions at high temperatures (80–94 °C) and the dimer structure of *A. aeolicus* TrmD is stabilized by inter-subunit disulfide bonds [165].

5.2. Other Environmental Stresses

Thermophiles often live in severe environments such as extreme pH and high pressure in addition to high temperatures. These environmental stresses may give rise to the diversity of tRNA modifications. At present, however, there are no data to support this viewpoint.

UV-stress is one such environmental stress and the s^4U modification in tRNA is a known UV-stress-resistance factor for *E. coli* [368] and *Salmonella typhimurium* [27]. Thus, the s^4U modification in tRNA is likely to work similarly to a UV-resistant factor in thermophiles. Interestingly, the genomes of *Archaeoglobus fulgidus* and *Methanocaldococcus janaschii,* which were isolated from the oil mines under the sea and deep sea, respectively, contain a *thiI* genes [369] (AF_RS04455 and MJ_RS04985, respectively) encoding tRNA s^4U_8 synthetase. Since sunlight does not reach the environments in which these thermophilic archaea live, the s^4U modification and/or ThiI may have an additional function (e.g., sulfur-metabolism) in these archaea. Furthermore, it was recently reported that the melting temperature of tRNA from an *E. coli thiI*-gene disruptant strain was decreased relative to the wild-type strain [33]. Therefore, the s^4U_8 modification may contribute to stabilize tRNA structure. Furthermore, UV-stress may have an effect on other tRNA modifications via the cross-linking of s^4U in tRNA. For example, the methylation speed of *T. thermophilus* TrmH is decreased when the substrate tRNA is cross-linked [30].

Lastly, the availability of nutrient-factors may have an effect on tRNA modifications in thermophiles. To test this idea, the extent of modifications in tRNA from *T. thermophilus* cells cultured in a nutrient-poor condition was investigated [227]. Contrary to expectation, the extent of the modification of all methylated nucleosides analyzed was normal, which demonstrates that the limited nutrients were preferentially consumed in the tRNA modification systems [227]. Thus, the findings indicated the importance of tRNA modifications for the survival of *T. thermophilus*.

6. Utilization of tRNA Modification Enzymes from Thermophiles

Given that proteins from thermophiles are heat-resistant and very stable, numerous tRNA modification enzymes have been used in biochemical and structural studies (Tables 1 and 2). In particular, crystal structural studies of thermostable enzymes provided significant information on catalytic mechanisms and RNA-protein interactions. Studies on the crystal structures of tRNA modification enzymes from thermophiles are summarized in Supplementary Table S2. It is anticipated that thermostable proteins will continue to contribute structural studies in the future. Thermostable tRNA modification enzymes can be a tool for molecular and cell biology. For example, *A. fulgidus* TiaS with agmatine analogues has been used for site-specific RNA-labeling in mammalian cells [315]. In addition, thermostable tRNA modification enzymes may be used for healthcare. For example, Gm_{18} modification in tRNA does not stimulate the Toll-like receptor 7 [287,288] and tRNA with Gm_{18} alleviates inflammation [288]. Since TrmH from *T. thermophilus* can methylate all tRNA species [72] and is very stable, it may be useful for preparing tRNAs with Gm_{18} modifications for tRNA therapy.

7. Perspective

Given that the temperature of ancient Earth was very high relative to that of present-day Earth, thermophiles may be remnants of ancient living organisms. Therefore, studies on tRNA modification enzymes and modified nucleosides in tRNA from thermophiles will contribute to the considerations of the evolutionary pathways of living organisms. Furthermore, such studies will continue to shed light on the variety and environmental adaptations of living organisms. Moreover, as outlined above, the thermostable enzymes may be useful as biotechnological and medical tools and may contribute toward the production of valuable materials.

Supplementary Materials: The following are available online at http://www.mdpi.com/2076-2607/6/4/110/s1, Table S1: Abbreviations of modified nucleosides, Table S2: Crystal structural studies on tRNA modification enzymes from thermophiles.

Author Contributions: All authors determined the concept of this review and collected the information from the references. H.H. wrote the manuscript and all authors revised and approved the manuscript.

Funding: This work was supported by a Grant-in-Aid for Scientific Research (16H04763 to H.H.) from the Japan Society for the Promotion of Science (JSPS).

Acknowledgments: We dedicate this review to Kimitsuna Watanabe who died in 2016. He discovered the $m^5s^2U_{54}$ modification in tRNA and encouraged our studies. We thank previous collaborators.

Conflicts of Interest: The authors declare no conflict of interest.

References

1. Crick, F.H. On protein synthesis. *Symp. Soc. Exp. Biol.* **1958**, *12*, 138–163. [PubMed]
2. Grunberger, D.; Weinstein, I.B.; Jacobson, K.B. Codon recognition by enzymatically mischarged valine transfer ribonucleic acid. *Science* **1969**, *166*, 1635–1637. [CrossRef] [PubMed]
3. Sprinzl, M.; Horn, C.; Brown, M.; Ioudovitch, A.; Steinberg, S. Compilation of tRNA sequences and sequences of tRNA genes. *Nucleic Acids Res.* **1998**, *26*, 148–153. [CrossRef] [PubMed]
4. Robertus, J.D.; Ladner, J.E.; Finch, J.T.; Rhodes, D.; Brown, R.S.; Clark, B.F.C.; Klug, A. Structure of yeast phenylalanine tRNA at 3 Å resolution. *Nature* **1974**, *250*, 546–551. [CrossRef] [PubMed]
5. Kim, S.H.; Suddath, F.L.; Quigley, G.J.; McPherson, A.; Sussan, J.L.; Wang, A.H.J.; Seeman, N.C.; Rich, A. Three-dimensional tertiary structure of yeast phenylalanine transfer RNA. *Science* **1974**, *185*, 435–440. [CrossRef] [PubMed]
6. Shi, H.; Moore, P.B. The crystal structure of yeast phenylalanine tRNA at 1.93 A resolution: A classic structure revisited. *RNA* **2000**, *6*, 1091–1105. [CrossRef] [PubMed]

7. Boccaletto, P.; Machnicka, M.A.; Purta, E.; Piatkowski, P.; Baginski, B.; Wirecki, T.K.; de Crécy-Lagard, V.; Ross, R.; Limbach, P.A.; Kotter, A.; et al. MODOMICS: A database of RNA modification pathways. 2017 update. *Nucleic Acids Res* **2018**, *46*, D303–D307. [CrossRef] [PubMed]

8. Juhling, F.; Morl, M.; Hartmann, R.K.; Sprinzl, M.; Stadler, P.F.; Putz, J. tRNAdb 2009: Compilation of tRNA sequences and tRNA genes. *Nucleic Acids Res.* **2009**, *37*, D159–D162. [CrossRef] [PubMed]

9. Constantinesco, F.; Motorin, Y.; Grosjean, H. Transfer RNA modification enzymes from *Pyrococcus furiosus*: Detection of the enzymatic activities in vitro. *Nucleic Acids Res.* **1999**, *27*, 1308–1315. [CrossRef] [PubMed]

10. Grawunder, U.; Schön, A.; Sprintzl, M. Sequence and base modifications of two phenylalanine-tRNAs from *Thermus thermophilus* HB8. *Nucleic Acids Res.* **1992**, *20*, 137. [CrossRef] [PubMed]

11. Tomikawa, C.; Yokogawa, T.; Kanai, T.; Hori, H. N^7-Methylguanine at position 46 (m^{7}G46) in tRNA from *Thermus thermophilus* is required for cell viability through a tRNA modification network. *Nucleic Acids Res.* **2010**, *38*, 942–957. [CrossRef] [PubMed]

12. Menezes, S.; Gaston, K.W.; Krivos, K.L.; Apolinario, E.E.; Reich, N.O.; Sowers, K.R.; Limbach, P.A.; Perona, J.J. Formation of m^{2}G6 in *Methanocaldococcus jannaschii* tRNA catalyzed by the novel methyltransferase Trm14. *Nucleic Acids Res.* **2011**, *39*, 7641–7655. [CrossRef] [PubMed]

13. Fislage, M.; Roovers, M.; Münnich, S.; Droogmans, L.; Versées, W. Crystallization and preliminary X-ray crystallographic analysis of putative tRNA-modification enzymes from *Pyrococcus furiosus* and *Thermus thermophilus*. *Acta Crystallogr. Sect. F Struct. Biol. Cryst. Commun.* **2011**, *67*, 1432–1435. [CrossRef] [PubMed]

14. Roovers, M.; Oudjama, Y.; Fislage, M.; Bujnicki, J.M.; Versées, W.; Droogmans, L. The open reading frame TTC1157 of *Thermus thermophilus* HB27 encodes the methyltransferase forming N^2-methylguanosine at position 6 in tRNA. *RNA* **2012**, *18*, 815–824. [CrossRef] [PubMed]

15. Fislage, M.; Roovers, M.; Tuszynska, I.; Bujnicki, J.M.; Droogmans, L.; Versées, W. Crystal structures of the tRNA:m^{2}G6 methyltransferase Trm14/TrmN from two domains of life. *Nucleic Acids Res.* **2012**, *40*, 5149–5161. [CrossRef] [PubMed]

16. Randau, L.; Stanley, B.J.; Kohlway, A.; Mechta, S.; Xiong, Y.; Söll, D. A cytidine deaminase edits C to U in transfer RNAs in Archaea. *Science* **2009**, *324*, 657–659. [CrossRef] [PubMed]

17. Su, A.A.; Tripp, V.; Randau, L. RNA-Seq analyses reveal the order of tRNA processing events and the maturation of C/D box and CRISPR RNAs in the hyperthermophile *Methanopyrus kandleri*. *Nucleic Acids Res.* **2013**, *41*, 6250–6258. [CrossRef] [PubMed]

18. Pixa, G.; Dirheimer, G.; Keith, G. Sequence of tRNA^{Leu} CmAA from *Bacillus stearothermophilus*. *Biochem. Biophys. Res. Commun.* **1983**, *112*, 578–585. [CrossRef]

19. Keith, G.; Guerrier-Takada, C.; Grosjean, H.; Dirheimer, G. A revised sequence for *Bacillus stearothermophilus* phenylalanine tRNA. *FEBS Lett.* **1977**, *84*, 241–243. [CrossRef]

20. Brown, R.S.; Rubin, J.R.; Rhodes, D.; Guilley, H.; Simoncsits, A.; Brownlee, G.G. The nucleoside sequence of tyrosine tRNA from *Bacillus stearothermophilus*. *Nucleic Acids Res.* **1978**, *5*, 23–36. [CrossRef] [PubMed]

21. Watanabe, K.; Kuchino, Y.; Yamaizumi, Z.; Kato, M.; Oshima, T.; Nishimura, S. Nucleotide sequence of formylmethionine tRNA from an extreme thermophile, *Thermus thermophilus* HB8. *J. Biochem.* **1979**, *86*, 893–905. [CrossRef] [PubMed]

22. Watanabe, K.; Oshima, T.; Hansske, F.; Ohta, T. Separation and comparison of 2-thioribothymidine-containing transfer ribonucleic acid and the ribothymidine-containing counterpart from cells of *Thermus thermophilus* HB 8. *Biochemistry* **1983**, *22*, 98–102. [CrossRef] [PubMed]

23. Keith, G.; Yusupov, M.; Briand, C.; Moras, D.; Kern, D.; Brion, C. Sequence of tRNA^{Asp} from *Thermus thermophilus* HB8. *Nucleic Acids Res.* **1993**, *21*, 4399. [CrossRef] [PubMed]

24. Awai, T.; Kimura, S.; Tomikawa, C.; Ochi, A.; Ihsanawati, I.S.; Bessho, Y.; Yokoyama, S.; Ohno, S.; Nishikawa, K.; Yokogawa, T.; et al. *Aquifex aeolicus* tRNA (N^2, N^2-guanine)-dimethyltransferase (Trm1) catalyzes transfer of methyl groups not only to guanine 26 but also to guanine 27 in tRNA. *J. Biol. Chem.* **2009**, *284*, 20467–20478. [CrossRef] [PubMed]

25. Kilpatrick, M.W.; Walker, R.T. The nucleotide sequence of the tRNA$_M$^{Met} from the archaebacterium *Thermoplasma acidophilum*. *Nucleic Acids Res.* **1981**, *9*, 4387–4390. [CrossRef] [PubMed]

26. Caldeira de Araujo, A.; Favre, A. Induction of size reduction in *Escherichia coli* by near-ultraviolet light. *Eur. J. Biochem.* **1985**, *146*, 605–610. [CrossRef] [PubMed]

27. Kramer, G.F.; Baker, J.C.; Ames, B.N. Near-UV stress in *Salmonella typhimurium*: 4-thiouridine in tRNA, ppGpp, and ApppGpp as components of an adaptive response. *J. Bacteriol.* **1988**, *170*, 2344–2351. [CrossRef] [PubMed]

28. Mueller, E.G.; Buck, C.J.; Palenchar, P.M.; Barnhart, L.E.; Paulson, J.L. Identification of a gene involved in the generation of 4-thiouridine in tRNA. *Nucleic Acids Res.* **1998**, *26*, 2060–2610. [CrossRef]

29. Kambampati, R.; Lauhon, C.T. IscS is a sulfurtransferase for the in vitro biosynthesis of 4-thiouridine in *Escherichia coli* tRNA. *Biochemistry* **1999**, *38*, 16561–16568. [CrossRef] [PubMed]

30. Hori, H.; Saneyoshi, M.; Kumagai, I.; Miura, K.; Watanabe, K. Effects of modification of 4-thiouridine in *E. coli* tRNA$^{Met}_f$ on its methyl acceptor activity by thermostable Gm-methylases. *J. Biochem.* **1989**, *106*, 798–802. [CrossRef] [PubMed]

31. Naumann, P.T.; Lauhon, C.T.; Ficner, R. Purification, crystallization and preliminary crystallographic analysis of a 4-thiouridine synthetase-RNA complex. *Acta Crystallogr. Sect. F Struct. Biol. Cryst. Commun.* **2013**, *69*, 421–424. [CrossRef] [PubMed]

32. Neumann, P.; Lakomek, K.; Naumann, P.T.; Erwin, W.M.; Lauhon, C.T.; Ficner, R. Crystal structure of a 4-thiouridine synthetase-RNA complex reveals specificity of tRNA U8 modification. *Nucleic Acids Res.* **2014**, *42*, 6673–6685. [CrossRef] [PubMed]

33. Nomura, Y.; Ohno, S.; Nishikawa, K.; Yokogawa, T. Correlation between the stability of tRNA tertiary structure and the catalytic efficiency of a tRNA-modifying enzyme, archaeal tRNA-guanine transglycosylase. *Genes Cells* **2016**, *21*, 41–52. [CrossRef] [PubMed]

34. Liu, Y.; Vinyard, D.J.; Reesbeck, M.E.; Suzuki, T.; Manakongtreecheep, K.; Holland, P.L.; Brudvig, G.W.; Söll, D. A [3Fe-4S] cluster is required for tRNA thiolation in archaea and eukaryotes. *Proc. Natl. Acad. Sci. USA* **2016**, *113*, 12703–12708. [CrossRef] [PubMed]

35. Čavužić, M.; Liu, Y. Biosynthesis of sulfur-containing tRNA modifications: A comparison of bacterial, archaeal, and eukaryotic pathways. *Biomolecules* **2017**, *7*, 27. [CrossRef] [PubMed]

36. Tomikawa, C.; Ohira, T.; Inoue, Y.; Kawamura, T.; Yamagishi, A.; Suzuki, T.; Hori, H. Distinct tRNA modifications in the thermo-acidophilic archaeon, *Thermoplasma acidophilum*. *FEBS Lett.* **2013**, *587*, 3537–3580. [CrossRef] [PubMed]

37. Kempenaers, M.; Roovers, M.; Oudjama, Y.; Tkaczuk, K.L.; Bujnicki, J.M.; Droogmans, L. New archaeal methyltransferases forming 1-methyladenosine or 1-methyladenosine and 1-methylguanosine at position 9 of tRNA. *Nucleic Acids Res.* **2010**, *38*, 6533–6543. [CrossRef] [PubMed]

38. Van Laer, B.; Roovers, M.; Wauters, L.; Kasprzak, J.M.; Dyzma, M.; Deyaert, E.; Kumar Singh, R.; Feller, A.; Bujnicki, J.M.; Droogmans, L.; et al. Structural and functional insights into tRNA binding and adenosine N^1-methylation by an archaeal Trm10 homologue. *Nucleic Acids Res.* **2016**, *44*, 940–953. [CrossRef] [PubMed]

39. Singh, R.K.; Feller, A.; Roovers, M.; Van Elder, D.; Wauters, L.; Droogmans, L.; Versées, W. Structural and biochemical analysis of the dual-specificity Trm10 enzyme from *Thermococcus kodakaraensis* prompts reconsideration of its catalytic mechanism. *RNA* **2018**, *24*, 1080–1092. [CrossRef] [PubMed]

40. Armengaud, J.; Urbonavicius, J.; Fernandez, B.; Chaussinand, G.; Bujnicki, J.M.; Grosjean, H. N^2-methylation of guanosine at position 10 in tRNA is catalyzed by a THUMP domain-containing, S-adenosylmethionine-dependent methyltransferase, conserved in Archaea and Eukaryota. *J. Biol. Chem.* **2004**, *279*, 37142–37152. [CrossRef] [PubMed]

41. Gabant, G.; Auxilien, S.; Tuszynska, I.; Locard, M.; Gajda, M.J.; Chaussinand, G.; Fernandez, B.; Dedieu, A.; Grosjean, H.; Golinelli-Pimpaneau, B.; et al. THUMP from archaeal tRNA:m$^2_2$G10 methyltransferase, a genuine autonomously folding domain. *Nucleic Acids Res.* **2006**, *34*, 2483–2494. [CrossRef] [PubMed]

42. Urbonavicius, J.; Armengaud, J.; Grosjean, H. Identity elements required for enzymatic formation of N^2,N^2-dimethylguanosine from N^2-monomethylated derivative and its possible role in avoiding alternative conformations in archaeal tRNA. *J. Mol. Biol.* **2006**, *357*, 387–399. [CrossRef] [PubMed]

43. Hirata, A.; Nishiyama, S.; Tamura, T.; Yamauchi, A.; Hori, H. Structural and functional analyses of the archaeal tRNA m^{2}G/m$^2_2$G10 methyltransferase aTrm11 provide mechanistic insights into site specificity of a tRNA methyltransferase that contains common RNA-binding modules. *Nucleic Acids Res.* **2016**, *44*, 6377–6390. [CrossRef] [PubMed]

44. Kuchino, Y.; Ihara, M.; Yabusaki, Y.; Nishimura, S. Initiator tRNAs from archaebacteria show common unique sequence characteristics. *Nature* **1982**, *298*, 684–685. [CrossRef] [PubMed]

45. Kaya, Y.; Ofengand, J. A novel unanticipated type of pseudouridine synthase with homologs in bacteria, archaea, and eukarya. *RNA* **2003**, *9*, 711–721. [CrossRef] [PubMed]

46. Muller, S.; Urban, A.; Hecker, A.; Leclerc, F.; Branlant, C.; Motorin, Y. Deficiency of the tRNATyr: Psi 35-synthase aPus7 in Archaea of the Sulfolobales order might be rescued by the H/ACA sRNA-guided machinery. *Nucleic Acids Res.* **2009**, *37*, 1308–1322. [CrossRef] [PubMed]

47. Kawamura, T.; Hirata, A.; Ohno, S.; Nomura, Y.; Nagano, T.; Nameki, N.; Yokogawa, T.; Hori, H. Multisite-specific archaeosine tRNA-guanine transglycosylase (ArcTGT) from *Thermoplasma acidophilum*, a thermo-acidophilic archaeon. *Nucleic Acids Res.* **2016**, *44*, 1894–1908. [CrossRef] [PubMed]

48. Gupta, R. *Halobacterium volcanii* tRNAs. Identification of 41 tRNAs covering all amino acids, and the sequences of 33 class I tRNAs. *J. Biol. Chem.* **1984**, *259*, 9461–9471. [PubMed]

49. Gregson, J.M.; Crain, P.F.; Edmonds, C.G.; Gupta, R.; Hashizume, T.; Phillipson, D.W.; McCloskey, J.A. Structure of the archaeal transfer RNA nucleoside G*-15 (2-amino-4,7-dihydro-4-oxo-7-beta-D-ribofuranosyl-1H-pyrrolo[2,3-d]pyrimidine-5-carboximi dam ide (archaeosine)). *J. Biol. Chem.* **1993**, *268*, 10076–10086. [PubMed]

50. Watanabe, M.; Matsuo, M.; Tanaka, S.; Akimoto, H.; Asahi, S.; Nishimura, S.; Katze, J.R.; Hashizume, T.; Crain, P.F.; McCloskey, J.A.; et al. Biosynthesis of archaeosine, a novel derivative of 7-deazaguanosine specific to archaeal tRNA, proceeds via a pathway involving base replacement on the tRNA polynucleotide chain. *J. Biol. Chem.* **1997**, *272*, 20146–20151. [CrossRef] [PubMed]

51. Bai, Y.; Fox, D.T.; Lacy, J.A.; Van Lanen, S.G.; Iwata-Reuyl, D. Hypermodification of tRNA in Thermophilic archaea. Cloning, overexpression, and characterization of tRNA-guanine transglycosylase from *Methanococcus jannaschii*. *J. Biol. Chem.* **2000**, *275*, 28731–28738. [CrossRef] [PubMed]

52. Watanabe, M.; Nameki, N.; Matsuo-Takasaki, M.; Nishimura, S.; Okada, N. tRNA recognition of tRNA-guanine transglycosylase from a hyperthermophilic archaeon, *Pyrococcus horikoshii*. *J. Biol. Chem.* **2001**, *276*, 2387–2394. [CrossRef] [PubMed]

53. Ishitani, R.; Nureki, O.; Kijimoto, T.; Watanabe, M.; Kondo, H.; Nameki, N.; Okada, N.; Nishimura, S.; Yokoyama, S. Crystallization and preliminary X-ray analysis of the archaeosine tRNA-guanine transglycosylase from *Pyrococcus horikoshii*. *Acta Crystallogr. D Biol. Crystallogr.* **2001**, *57*, 1659–1662. [CrossRef] [PubMed]

54. Ishitani, R.; Nureki, O.; Fukai, S.; Kijimoto, T.; Nameki, N.; Watanabe, M.; Kondo, H.; Sekine, M.; Okada, N.; Nishimura, S.; et al. Crystal structure of archaeosine tRNA-guanine transglycosylase. *J. Mol. Biol.* **2002**, *318*, 665–677. [CrossRef]

55. Ishitani, R.; Nureki, O.; Fukai, S.; Kijimoto, T.; Nameki, N.; Watanabe, M.; Kondo, H.; Sekine, M.; Okada, N.; Nishimura, S.; et al. Alternative tertiary structure of tRNA for recognition by a posttranscriptional modification enzyme. *Cell* **2003**, *113*, 383–394. [CrossRef]

56. Noon, K.R.; Guymon, R.; Crain, P.F.; McCloskey, J.A.; Thomm, M.; Lim, J.; Cavicchioli, R. Influence of temperature on tRNA modification in archaea: *Methanococcoides burtonii* (optimum growth temperature [Topt], 23 degrees C) and *Stetteria hydrogenophila* (Topt, 95 degrees C). *J. Bacteriol.* **2003**, *185*, 5483–5490. [CrossRef] [PubMed]

57. Sabina, J.; Söll, D. The RNA-binding PUA domain of archaeal tRNA-guanine transglycosylase is not required for archaeosine formation. *J. Biol. Chem.* **2006**, *281*, 6993–7001. [CrossRef] [PubMed]

58. Oliva, R.; Tramontano, A.; Cavallo, L. Mg^{2+} binding and archaeosine modification stabilize the G15 C48 Levitt base pair in tRNAs. *RNA* **2007**, *13*, 1427–1436. [CrossRef] [PubMed]

59. Phillips, G.; Chikwana, V.M.; Maxwell, A.; El-Yacoubi, B.; Swairjo, M.A.; Iwata-Reuyl, D.; de Crécy-Lagard, V. Discovery and characterization of an amidinotransferase involved in the modification of archaeal tRNA. *J. Biol. Chem.* **2010**, *285*, 12706–12713. [CrossRef] [PubMed]

60. Phillips, G.; Swairjo, M.A.; Gaston, K.W.; Bailly, M.; Limbach, P.A.; Iwata-Reuyl, D.; de Crécy-Lagard, V. Diversity of archaeosine synthesis in crenarchaeota. *ACS Chem. Biol.* **2012**, *7*, 300–305. [CrossRef] [PubMed]

61. Nomura, Y.; Onda, Y.; Ohno, S.; Taniguchi, H.; Ando, K.; Oka, N.; Nishikawa, K.; Yokogawa, T. Purification and comparison of native and recombinant tRNA-guanine transglycosylases from Methanosarcina acetivorans. *Protein Expr. Purif.* **2013**, *88*, 13–19. [CrossRef] [PubMed]

62. Mei, X.; Alvarez, J.; Bon Ramos, A.; Samanta, U.; Iwata-Reuyl, D.; Swairjo, M.A. Crystal structure of the archaeosine synthase QueF-like-Insights into amidino transfer and tRNA recognition by the tunnel fold. *Proteins* **2017**, *85*, 103–116. [CrossRef] [PubMed]

63. Bon Ramos, A.; Bao, L.; Turner, B.; de Crécy-Lagard, V.; Iwata-Reuyl, D. QueF-Like, a Non-homologous archaeosine synthase from the crenarchaeota. *Biomolecules* **2017**, *7*, 36. [CrossRef] [PubMed]

64. Takada-Guerrier, C.; Grosjean, H.; Dirheimer, G.; Keith, G. The primary structure of tRNA$_2^{Val}$ from *Bacillus stearothermophilus*. *FEBS Lett.* **1976**, *62*, 1–3. [CrossRef]

65. Bishop, A.C.; Xu, J.; Johnson, R.C.; Schimmel, P.; de Crécy-Lagard, V. Identification of the tRNA-dihydrouridine synthase family. *J. Biol. Chem.* **2002**, *277*, 25090–25095. [CrossRef] [PubMed]

66. Bou-Nader, C.; Montémont, H.; Guérineau, V.; Jean-Jean, O.; Brégeon, D.; Hamdane, D. Unveiling structural and functional divergences of bacterial tRNA dihydrouridine synthases: Perspectives on the evolution scenario. *Nucleic Acids Res.* **2018**, *46*, 1386–1394. [CrossRef] [PubMed]

67. Horie, N.; Hara-Yokoyama, M.; Yokoyama, S.; Watanabe, K.; Kuchino, Y.; Nishimura, S.; Miyazawa, T. Two tRNA^{Ile}1 species from an extreme thermophile, *Thermus thermophilus* HB8: Effect of 2-thiolation of ribothymidine on the thermostability of tRNA. *Biochemistry* **1985**, *24*, 5711–5715. [CrossRef] [PubMed]

68. Persson, B.C.; Jäger, G.; Gustafsson, C. The spoU gene of *Escherichia coli*, the fourth gene of the spoT operon, is essential for tRNA (Gm18) 2'-*O*-methyltransferase activity. *Nucleic Acids Res.* **1997**, *25*, 4093–4097. [CrossRef] [PubMed]

69. Kumagai, I.; Watanabe, K.; Oshima, T. A thermostable tRNA (guanosine-2')-methyltransferase from *Thermus. thermophilus* HB27 and the effect of ribose methylation on the conformational stability of tRNA. *J. Biol. Chem.* **1982**, *257*, 7388–7395. [PubMed]

70. Matsumoto, T.; Ohta, T.; Kumagai, I.; Oshima, T.; Murao, K.; Hasegawa, T.; Ishikura, H.; Watanabe, K. A thermostable Gm-methylase recognizes the tertiary structure of tRNA. *J. Biochem.* **1987**, *101*, 1191–1198. [CrossRef] [PubMed]

71. Matsumoto, T.; Nishikawa, K.; Hori, H.; Ohta, T.; Miura, K.; Watanabe, K. Recognition sites of tRNA by a thermostable tRNA(guanosine-2'-)-methyltransferase from *Thermus thermophilus* HB27. *J. Biochem.* **1990**, *107*, 331–338. [CrossRef] [PubMed]

72. Hori, H.; Yamazaki, N.; Matsumoto, T.; Watanabe, Y.; Ueda, T.; Nishikawa, K.; Kumagai, I.; Watanabe, K. Substrate recognition of tRNA (Guanosine-2'-)-methyltransferase from *Thermus thermophilus* HB27. *J. Biol. Chem.* **1998**, *273*, 25721–25727. [CrossRef] [PubMed]

73. Hori, H.; Suzuki, T.; Sugawara, K.; Inoue, Y.; Shibata, T.; Kuramitsu, S.; Yokoyama, S.; Oshima, T.; Watanabe, K. Identification and characterization of tRNA (Gm18) methyltransferase from *Thermus thermophilus* HB8: Domain structure and conserved amino acid sequence motifs. *Genes Cells* **2002**, *7*, 259–272. [CrossRef] [PubMed]

74. Hori, H.; Kubota, S.; Watanabe, K.; Kim, J.M.; Ogasawara, T.; Sawasaki, T.; Endo, Y. *Aquifex aeolicus* tRNA (Gm18) methyltransferase has unique substrate specificity: tRNA recognition mechanism of the enzyme. *J. Biol. Chem.* **2003**, *278*, 25081–25090. [CrossRef] [PubMed]

75. Nureki, O.; Watanabe, K.; Fukai, S.; Ishii, R.; Endo, Y.; Hori, H.; Yokoyama, S. Deep knot structure for construction of active site and cofactor binding site of tRNA modification enzyme. *Structure* **2004**, *12*, 593–604. [CrossRef] [PubMed]

76. Watanabe, K.; Nureki, O.; Fukai, S.; Ishii, R.; Okamoto, H.; Yokoyama, S.; Endo, Y.; Hori, H. Roles of conserved amino acid sequence motifs in the SpoU (TrmH) RNA methyltransferase family. *J. Biol. Chem.* **2005**, *280*, 10368–10377. [CrossRef] [PubMed]

77. Pleshe, E.; Truesdell, J.; Batey, R.T. Structure of a class II TrmH tRNA-modifying enzyme from *Aquifex aeolicus*. *Acta Crystallogr. Sect. F Struct. Biol. Cryst. Commun.* **2005**, *61*, 722–728. [CrossRef] [PubMed]

78. Watanabe, K.; Nureki, O.; Fukai, S.; Endo, Y.; Hori, H. Functional categorization of the conserved basic amino acid residues in TrmH (tRNA (Gm18) methyltransferase) enzymes. *J. Biol. Chem.* **2006**, *281*, 34630–34639. [CrossRef] [PubMed]

79. Ochi, A.; Makabe, K.; Kuwajima, K.; Hori, H. Flexible recognition of the tRNA G18 methylation target site by TrmH methyltransferase through first binding and induced fit processes. *J. Biol. Chem.* **2010**, *285*, 9018–9029. [CrossRef] [PubMed]

80. Ochi, A.; Makabe, K.; Yamagami, R.; Hirata, A.; Sakaguchi, R.; Hou, Y.M.; Watanabe, K.; Nureki, O.; Kuwajima, K.; Hori, H. The catalytic domain of topological knot tRNA methyltransferase (TrmH) discriminates between substrate tRNA and nonsubstrate tRNA via an induced-fit process. *J. Biol. Chem.* **2013**, *288*, 25562–25574. [CrossRef] [PubMed]

81. Hori, H.; Terui, Y.; Nakamoto, C.; Iwashita, C.; Ochi, A.; Watanabe, K.; Oshima, T. Effects of polyamines from *Thermus thermophilus*, an extreme-thermophilic eubacterium, on tRNA methylation by tRNA (Gm18) methyltransferase (TrmH). *J. Biochem.* **2016**, *159*, 509–517. [CrossRef] [PubMed]
82. Savage, D.F.; de Crécy-Lagard, V.; Bishop, A.C. Molecular determinants of dihydrouridine synthase activity. *FEBS Lett.* **2006**, *580*, 5198–5202. [CrossRef] [PubMed]
83. Yu, F.; Tanaka, Y.; Yamamoto, S.; Nakamura, A.; Kita, S.; Hirano, N.; Tanaka, I.; Yao, M. Crystallization and preliminary X-ray crystallographic analysis of dihydrouridine synthase from *Thermus thermophilus* and its complex with tRNA. *Acta Crystallogr. Sect. F Struct. Biol. Cryst. Commun.* **2011**, *67*, 685–688. [CrossRef] [PubMed]
84. Yu, F.; Tanaka, Y.; Yamashita, K.; Suzuki, T.; Nakamura, A.; Hirano, N.; Suzuki, T.; Yao, M.; Tanaka, I. Molecular basis of dihydrouridine formation on tRNA. *Proc. Natl. Acad. Sci. USA* **2011**, *108*, 19593–19598. [CrossRef] [PubMed]
85. Kusuba, H.; Yoshida, T.; Iwasaki, E.; Awai, T.; Kazayama, A.; Hirata, A.; Tomikawa, C.; Yamagami, R.; Hori, H. In vitro dihydrouridine formation by tRNA dihydrouridine synthase from *Thermus thermophilus*, an extreme-thermophilic eubacterium. *J. Biochem.* **2015**, *158*, 513–521. [PubMed]
86. Roovers, M.; Kaminska, K.H.; Tkaczuk, K.L.; Gigot, D.; Droogmans, L.; Bujnicki, J.M. The YqfN protein of *Bacillus subtilis* is the tRNA: M^1A22 methyltransferase (TrmK). *Nucleic Acids Res.* **2008**, *36*, 3252–3262. [CrossRef] [PubMed]
87. Blaby, I.K.; Majumder, M.; Chatterjee, K.; Jana, S.; Grosjean, H.; de Crécy-Lagard, V.; Gupta, R. Pseudouridine formation in archaeal RNAs: The case of *Haloferax volcanii*. *RNA* **2011**, *17*, 1367–1380. [CrossRef] [PubMed]
88. Lorenz, C.; Lünse, C.E.; Mörl, M. tRNA modifications: Impact on structure and thermal adaptation. *Biomolecules* **2017**, *7*, 35. [CrossRef] [PubMed]
89. Kawamura, T.; Anraku, R.; Hasegawa, T.; Tomikawa, C.; Hori, H. Transfer RNA methyltransferases from *Thermoplasma acidophilum*, a thermoacidophilic archaeon. *Int. J. Mol. Sci.* **2014**, *16*, 91–113. [CrossRef] [PubMed]
90. Phillips, J.H.; Kjellin-Straby, K. Studies on microbial ribonucleic acid. IV. Two mutants of *Saccharomyces cerevisiae* lacking N-2-dimethylguanine in soluble ribonucleic acid. *J. Mol. Biol.* **1967**, *26*, 509–518. [CrossRef]
91. Constantinesco, F.; Benachenhou, N.; Motorin, Y.; Grosjean, H. The tRNA(guanine-26,N^2-N^2) methyltransferase (Trm1) from the hyperthermophilic archaeon *Pyrococcus furiosus*: Cloning, sequencing of the gene and its expression in *Escherichia coli*. *Nucleic Acids Res.* **1998**, *26*, 3753–3761. [CrossRef] [PubMed]
92. Constantinesco, F.; Motorin, Y.; Grosjean, H. Characterisation and enzymatic properties of tRNA (guanine 26, N (2), N (2))-dimethyltransferase (Trm1p) from *Pyrococcus furiosus*. *J. Mol. Biol.* **1999**, *291*, 375–392. [CrossRef] [PubMed]
93. Ihsanawati; Nishimoto, M.; Higashijima, K.; Shirouzu, M.; Grosjean, H.; Bessho, Y.; Yokoyama, S. Crystal structure of tRNA N^2,N^2-guanosine dimethyltransferase Trm1 from *Pyrococcus horikoshii*. *J. Mol. Biol.* **2008**, *383*, 871–884. [CrossRef] [PubMed]
94. Vakiloroayaei, A.; Shah, N.S.; Oeffinger, M.; Bayfield, M.A. The RNA chaperone La promotes pre-tRNA maturation via indiscriminate binding of both native and misfolded targets. *Nucleic Acids Res.* **2017**, *45*, 11341–11355. [CrossRef] [PubMed]
95. Awai, T.; Ochi, A.; Ihsanawati; Sengoku, T.; Hirata, A.; Bessho, Y.; Yokoyama, S.; Hori, H. Substrate tRNA recognition mechanism of a multisite-specific tRNA methyltransferase, *Aquifex aeolicus* Trm1, based on the X-ray crystal structure. *J. Biol. Chem.* **2011**, *286*, 35236–35246. [CrossRef] [PubMed]
96. Somme, J.; Van Laer, B.; Roovers, M.; Steyaert, J.; Versées, W.; Droogmans, L. Characterization of two homologous 2′-O-methyltransferases showing different specificities for their tRNA substrates. *RNA* **2014**, *20*, 1257–1271. [CrossRef] [PubMed]
97. Edmonds, C.G.; Crain, P.F.; Gupta, R.; Hashizume, T.; Hocart, C.H.; Kowalak, J.A.; Pomerantz, S.C.; Stetter, K.O.; McCloskey, J.A. Posttranscriptional modification of tRNA in *Thermophilic archaea* (Archaebacteria). *J. Bacteriol.* **1991**, *173*, 3138–3148. [CrossRef] [PubMed]
98. Kowalak, J.A.; Dalluge, J.J.; McCloskey, J.A.; Stetter, K.O. The role of posttranscriptional modification in stabilization of transfer RNA from *Hyperthermophiles*. *Biochemistry* **1994**, *33*, 7869–7876. [CrossRef] [PubMed]

99. McCloskey, J.A.; Graham, D.E.; Zou, S.; Crain, P.F.; Ibba, M.; Konisky, J.; Soll, D.; Olsen, G.J. Post-transcriptional modification in archaeal tRNAs: Identities and phylogenetic relations of nucleotides from mesophilic and hyperthermophilic Methanococcales. *Nucleic Acids Res.* **2001**, *29*, 4299–4706. [CrossRef]

100. Sauerwald, A.; Sitaramaiah, D.; McCloskey, J.A.; Söll, D.; Crain, P.F. N^6-Acetyladenosine: A new modified nucleoside from *Methanopyrus kandleri* tRNA. *FEBS Lett.* **2005**, *579*, 2807–2810. [CrossRef] [PubMed]

101. Purta, E.; van Vliet, F.; Tkaczuk, K.L.; Dunin-Horkawicz, S.; Mori, H.; Droogmans, L.; Bujnicki, J.M. The *yfhQ* gene of *Escherichia coli* encodes a tRNA: Cm32/Um32 methyltransferase. *BMC Mol. Biol.* **2006**, *7*, 23. [CrossRef] [PubMed]

102. Jaroensuk, J.; Atichartpongkul, S.; Chionh, Y.H.; Wong, Y.H.; Liew, C.W.; McBee, M.E.; Thongdee, N.; Prestwich, E.G.; DeMott, M.S.; Mongkolsuk, S.; et al. Methylation at position 32 of tRNA catalyzed by TrmJ alters oxidative stress response in *Pseudomonas aeruginosa*. *Nucleic Acids Res.* **2016**, *44*, 10834–10848. [CrossRef] [PubMed]

103. Wolf, J.; Gerber, A.P.; Keller, W. *tadA*, an essential tRNA-specific adenosine deaminase from *Escherichia coli*. *EMBO J.* **2002**, *21*, 3841–3851. [CrossRef] [PubMed]

104. Elias, Y.; Huang, R.H. Biochemical and structural studies of A-to-I editing by tRNA: A34 deaminases at the wobble position of transfer RNA. *Biochemistry* **2005**, *44*, 12057–12065. [CrossRef] [PubMed]

105. Kuratani, M.; Ishii, R.; Bessho, Y.; Fukunaga, R.; Sengoku, T.; Shirouzu, M.; Sekine, S.; Yokoyama, S. Crystal structure of tRNA adenosine deaminase (TadA) from *Aquifex aeolicus*. *J. Biol. Chem.* **2005**, *280*, 16002–16008. [CrossRef] [PubMed]

106. Yokobori, S.; Kitamura, A.; Grosjean, H.; Bessho, Y. Life without tRNAArg-adenosine deaminase TadA: Evolutionary consequences of decoding the four CGN codons as arginine in Mycoplasmas and other Mollicutes. *Nucleic Acids Res.* **2013**, *41*, 6531–6543. [CrossRef] [PubMed]

107. Torres, A.G.; Piñeyro, D.; Filonava, L.; Stracker, T.H.; Batlle, E.; Ribas de Pouplana, L. A-to-I editing on tRNAs: Biochemical, biological and evolutionary implications. *FEBS Lett.* **2014**, *588*, 4279–4286. [CrossRef] [PubMed]

108. Muramatsu, T.; Yokoyama, S.; Horie, N.; Matsuda, A.; Ueda, T.; Yamaizumi, Z.; Kuchino, Y.; Nishimura, S.; Miyazawa, T. A novel lysine-substituted nucleoside in the first position of the anticodon of minor isoleucine tRNA from *Escherichia coli*. *J. Biol. Chem.* **1988**, *263*, 9261–9267. [PubMed]

109. Muramatsu, T.; Nishikawa, K.; Nemoto, F.; Kuchino, Y.; Nishimura, S.; Miyazawa, T.; Muramatsu, T.; Nishikawa, K.; Nemoto, F.; Kuchino, Y.; et al. Codon and amino-acid specificities of a transfer RNA are both converted by a single post-transcriptional modification. *Nature* **1988**, *336*, 179–181. [CrossRef] [PubMed]

110. Soma, A.; Ikeuchi, Y.; Kanemasa, S.; Kobayashi, K.; Ogasawara, N.; Ote, T.; Kato, J.; Watanabe, K.; Sekine, Y.; Suzuki, T. An RNA-modifying enzyme that governs both the codon and amino acid specificities of isoleucine tRNA. *Mol. Cell* **2003**, *12*, 689–698. [CrossRef]

111. Nakanishi, K.; Fukai, S.; Ikeuchi, Y.; Soma, A.; Sekine, Y.; Suzuki, T.; Nureki, O. Structural basis for lysidine formation by ATP pyrophosphatase accompanied by a lysine-specific loop and a tRNA-recognition domain. *Proc. Natl. Acad. Sci. USA* **2005**, *102*, 7487–7492. [CrossRef] [PubMed]

112. Kuratani, M.; Yoshikawa, Y.; Bessho, Y.; Higashijima, K.; Ishii, T.; Shibata, R.; Takahashi, S.; Yutani, K.; Yokoyama, S. Structural basis of the initial binding of tRNA(Ile) lysidine synthetase TilS with ATP and L-lysine. *Structure* **2007**, *15*, 1642–1653. [CrossRef] [PubMed]

113. Nakanishi, K.; Bonnefond, L.; Kimura, S.; Suzuki, T.; Ishitani, R.; Nureki, O. Structural basis for translational fidelity ensured by transfer RNA lysidine synthetase. *Nature* **2009**, *461*, 1144–1148. [CrossRef] [PubMed]

114. Suzuki, T.; Numata, T. Convergent evolution of AUA decoding in bacteria and archaea. *RNA Biol.* **2014**, *11*, 1586–1596. [CrossRef] [PubMed]

115. Numata, T. Mechanisms of the tRNA wobble cytidine modification essential for AUA codon decoding in prokaryotes. *Biosci. Biotechnol. Biochem.* **2015**, *79*, 347–353. [CrossRef] [PubMed]

116. Ikeuchi, Y.; Kimura, S.; Numata, T.; Nakamura, D.; Yokogawa, T.; Ogata, T.; Wada, T.; Suzuki, T.; Suzuki, T. Agmatine-conjugated cytidine in a tRNA anticodon is essential for AUA decoding in archaea. *Nat. Chem. Biol.* **2010**, *6*, 277–282. [CrossRef] [PubMed]

117. Mandal, D.; Köhrer, C.; Su, D.; Russell, S.P.; Krivos, K.; Castleberry, C.M.; Blum, P.; Limbach, P.A.; Söll, D.; RajBhandary, U.L. Agmatidine, a modified cytidine in the anticodon of archaeal tRNA(Ile), base pairs with adenosine but not with guanosine. *Proc. Natl. Acad. Sci. USA* **2010**, *107*, 2872–2877. [CrossRef] [PubMed]

118. Terasaka, N.; Kimura, S.; Osawa, T.; Numata, T.; Suzuki, T. Biogenesis of 2-agmatinylcytidine catalyzed by the dual protein and RNA kinase TiaS. *Nat. Struct. Mol. Biol.* **2011**, *18*, 1268–1274. [CrossRef] [PubMed]

119. Osawa, T.; Kimura, S.; Terasaka, N.; Inanaga, H.; Suzuki, T.; Numata, T. Structural basis of tRNA agmatinylation essential for AUA codon decoding. *Nat. Struct. Mol. Biol.* **2011**, *18*, 1275–1280. [CrossRef] [PubMed]

120. Voorhees, R.M.; Mandal, D.; Neubauer, C.; Köhrer, C.; RajBhandary, U.L.; Ramakrishnan, V. The structural basis for specific decoding of AUA by isoleucine tRNA on the ribosome. *Nat. Struct. Mol. Biol.* **2013**, *20*, 641–643. [CrossRef] [PubMed]

121. Yamanaka, K.; Hwang, J.; Inouye, M. Characterization of GTPase activity of TrmE, a member of a novel GTPase superfamily, from *Thermotoga maritima*. *J. Bacteriol.* **2000**, *182*, 7078–7082. [CrossRef] [PubMed]

122. Urbonavicius, J.; Qian, Q.; Durand, J.M.; Hagervall, T.G.; Björk, G.R. Improvement of reading frame maintenance is a common function for several tRNA modifications. *EMBO J.* **2001**, *20*, 4863–4873. [CrossRef] [PubMed]

123. Urbonavicius, J.; Stahl, G.; Durand, J.M.; Ben Salem, S.N.; Qian, Q.; Farabaugh, P.J.; Björk, G.R. Transfer RNA modifications that alter +1 frameshifting in general fail to affect -1 frameshifting. *RNA* **2003**, *9*, 760–768. [CrossRef] [PubMed]

124. Takai, K.; Yokoyama, S. Roles of 5-substituents of tRNA wobble uridines in the recognition of purine-ending codons. *Nucleic Acids Res.* **2003**, *31*, 6383–6391. [CrossRef] [PubMed]

125. Bujnicki, J.M.; Oudjama, Y.; Roovers, M.; Owczarek, S.; Caillet, J.; Droogmans, L. Identification of a bifunctional enzyme MnmC involved in the biosynthesis of a hypermodified uridine in the wobble position of tRNA. *RNA* **2004**, *18*, 1236–1242. [CrossRef] [PubMed]

126. Armengod, M.E.; Moukadiri, I.; Prado, S.; Ruiz-Partida, R.; Benítez-Páez, A.; Villarroya, M.; Lomas, R.; Garzón, M.J.; Martínez-Zamora, A.; Meseguer, S.; Navarro-González, C. Enzymology of tRNA modification in the bacterial MnmEG pathway. *Biochimie* **2012**, *94*, 1510–1520. [CrossRef] [PubMed]

127. Moukadiri, I.; Garzón, M.J.; Björk, G.R.; Armengod, M.E. The output of the tRNA modification pathways controlled by the *Escherichia coli* MnmEG and MnmC enzymes depends on the growth conditions and the tRNA species. *Nucleic Acids Res.* **2014**, *42*, 2602–2623. [CrossRef] [PubMed]

128. Armengod, M.E.; Meseguer, S.; Villarroya, M.; Prado, S.; Moukadiri, I.; Ruiz-Partida, R.; Garzón, M.J.; Navarro-González, C.; Martínez-Zamora, A. Modification of the wobble uridine in bacterial and mitochondrial tRNAs reading NNA/NNG triplets of 2-codon boxes. *RNA Biol.* **2014**, *11*, 1495–1507. [CrossRef] [PubMed]

129. Kalhor, H.R.; Clarke, S. Novel methyltransferase for modified uridine residues at the wobble position of tRNA. *Mol. Cell Biol.* **2003**, *23*, 9283–9292. [CrossRef] [PubMed]

130. Huang, B.; Johansson, M.J.; Byström, A.S. An early step in wobble uridine tRNA modification requires the Elongator complex. *RNA* **2005**, *11*, 424–436. [CrossRef] [PubMed]

131. Begley, U.; Dyavaiah, M.; Patil, A.; Rooney, J.P.; DiRenzo, D.; Young, C.M.; Conklin, D.S.; Zitomer, R.S.; Begley, T.J. Trm9-catalyzed tRNA modifications link translation to the DNA damage response. *Mol. Cell* **2007**, *28*, 860–870. [CrossRef] [PubMed]

132. Mazauric, M.H.; Dirick, L.; Purushothaman, S.K.; Björk, G.R.; Lapeyre, B. Trm112p is a 15-kDa zinc finger protein essential for the activity of two tRNA and one protein methyltransferases in yeast. *J. Biol. Chem.* **2010**, *285*, 18505–18515. [CrossRef] [PubMed]

133. Chavarria, N.E.; Hwang, S.; Cao, S.; Fu, X.; Holman, M.; Elbanna, D.; Rodriguez, S.; Arrington, D.; Englert, M.; Uthandi, S.; et al. Archaeal Tuc1/Ncs6 homolog required for wobble uridine tRNA thiolation is associated with ubiquitin-proteasome, translation, and RNA processing system homologs. *PLoS ONE* **2014**, *9*, e99104. [CrossRef] [PubMed]

134. Arragain, S.; Bimai, O.; Legrand, P.; Caillat, S.; Ravanat, J.L.; Touati, N.; Binet, L.; Atta, M.; Fontecave, M.; Golinelli-Pimpaneau, B. Nonredox thiolation in tRNA occurring via sulfur activation by a [4Fe-4S] cluster. *Proc. Natl. Acad. Sci. USA* **2017**, *114*, 7355–7360. [CrossRef] [PubMed]

135. Black, K.A.; Dos Satos, P.C. Abbreviated Pathway for Biosynthesis of 2-Thiouridine in *Bacillus subtilis*. *J. Bacteriol.* **2015**, *197*, 1952–1962. [CrossRef] [PubMed]

136. Kitamura, A.; Nishimoto, M.; Sengoku, T.; Shibata, R.; Jäger, G.; Björk, G.R.; Grosjean, H.; Yokoyama, S.; Bessho, Y. Characterization and structure of the *Aquifex aeolicus* protein DUF752: A bacterial tRNA-methyltransferase (MnmC2) functioning without the usually fused oxidase domain (MnmC1). *J. Biol. Chem.* **2012**, *287*, 43950–43960. [CrossRef] [PubMed]

137. Mandal, D.; Köhrer, C.; Su, D.; Babu, I.R.; Chan, C.T.; Liu, Y.; Söll, D.; Blum, P.; Kuwahara, M.; Dedon, P.C.; Rajbhandary, U.L. Identification and codon reading properties of 5-cyanomethyl uridine, a new modified nucleoside found in the anticodon wobble position of mutant haloarchaeal isoleucine tRNAs. *RNA* **2014**, *20*, 177–188. [CrossRef] [PubMed]

138. Scrima, A.; Vetter, I.R.; Armengod, M.E.; Wittinghofer, A. The structure of the TrmE GTP-binding protein and its implications for tRNA modification. *EMBO J.* **2005**, *24*, 23–33. [CrossRef] [PubMed]

139. Scrima, A.; Wittinghofer, A. Dimerisation-dependent GTPase reaction of MnmE: How potassium acts as GTPase-activating element. *EMBO J.* **2006**, *25*, 2940–2951. [CrossRef] [PubMed]

140. Osawa, T.; Inanaga, H.; Numata, T. Crystallization and preliminary X-ray diffraction analysis of the tRNA-modification enzyme GidA from *Aquifex aeolicus. Acta Crystallogr. Sect. F Struct. Biol. Cryst. Commun.* **2009**, *65*, 508–511. [CrossRef] [PubMed]

141. Osawa, T.; Ito, K.; Inanaga, H.; Nureki, O.; Tomita, K.; Numata, T. Conserved cysteine residues of GidA are essential for biogenesis of 5-carboxymethylaminomethyluridine at tRNA anticodon. *Structure* **2009**, *17*, 713–724. [CrossRef] [PubMed]

142. Selcadurai, K.; Wang, P.; Seimetz, J.; Huang, R.H. Archaeal Elp3 catalyzes tRNA wobble uridine modification at C5 via a radical mechanism. *Nat. Chem. Biol.* **2014**, *10*, 810–812. [CrossRef] [PubMed]

143. Benítez-Páez, A.; Villarroya, M.; Douthwaite, S.; Gabaldón, T.; Armengod, M.E. YibK is the 2′-O-methyltransferase TrmL that modifies the wobble nucleotide in *Escherichia coli* tRNA(Leu) isoacceptors. *RNA* **2010**, *16*, 2131–2143. [CrossRef] [PubMed]

144. Pang, P.; Deng, X.; Wang, Z.; Xie, W. Structural and biochemical insights into the 2′-O-methylation of pyrimidines 34 in tRNA. *FEBS J.* **2017**, *284*, 2251–2263. [CrossRef] [PubMed]

145. Okada, N.; Noguchi, S.; Nishimura, S.; Ohgi, T.; Goto, T.; Crain, P.F.; McCloskey, J.A. Structure determination of a nucleoside Q precursor isolated from *E. coli* tRNA: 7-(aminomethyl)-7-deazaguanosine. *Nucleic Acids Res.* **1978**, *5*, 2289–2296. [CrossRef] [PubMed]

146. Okada, N.; Noguchi, S.; Kasai, H.; Shindo-Okada, N.; Ohgi, T.; Goto, T.; Nishimura, S. Novel mechanism of post-transcriptional modification of tRNA. Insertion of bases of Q precursors into tRNA by a specific tRNA transglycosylase reaction. *J. Biol. Chem.* **1979**, *254*, 3067–3073. [PubMed]

147. Nakanishi, S.; Ueda, T.; Hori, H.; Yamazaki, N.; Okada, N.; Watanabe, K. A UGU sequence in the anticodon loop is a minimum requirement for recognition by Escherichia coli tRNA-guanine transglycosylase. *J. Biol. Chem.* **1994**, *269*, 32221–32225. [PubMed]

148. Slany, R.K.; Bösl, M.; Kersten, H. Transfer and isomerization of the ribose moiety of AdoMet during the biosynthesis of queuosine tRNAs, a new unique reaction catalyzed by the QueA protein from *Escherichia coli. Biochimie* **1994**, *76*, 389–393. [CrossRef]

149. Van Lanen, S.G.; Kinzie, S.D.; Matthieu, S.; Link, T.; Culp, J.; Iwata-Reuyl, D. tRNA modification by S-adenosylmethionine: tRNA ribosyltransferase-isomerase. Assay development and characterization of the recombinant enzyme. *J. Biol. Chem.* **2003**, *278*, 10491–10499. [CrossRef] [PubMed]

150. Miles, Z.D.; Myers, W.K.; Kincannon, W.M.; Britt, R.D.; Bandarian, V. Biochemical and Spectroscopic Studies of Epoxyqueuosine Reductase: A Novel Iron-Sulfur Cluster- and Cobalamin-Containing Protein Involved in the Biosynthesis of Queuosine. *Biochemistry* **2015**, *54*, 4927–4935. [CrossRef] [PubMed]

151. Mathews, I.; Schwarzenbacher, R.; McMullan, D.; Abdubek, P.; Ambing, E.; Axelrod, H.; Biorac, T.; Canaves, J.M.; Chiu, H.J.; Deacon, A.M.; et al. Crystal structure of S-adenosylmethionine: tRNA ribosyltransferase-isomerase (QueA) from *Thermotoga maritima* at 2.0 A resolution reveals a new fold. *Proteins* **2005**, *59*, 869–974. [CrossRef] [PubMed]

152. Vinayak, M.; Pathak, C. Queuosine modification of tRNA: Its divergent role in cellular machinery. *Biosci. Rep.* **2009**, *30*, 135–148. [CrossRef] [PubMed]

153. Hutinet, G.; Swarjo, M.A.; de Crécy-Lagard, V. Deazaguanine derivatives, examples of crosstalk between RNA and DNA modification pathways. *RNA Biol.* **2017**, *14*, 1175–1184. [CrossRef] [PubMed]

154. Clouet d'Orval, B.; Bortolin, M.L.; Gaspin, C.; Bachellerie, J.P. Box C/D RNA guides for the ribose methylation of archaeal tRNAs. The tRNA^{Trp} intron guides the formation of two ribose-methylated nucleosides in the mature tRNA^{Trp}. *Nucleic Acids Res.* **2001**, *29*, 4518–4529. [CrossRef] [PubMed]

155. Bortolin, M.L.; Bachellerie, J.P.; Clouet-d'Orval, B. In vitro RNP assembly and methylation guide activity of an unusual box C/D RNA, cis-acting archaeal pre-tRNA^{Trp}. *Nucleic Acids Res.* **2003**, *31*, 6524–6535. [CrossRef] [PubMed]

156. Byström, A.S.; Björk, G.R. Chromosomal location and cloning of the gene (*trmD*) responsible for the synthesis of tRNA (m^{1}G) methyltransferase in *Escherichia coli* K-12. *Mol. Gen. Genet.* **1982**, *188*, 440–446. [CrossRef] [PubMed]

157. Björk, G.R.; Wikstrom, P.M.; Byström, A.S. Prevention of translational frameshifting by the modified nucleoside 1-methylguanosine. *Science* **1989**, *244*, 986–989. [CrossRef] [PubMed]

158. Farabaugh, P.J.; Björk, G.R. How translational accuracy influences reading frame maintenance. *EMBO J.* **1999**, *18*, 1427–1434. [CrossRef] [PubMed]

159. Björk, G.R.; Jacobsson, K.; Nilsson, K.; Johansson, M.J.; Byström, A.S.; Persson, O.P. A primordial tRNA modification required for the evolution of life? *EMBO J.* **2001**, *20*, 231–239. [CrossRef] [PubMed]

160. Liu, J.; Wang, W.; Shin, D.H.; Yokota, H.; Kim, R.; Kim, S.H. Crystal structure of tRNA (m^{1}G37) methyltransferase from *Aquifex aeolicus* at 2.6 A resolution: A novel methyltransferase fold. *Proteins* **2003**, *53*, 326–328. [CrossRef] [PubMed]

161. Christian, T.; Evilia, C.; Williams, S.; Hou, Y.M. Distinct origins of tRNA(m^{1}G37) methyltransferase. *J. Mol. Biol.* **2004**, *339*, 707–719. [CrossRef] [PubMed]

162. Takeda, H.; Toyooka, T.; Ikeuchi, Y.; Yokobori, S.; Okadome, K.; Takano, F.; Oshima, T.; Suzuki, T.; Endo, Y.; Hori, H. The substrate specificity of tRNA (m^{1}G37) methyltransferase (TrmD) from *Aquifex aeolicus*. *Genes Cells* **2006**, *11*, 1353–1365. [CrossRef] [PubMed]

163. Christian, T.; Evilia, C.; Hou, Y.M. Catalysis by the second class of tRNA(m^{1}G37) methyl transferase requires a conserved proline. *Biochemistry* **2006**, *45*, 7463–7473. [CrossRef] [PubMed]

164. Christian, T.; Hou, Y.M. Distinct determinants of tRNA recognition by the TrmD and Trm5 methyl transferases. *J. Mol. Biol.* **2007**, *373*, 623–632. [CrossRef] [PubMed]

165. Toyooka, T.; Awai, T.; Kanai, T.; Imanaka, T.; Hori, H. Stabilization of tRNA (m^{1}G37) methyltransferase [TrmD] from *Aquifex aeolicus* by an intersubunit disulfide bond formation. *Genes Cells* **2008**, *13*, 807–816. [CrossRef] [PubMed]

166. Goto-Ito, S.; Ito, T.; Ishii, R.; Muto, Y.; Bessho, Y.; Yokoyama, S. Crystal structure of archaeal tRNA(m(1)G37)methyltransferase aTrm5. *Proteins* **2008**, *72*, 1274–1289. [CrossRef] [PubMed]

167. Goto-Ito, S.; Ito, T.; Kuratani, M.; Bessho, Y.; Yokoyama, S. Tertiary structure checkpoint at anticodon loop modification in tRNA functional maturation. *Nat. Struct. Mol. Biol.* **2009**, *16*, 1109–1115. [CrossRef] [PubMed]

168. Lahoud, G.; Goto-Ito, S.; Yoshida, K.; Ito, T.; Yokoyama, S.; Hou, Y.M. Differentiating analogous tRNA methyltransferases by fragments of the methyl donor. *RNA* **2011**, *17*, 1236–1246. [CrossRef] [PubMed]

169. Sakaguchi, R.; Giessing, A.; Dai, Q.; Lahoud, G.; Liutkeviciute, Z.; Klimasauskas, S.; Piccirilli, J.; Kirpekar, F.; Hou, Y.M. Recognition of guanosine by dissimilar tRNA methyltransferases. *RNA* **2012**, *18*, 1687–1701. [CrossRef] [PubMed]

170. Christian, T.; Gamper, H.; Hou, Y.M. Conservation of structure and mechanism by Trm5 enzymes. *RNA* **2013**, *19*, 1192–1199. [CrossRef] [PubMed]

171. Ito, T.; Masuda, I.; Yoshida, K.; Goto-Ito, S.; Sekine, S.; Suh, S.W.; Hou, Y.M.; Yokoyama, S. Structural basis for methyl-donor-dependent and sequence-specific binding to tRNA substrates by knotted methyltransferase TrmD. *Proc. Natl. Acad. Sci. USA* **2015**, *112*, E4197–4205. [CrossRef] [PubMed]

172. Goto-Ito, S.; Ishii, R.; Ito, T.; Shibata, R.; Fusatomi, E.; Sekine, S.I.; Bessho, Y.; Yokoyama, S. Structure of an archaeal TYW1, the enzyme catalyzing the second step of wye-base biosynthesis. *Acta Crystallogr. D Biol. Crystallogr.* **2007**, *63*, 1059–1068. [CrossRef] [PubMed]

173. Suzuki, Y.; Noma, A.; Suzuki, T.; Senda, M.; Senda, T.; Ishitani, R.; Nureki, O. Crystal structure of the radical SAM enzyme catalyzing tricyclic modified base formation in tRNA. *J. Mol. Biol.* **2007**, *372*, 1204–1214. [CrossRef] [PubMed]

174. Umitsu, M.; Nishimasu, H.; Noma, A.; Suzuki, T.; Ishitani, R.; Nureki, O. Structural basis of AdoMet-dependent aminocarboxypropyl transfer reaction catalyzed by tRNA-wybutosine synthesizing enzyme, TYW2. *Proc. Natl. Acad. Sci. USA* **2009**, *106*, 15616–15621. [CrossRef] [PubMed]

175. Perche-Letuvée, P.; Kathirvelu, V.; Berggren, G.; Clemancey, M.; Latour, J.M.; Maurel, V.; Douki, T.; Armengaud, J.; Mulliez, E.; Fontecave, M.; et al. 4-Demethylwyosine synthase from *Pyrococcus abyssi* is a radical-S-adenosyl-L-methionine enzyme with an additional [4Fe-4S](+2) cluster that interacts with the pyruvate co-substrate. *J. Biol. Chem.* **2012**, *287*, 41174–41185. [CrossRef] [PubMed]

176. Urbonavičius, J.; Rutkienė, R.; Lopato, A.; Tauraitė, D.; Stankevičiūtė, J.; Aučynaitėm, A.; Kaliniene, L.; van Tilbeurgh, H.; Meškys, R. Evolution of tRNA[Phe]: imG2 methyltransferases involved in the biosynthesis of wyosine derivatives in Archaea. *RNA* **2016**, *22*, 1871–1883.

177. Wang, C.; Jia, Q.; Chen, R.; Wei, Y.; Li, J.; Ma, J.; Xie, W. Crystal structures of the bifunctional tRNA methyltransferase Trm5a. *Sci. Rep.* **2016**, *6*, 33553. [CrossRef] [PubMed]

178. Currie, M.A.; Brown, G.; Wong, A.; Ohira, T.; Sugiyama, K.; Suzuki, T.; Yakunin, A.F.; Jia, Z. Structural and functional characterization of the TYW3/Taw3 class of SAM-dependent methyltransferases. *RNA* **2017**, *23*, 346–354. [CrossRef] [PubMed]

179. Wang, C.; Jia, Q.; Zeng, J.; Chen, R.; Xie, W. Structural insight into the methyltransfer mechanism of the bifunctional Trm5. *Sci. Adv.* **2017**, *3*, e1700195. [CrossRef] [PubMed]

180. Wu, J.; Jia, Q.; Wu, S.; Zeng, H.; Sun, Y.; Wang, C.; Ge, R.; Xie, W. The crystal structure of the *Pyrococcus abyssi* mono-functional methyltransferase PaTrm5b. *Biochem. Biophys. Res. Commun.* **2017**, *493*, 240–245. [CrossRef] [PubMed]

181. de Crécy-Lagard, V.; Brochier-Armanet, C.; Urbonavicius, J.; Fernandez, B.; Phillips, G.; Lyons, B.; Noma, A.; Alvarez, S.; Droogmans, L.; Armengaud, J.; et al. Biosynthesis of wyosine derivatives in tRNA: An ancient and highly diverse pathway in Archaea. *Mol. Biol. Evol.* **2010**, *27*, 2062–2077. [CrossRef] [PubMed]

182. Urbonavičius, J.; Meškys, R.; Grosjean, H. Biosynthesis of wyosine derivatives in tRNA(Phe) of Archaea: Role of a remarkable bifunctional tRNA(Phe):m1G/imG2 methyltransferase. *RNA* **2014**, *20*, 747–753. [CrossRef] [PubMed]

183. Perche-Letuvée, P.; Molle, T.; Forouhar, F.; Mulliez, E.; Atta, M. Wybutosine biosynthesis: Structural and mechanistic overview. *RNA Biol.* **2014**, *11*, 1508–1518. [CrossRef] [PubMed]

184. Mao, D.Y.; Neculai, D.; Downey, M.; Orlicky, S.; Haffani, Y.Z.; Ceccarelli, D.F.; Ho, J.S.; Szilard, R.K.; Zhang, W.; Ho, C.S.; et al. Atomic structure of the KEOPS complex: An ancient protein kinase-containing molecular machine. *Mol. Cell* **2008**, *32*, 259–275. [CrossRef] [PubMed]

185. Hecker, A.; Lopreiato, R.; Graille, M.; Collinet, B.; Forterre, P.; Libri, D.; van Tilbeurgh, H. Structure of the archaeal Kae1/Bud32 fusion protein MJ1130: A model for the eukaryotic EKC/KEOPS subcomplex. *EMBO J.* **2008**, *27*, 2340–2351. [CrossRef] [PubMed]

186. Hecker, A.; Graille, M.; Madec, E.; Gadelle, D.; Le Cam, E.; van Tilbergh, H.; Forterre, P. The universal Kae1 protein and the associated Bud32 kinase (PRPK), a mysterious protein couple probably essential for genome maintenance in Archaea and Eukarya. *Biochem. Soc. Trans.* **2009**, *37*, 29–35. [CrossRef] [PubMed]

187. Perrochia, L.; Crozat, E.; Hecker, A.; Zhang, W.; Bareille, J.; Collinet, B.; van Tilbeurgh, H.; Forterre, P.; Basta, T. In vitro biosynthesis of a universal t[6]A tRNA modification in Archaea and Eukarya. *Nucleic Acids Res.* **2013**, *41*, 1953–1964. [CrossRef] [PubMed]

188. Wan, L.C.; Pillon, M.C.; Thevakumaran, N.; Sun, Y.; Chakrabartty, A.; Guarné, A.; Kurinov, I.; Durocher, D.; Sicheri, F. Structural and functional characterization of KEOPS dimerization by Pcc1 and its role in t[6]A biosynthesis. *Nucleic Acids Res.* **2016**, *44*, 6971–6980. [CrossRef] [PubMed]

189. Pichard-Kostuch, A.; Zhang, W.; Liger, D.; Daugeron, M.C.; Letoquart, J.; Li de la Sierra-Gallay, I.; Forterre, P.; Collinet, B.; van Tilbeurgh, H.; Basta, T. Structure-function analysis of Sua5 protein reveals novel functional motifs required for the biosynthesis of the universal t[6]A tRNA modification. *RNA* **2018**, *24*, 926–938. [CrossRef] [PubMed]

190. Luthra, A.; Swinehart, W.; Bayooz, S.; Phan, P.; Stec, B.; Iwata-Reuyl, D.; Swairjo, M.A. Structure and mechanism of a bacterial t[6]A biosynthesis system. *Nucleic Acids Res.* **2018**, *46*, 1395–1411. [CrossRef] [PubMed]

191. Thiaville, P.C.; Iwata-Reuyl, D.; de Crécy-Lagard, V. Diversity of the biosynthesis pathway for threonylcarbamoyladenosine (t[6]A), a universal modification of tRNA. *RNA Biol.* **2014**, *11*, 1529–1539. [CrossRef] [PubMed]

192. Caillet, J.; Droogmans, L. Molecular cloning of the *Escherichia coli miaA* gene involved in the formation of delta 2-isopentenyl adenosine in tRNA. *J. Bacteriol.* **1988**, *170*, 4147–4152. [CrossRef] [PubMed]

193. Esberg, B.; Leung, H.C.; Tsui, H.C.; Björk, G.R.; Winkler, M.E. Identification of the *miaB* gene, involved in methylthiolation of isopentenylated A37 derivatives in the tRNA of *Salmonella typhimurium* and *Escherichia coli*. *J. Bacteriol.* **1999**, *181*, 7256–7265. [PubMed]

194. Pierrel, F.; Hernandez, H.L.; Johnson, M.K.; Fontecave, M.; Atta, M. MiaB protein from *Thermotoga maritima*. Characterization of an extremely thermophilic tRNA-methylthiotransferase. *J. Biol. Chem.* **2003**, *278*, 29515–29524. [CrossRef] [PubMed]

195. Pierrel, F.; Douki, T.; Fontecave, M.; Atta, M. MiaB protein is a bifunctional radical-S-adenosylmethionine enzyme involved in thiolation and methylation of tRNA. *J. Biol. Chem.* **2004**, *279*, 47555–47563. [CrossRef] [PubMed]

196. Hernández, H.L.; Pierrel, F.; Elleingand, E.; García-Serres, R.; Huynh, B.H.; Johnson, M.K.; Fontecave, M.; Atta, M. MiaB, a bifunctional radical-S-adenosylmethionine enzyme involved in the thiolation and methylation of tRNA, contains two essential [4Fe-4S] clusters. *Biochemistry* **2007**, *46*, 5140–5147. [CrossRef] [PubMed]

197. Schweizer, U.; Bohleber, S.; Fradejas-Villar, N. The modified base isopentenyladenosine and its derivatives in tRNA. *RNA Biol.* **2017**, *14*, 1197–1208. [CrossRef] [PubMed]

198. Golovina, A.Y.; Sergiev, P.V.; Golovin, A.V.; Serebryakova, M.V.; Demina, I.; Govorun, V.M.; Dontsova, O.A. The *yfiC* gene of *E. coli* encodes an adenine-N^6 methyltransferase that specifically modifies A37 of tRNA$_1$^{Val}(cmo^{5}UAC). *RNA* **2009**, *15*, 1134–1141. [CrossRef] [PubMed]

199. Allaudeen, H.S.; Yang, S.K.; Söll, D. Leucine tRNA(1) from HisT mutant of *Salmonella typhimurium* lacks two pseudouridines. *FEBS Lett.* **1972**, *28*, 205–208. [CrossRef]

200. Kammen, H.O.; Marvel, C.C.; Hardy, L.; Penhoet, E.E. Purification, structure, and properties of *Escherichia coli* tRNA pseudouridine synthase I. *J. Biol. Chem.* **1988**, *263*, 2255–2263. [PubMed]

201. Lecointe, F.; Simos, G.; Sauer, A.; Hurt, E.C.; Motorin, Y.; Grosjean, H. Characterization of yeast protein Deg1 as pseudouridine synthase (Pus3) catalyzing the formation of psi 38 and psi 39 in tRNA anticodon loop. *J. Biol. Chem.* **1998**, *273*, 1316–1323. [CrossRef] [PubMed]

202. Dong, X.; Bessho, Y.; Shibata, R.; Nishimoto, M.; Shirouzu, M.; Kuramitsu, S.; Yokoyama, S. Crystal structure of tRNA pseudouridine synthase TruA from *Thermus thermophilus* HB8. *RNA Biol.* **2006**, *3*, 115–122. [CrossRef] [PubMed]

203. Spenkuch, F.; Motorin, Y.; Helm, M. Pseudouridine: Still mysterious, but never a fake (uridine)! *RNA Biol.* **2014**, *11*, 1540–1554. [CrossRef] [PubMed]

204. Morozov, I.A.; Gambaryan, A.S.; Lvova, T.N.; Nedospasov, A.A.; Venkstern, T.V. Purification and characterization of tRNA (adenine-1-)-methyltransferase from *Thermus flavus* strain 71. *Eur. J. Biochem.* **1982**, *129*, 429–436. [CrossRef] [PubMed]

205. De Bie, L.G.; Roovers, M.; Oudjama, Y.; Wattiez, R.; Tricot, C.; Stalon, V.; Droogmans, L.; Bujnicki, J.M. The *yggH* gene of *Escherichia coli* encodes a tRNA (m^{7}G46) methyltransferase. *J. Bacteriol.* **2003**, *185*, 3238–3243. [CrossRef] [PubMed]

206. Okamoto, H.; Watanabe, K.; Ikeuchi, Y.; Suzuki, T.; Endo, Y.; Hori, H. Substrate tRNA recognition mechanism of tRNA (m^{7}G46) methyltransferase from *Aquifex aeolicus*. *J. Biol. Chem.* **2004**, *279*, 49151–49159. [CrossRef] [PubMed]

207. Tomikawa, C.; Ochi, A.; Hori, H. The C-terminal region of thermophilic tRNA (m^{7}G46) methyltransferase (TrmB) stabilizes the dimer structure and enhances fidelity of methylation. *Proteins* **2008**, *71*, 1400–1408. [CrossRef] [PubMed]

208. Tomikawa, C.; Takai, K.; Hori, H. Kinetic characterization of substrate-binding sites of thermostable tRNA methyltransferase (TrmB). *J. Biochem.* **2018**, *163*, 133–142. [CrossRef] [PubMed]

209. Auxilien, S.; El Khadali, F.; Rasmussen, A.; Douthwaite, S.; Grosjean, H. Archease from *Pyrococcus abyssi* improves substrate specificity and solubility of a tRNA m^{5}C methyltransferase. *J. Biol. Chem.* **2007**, *282*, 18711–18721. [CrossRef] [PubMed]

210. Kuratani, M.; Hirano, M.; Goto-Ito, S.; Itoh, Y.; Hikida, Y.; Nishimoto, M.; Sekine, S.; Bessho, Y.; Ito, T.; Grosjean, H.; et al. Crystal structure of *Methanocaldococcus jannaschii* Trm4 complexed with sinefungin. *J. Mol. Biol.* **2010**, *401*, 323–333. [CrossRef] [PubMed]

211. Roovers, M.; Hale, C.; Tricot, C.; Terns, M.P.; Terns, R.M.; Grosjean, H.; Droogmans, L. Formation of the conserved pseudouridine at position 55 in archaeal tRNA. *Nucleic Acids Res.* **2006**, *34*, 4293–4301. [CrossRef] [PubMed]

212. Gurha, P.; Gupta, R. Archaeal Pus10 proteins can produce both pseudouridine 54 and 55 in tRNA. *RNA* **2008**, *14*, 2521–2527. [CrossRef] [PubMed]

213. Joardar, A.; Jana, S.; Fitzek, E.; Gurha, P.; Majumder, M.; Chatterjee, K.; Geisler, M.; Gupta, R. Role of forefinger and thumb loops in production of Ψ54 and Ψ55 in tRNAs by archaeal Pus10. *RNA* **2013**, *19*, 1279–1294. [CrossRef] [PubMed]

214. Kamalampeta, R.; Keffer-Wilkes, L.C.; Kothe, U. tRNA binding, positioning, and modification by the pseudouridine synthase Pus10. *J. Mol. Biol.* **2013**, *425*, 3863–3874. [CrossRef] [PubMed]

215. Chen, H.Y.; Yuan, Y.A. Crystal structure of Mj1640/DUF358 protein reveals a putative SPOUT-class RNA methyltransferase. *J. Mol. Cell. Biol.* **2010**, *2*, 366–374. [CrossRef] [PubMed]

216. Wurm, J.P.; Griese, M.; Bahr, U.; Held, M.; Heckel, A.; Karas, M.; Soppa, J.; Wöhnert, J. Identification of the enzyme responsible for N^1-methylation of pseudouridine 54 in archaeal tRNAs. *RNA* **2012**, *18*, 412–420. [CrossRef] [PubMed]

217. Chatterjee, K.; Blaby, I.K.; Thiaville, P.C.; Majumder, M.; Grosjean, H.; Yuan, Y.A.; Gupta, R.; de Crécy-Lagard, V. The archaeal COG1901/DUF358 SPOUT-methyltransferase members, together with pseudouridine synthase Pus10, catalyze the formation of 1-methylpseudouridine at position 54 of tRNA. *RNA* **2012**, *18*, 421–433. [CrossRef] [PubMed]

218. Davanloo, P.; Sprinzl, M.; Watanabe, K.; Albani, M.; Kersten, H. Role of ribothymidine in the thermal stability of transfer RNA as monitored by proton magnetic resonance. *Nucleic Acids Res.* **1979**, *6*, 1571–1581. [CrossRef] [PubMed]

219. Watanabe, K.; Himeno, H.; Ohta, T. Selective utilization of 2-thioribothymidine- and ribothymidine-containing tRNAs by the protein synthetic systems of *Thermus thermophilus* HB 8 depending on the environmental temperature. *J. Biochem.* **1984**, *96*, 1625–1632. [CrossRef] [PubMed]

220. Yokoyama, S.; Watanabe, K.; Miyazawa, T. Dynamic structures and functions of transfer ribonucleic acids from extreme thermophiles. *Adv. Biophys.* **1987**, *23*, 115–147. [CrossRef]

221. Urbonavicius, J.; Skouloubris, S.; Myllykallio, H.; Grosjean, H. Identification of a novel gene encoding a flavin-dependent tRNA:m^{5}U methyltransferase in bacteria—evolutionary implications. *Nucleic Acids Res.* **2005**, *33*, 3955–3964. [CrossRef] [PubMed]

222. Cicmil, N. Crystallization and preliminary X-ray crystallographic characterization of TrmFO, a folate-dependent tRNA methyltransferase from *Thermotoga maritima. Acta Crystallogr. Sect. F Struct. Biol. Cryst. Commun.* **2008**, *64*, 193–195. [CrossRef] [PubMed]

223. Nishimasu, H.; Ishitani, R.; Yamashita, K.; Iwashita, C.; Hirata, A.; Hori, H.; Nureki, O. Atomic structure of a folate/FAD-dependent tRNA T54 methyltransferase. *Proc. Natl. Acad. Sci. USA* **2009**, *106*, 8180–8185. [CrossRef] [PubMed]

224. Yamagami, R.; Yamashita, K.; Nishimasu, H.; Tomikawa, C.; Ochi, A.; Iwashita, C.; Hirata, A.; Ishitani, R.; Nureki, O.; Hori, H. The tRNA recognition mechanism of folate/FAD-dependent tRNA methyltransferase (TrmFO). *J. Biol. Chem.* **2012**, *287*, 42480–42494. [CrossRef] [PubMed]

225. Yamagami, R.; Tomikawa, C.; Shigi, N.; Kazayama, A.; Asai, S.; Takuma, H.; Hirata, A.; Fourmy, D.; Asahara, H.; Watanabe, K.; et al. The folate/FAD-dependent tRNA methyltransferase (TrmFO) from *Thermus thermophilus* regulates the other modifications in tRNA at low temperatures. *Genes Cells* **2016**, *21*, 740–754. [CrossRef] [PubMed]

226. Hamdane, D.; Grosjean, H.; Fontecave, M. Flavin-Dependent Methylation of RNAs: Complex Chemistry for a Simple Modification. *J. Mol. Biol.* **2016**, *428*, 4867–4881. [CrossRef] [PubMed]

227. Yamagami, R.; Miyake, R.; Fukumoto, A.; Nakashima, M.; Hori, H. Consumption of N^5, N^{10}-methylenetetrahydrofolate in *Thermus thermophilus* under nutrient-poor condition. *J. Biochem.* **2018**, *164*, 141–152. [CrossRef] [PubMed]

228. Shigi, N.; Suzuki, T.; Tamakoshi, M.; Oshima, T.; Watanabe, K. Conserved bases in the TΨsi C loop of tRNA are determinants for thermophile-specific 2-thiouridylation at position 54. *J. Biol. Chem.* **2002**, *277*, 39128–39135. [CrossRef] [PubMed]

229. Shigi, N.; Suzuki, T.; Terada, T.; Shirouzu, M.; Yokoyama, S.; Watanabe, K. Temperature-dependent biosynthesis of 2-thioribothymidine of *Thermus thermophilus* tRNA. *J. Biol. Chem.* **2006**, *281*, 2104–2113. [CrossRef] [PubMed]

230. Shigi, N.; Sakaguchi, Y.; Suzuki, T.; Watanabe, K. Identification of two tRNA thiolation genes required for cell growth at extremely high temperatures. *J. Biol. Chem.* **2006**, *281*, 14296–14306. [CrossRef] [PubMed]

231. Shigi, N.; Sakaguchi, Y.; Asai, S.; Suzuki, T.; Watanabe, K. Common thiolation mechanism in the biosynthesis of tRNA thiouridine and sulphur-containing cofactors. *EMBO J.* **2008**, *27*, 3267–3278. [CrossRef] [PubMed]

232. Shigi, N. Posttranslational modification of cellular proteins by a ubiquitin-like protein in bacteria. *J. Biol. Chem.* **2012**, *287*, 17568–17577. [CrossRef] [PubMed]

233. Nakagawa, H.; Kuratani, M.; Goto-Ito, S.; Ito, T.; Katsura, K.; Terada, T.; Shirouzu, M.; Sekine, S.; Shigi, N.; Yokoyama, S. Crystallographic and mutational studies on the tRNA thiouridine synthetase TtuA. *Proteins* **2013**, *81*, 1232–1244. [CrossRef] [PubMed]

234. Chen, M.; Narai, S.; Omura, N.; Shigi, N.; Chimnaronk, S.; Tanaka, Y.; Yao, M. Crystallographic study of the 2-thioribothymidine-synthetic complex TtuA-TtuB from *Thermus thermophilus*. *Acta Crystallogr. F Struct. Biol. Commun.* **2016**, *72*, 777–781. [CrossRef] [PubMed]

235. Shigi, N.; Asai, S.-I.; Watanabe, K. Identification of a rhodanese-like protein involved in thiouridine biosynthesis in *Thermus thermophilus* tRNA. *FEBS Lett.* **2016**, *590*, 4628–4637. [CrossRef] [PubMed]

236. Chen, M.; Asai, S.-I.; Narai, S.; Nambu, S.; Omura, N.; Sakaguchi, Y.; Suzuki, T.; Ikeda-Saito, M.; Watanabe, K.; Yao, M.; et al. Biochemical and structural characterization of oxygen-sensitive 2-thiouridine synthesis catalyzed by an iron-sulfur protein TtuA. *Proc. Natl. Acad. Sci. USA* **2017**, *114*, 4954–4959. [CrossRef] [PubMed]

237. Urbonavicius, J.; Auxilien, S.; Walbott, H.; Trachana, K.; Golinelli-Pimpaneau, B.; Brochier-Armanet, C.; Grosjean, H. Acquisition of a bacterial RumA-type tRNA(uracil-54, C5)-methyltransferase by Archaea through an ancient horizontal gene transfer. *Mol. Microbiol.* **2008**, *67*, 323–335. [CrossRef] [PubMed]

238. Walbott, H.; Leulliot, N.; Grosjean, H.; Golinelli-Pimpaneau, B. The crystal structure of *Pyrococcus abyssi* tRNA (uracil-54, C5)-methyltransferase provides insights into its tRNA specificity. *Nucleic Acids Res.* **2008**, *36*, 4929–4940. [CrossRef] [PubMed]

239. Shigi, N. Biosynthesis and functions of sulfur modifications in tRNA. *Front. Genet.* **2014**, *5*, 67. [CrossRef] [PubMed]

240. Gurha, P.; Joardar, A.; Chaurasia, P.; Gupta, R. Differential roles of archaeal box H/ACA proteins in guide RNA-dependent and independent pseudouridine formation. *RNA Biol.* **2007**, *4*, 101–109. [CrossRef] [PubMed]

241. Muller, S.; Fourmann, J.B.; Loegler, C.; Charpentier, B.; Branlant, C. Identification of determinants in the protein partners aCBF5 and aNOP10 necessary for the tRNA: Psi55-synthase and RNA-guided RNA: Psi-synthase activities. *Nucleic Acids Res.* **2007**, *35*, 5610–5624. [CrossRef] [PubMed]

242. Kamalampeta, R.; Kothe, U. Archaeal proteins Nop10 and Gar1 increase the catalytic activity of Cbf5 in pseudouridylating tRNA. *Sci. Rep.* **2012**, *2*, 663. [CrossRef] [PubMed]

243. Nurse, K.; Wrzesinski, J.; Bakin, A.; Lane, B.G.; Ofengand, J. Purification, cloning, and properties of the tRNA psi 55 synthase from *Escherichia coli*. *RNA* **1995**, *1*, 102–112. [PubMed]

244. Pan, H.; Agarwalla, S.; Moustakas, D.T.; Finer-Moore, J.; Stroud, R.M. Structure of tRNA pseudouridine synthase TruB and its RNA complex: RNA recognition through a combination of rigid docking and induced fit. *Proc. Natl. Acad. Sci. USA* **2003**, *100*, 12648–12653. [CrossRef] [PubMed]

245. Wouters, J.; Tricot, C.; Durbecq, V.; Roovers, M.; Stalon, V.; Droogmans, L. Preliminary X-ray crystallographic analysis of tRNA pseudouridine 55 synthase from the thermophilic eubacterium *Thermotoga maritima*. *Acta Crystallogr. D Biol. Crystallogr.* **2003**, *59*, 152–154. [CrossRef] [PubMed]

246. Phannachet, K.; Huang, R.H. Conformational change of pseudouridine 55 synthase upon its association with RNA substrate. *Nucleic Acids Res.* **2004**, *32*, 1422–1429. [CrossRef] [PubMed]

247. Phannachet, K.; Elias, Y.; Huang, R.H. Dissecting the roles of a strictly conserved tyrosine in substrate recognition and catalysis by pseudouridine 55 synthase. *Biochemistry* **2005**, *44*, 15488–15494. [CrossRef] [PubMed]

248. Ishida, K.; Kunibayashi, T.; Tomikawa, C.; Ochi, A.; Kanai, T.; Hirata, A.; Iwashita, C.; Hori, H. Pseudouridine at position 55 in tRNA controls the contents of other modified nucleotides for low-temperature adaptation in the extreme-thermophilic eubacterium *Thermus thermophilus. Nucleic Acids Res.* **2011**, *39*, 2304–2318. [CrossRef] [PubMed]

249. Renalier, M.H.; Joseph, N.; Gaspin, C.; Thebault, P.; Mougin, A. The Cm56 tRNA modification in archaea is catalyzed either by a specific 2′-O-methylase, or a C/D sRNP. *RNA* **2005**, *11*, 1051–1063. [CrossRef] [PubMed]

250. Tkaczuk, K.L.; Dunin-Horkawicz, S.; Purta, E.; Bujnicki, J.M. Structural and evolutionary bioinformatics of the SPOUT superfamily of methyltransferases. *BMC Bioinfor.* **2007**, *8*, 73. [CrossRef] [PubMed]

251. Kuratani, M.; Bessho, Y.; Nishimoto, M.; Grosjean, H.; Yokoyama, S. Crystal structure and mutational study of a unique SpoU family archaeal methylase that forms 2′-O-methylcytidine at position 56 of tRNA. *J. Mol. Biol.* **2008**, *375*, 1064–1075. [CrossRef] [PubMed]

252. Walker, R.T. Mycoplasma evolution: A review of the use of ribosomal and transfer RNA nucleotide sequences in the determination of phylogenetic relationships. *Yale J. Biol. Med.* **1983**, *56*, 367–372. [PubMed]

253. Yamaizumi, Z.; Ihara, M.; Kuchino, Y.; Gupta, R.; Woese, C.R.; Nishimura, S. Archaebacterial tRNA contains 1-methylinosine at residue 57 in T psi C-loop. *Nucleic Acids Symp. Ser.* **1982**, *11*, 209–213.

254. Grosjean, H.; Constantinesco, F.; Foiret, D.; Benachenhou, N. A novel enzymatic pathway leading to 1-methylinosine modification in *Haloferax volcanii* tRNA. *Nucleic Acids Res.* **1995**, *23*, 4312–4319. [CrossRef] [PubMed]

255. Roovers, M.; Wouters, J.; Bujnicki, J.M.; Tricot, C.; Stalon, V.; Grosjean, H.; Droogmans, L. A primordial RNA modification enzyme: The case of tRNA (m^{1}A) methyltransferase. *Nucleic Acids Res.* **2004**, *32*, 465–476. [CrossRef] [PubMed]

256. Guelorget, A.; Roovers, M.; Guérineau, V.; Barbey, C.; Li, X.; Golinelli-Pimpaneau, B. Insights into the hyperthermostability and unusual region-specificity of archaeal *Pyrococcus abyssi* tRNA m^{1}A57/58 methyltransferase. *Nucleic Acids Res.* **2010**, *38*, 6206–6218. [CrossRef] [PubMed]

257. Guelorget, A.; Barraud, P.; Tisné, C.; Golinelli-Pimpaneau, B. Structural comparison of tRNA m(1)A58 methyltransferases revealed different molecular strategies to maintain their oligomeric architecture under extreme conditions. *BMC Struct. Biol.* **2011**, *11*, 48. [CrossRef] [PubMed]

258. Hamdane, D.; Guelorget, A.; Guérineau, V.; Golinelli-Pimpaneau, B. Dynamics of RNA modification by a multi-site-specific tRNA methyltransferase. *Nucleic Acids Res.* **2014**, *42*, 11697–11706. [CrossRef] [PubMed]

259. Biou, V.; Yaremchuk, A.; Tukalo, M.; Cusack, S. The 2.9 A crystal structure of *T. thermophilus* seryl-tRNA synthetase complexed with tRNA(Ser). *Science* **1994**, *263*, 1404–1410. [CrossRef] [PubMed]

260. Droogmans, L.; Roovers, M.; Bujnicki, J.M.; Tricot, C.; Hartsch, T.; Stalon, V.; Grosjean, H. Cloning and characterization of tRNA (m^{1}A58) methyltransferase (TrmI) from *Thermus thermophilus* HB27, a protein required for cell growth at extreme temperatures. *Nucleic Acids Res.* **2003**, *31*, 2148–2156. [CrossRef] [PubMed]

261. Barraud, P.; Golinelli-Pimpaneau, B.; Atmanene, C.; Sanglier, S.; Van Dorsselaer, A.; Droogmans, L.; Dardel, F.; Tisné, C. Crystal structure of *Thermus thermophilus* tRNA m^{1}A58 methyltransferase and biophysical characterization of its interaction with tRNA. *J. Mol. Biol.* **2008**, *377*, 535–550. [CrossRef] [PubMed]

262. Kuratani, M.; Yanagisawa, T.; Ishii, R.; Matsuno, M.; Si, S.Y.; Katsura, K.; Ushikoshi-Nakayama, R.; Shibata, R.; Shirouzu, M.; Bessho, Y.; et al. Crystal structure of tRNA m^{1}A58 methyltransferase TrmI from *Aquifex aeolicus* in complex with S-adenosyl-L-methionine. *J. Struct. Funct. Genomics* **2014**, *15*, 173–180. [CrossRef] [PubMed]

263. Takuma, H.; Ushio, N.; Minoji, M.; Kazayama, A.; Shigi, N.; Hirata, A.; Ochi, A.; Hori, H. Substrate tRNA recognition mechanism of eubacterial tRNA (m^{1}A58) methyltransferase (TrmI). *J. Biol. Chem.* **2015**, *290*, 5912–5925. [CrossRef] [PubMed]

264. Dégut, C.; Ponchon, L.; Folly-Klan, M.; Barraud, P.; Tisné, C. The m^{1}A58 modification in eubacterial tRNA: An overview of tRNA recognition and mechanism of catalysis by TrmI. *Biophys. Chem.* **2016**, *210*, 27–34. [CrossRef] [PubMed]

265. Väre, V.Y.; Eruysal, E.R.; Narendran, A.; Sarachan, K.L.; Agris, P.F. Chemical and Conformational Diversity of Modified Nucleosides Affects tRNA Structure and Function. *Biomolecules* **2017**, *7*, 29. [CrossRef] [PubMed]

266. Motorin, Y.; Helm, M. tRNA stabilization by modified nucleotides. *Biochemistry* **2010**, *49*, 4934–4944. [CrossRef] [PubMed]

267. El Yacoubi, B.; Bailly, M.; de Crécy-Lagard, V. Biosynthesis and function of posttranscriptional modifications of transfer RNAs. *Annu. Rev. Genet.* **2012**, *46*, 69–95. [CrossRef] [PubMed]

268. Manickam, N.; Joshi, K.; Bhatt, M.J.; Farabaugh, P.J. Effects of tRNA modification on translational accuracy depend on intrinsic codon-anticodon strength. *Nucleic Acids Res.* **2016**, *44*, 1871–1881. [CrossRef] [PubMed]

269. Grosjean, H.; Westhof, E. An integrated, structure- and energy-based view of the genetic code. *Nucleic Acids Res.* **2016**, *44*, 8020–8040. [CrossRef] [PubMed]

270. Agris, P.F.; Eruysal, E.R.; Narendran, A.; Väre, V.Y.P.; Vangaveti, S.; Ranganathan, S.V. Celebrating wobble decoding: Half a century and still much is new. *RNA Biol.* **2017**, *16*, 1–17. [CrossRef] [PubMed]

271. Perret, V.; Garcia, A.; Grosjean, H.; Ebel, J.P.; Florentz, C.; Giegé, R. Relaxation of a transfer RNA specificity by removal of modified nucleotides. *Nature* **1990**, *344*, 787–789. [CrossRef] [PubMed]

272. Aström, S.U.; Byström, A.S. Rit1, a tRNA backbone-modifying enzyme that mediates initiator and elongator tRNA discrimination. *Cell* **1994**, *79*, 535–546. [CrossRef]

273. Kaneko, T.; Suzuki, T.; Kapushoc, S.T.; Rubio, M.A.; Ghazvini, J.; Watanabe, K.; Simpson, L.; Suzuki, T. Wobble modification differences and subcellular localization of tRNAs in *Leishmania tarentolae*: Implication for tRNA sorting mechanism. *EMBO J.* **2003**, *22*, 657–667. [CrossRef] [PubMed]

274. Anderson, J.; Phan, L.; Cuesta, R.; Carison, B.A.; Pak, M.; Asano, K.; Björk, G.R.; Tamame, M.; Hinnebusch, A.G. The essential Gcd10p-Gcd14p nuclear complex is required for 1-methyladenosine modification and maturation of initiator methionyl-tRNA. *Genes Dev.* **1998**, *12*, 3650–3652. [CrossRef] [PubMed]

275. Anderson, J.; Phan, L.; Hinnebusch, A.G. The Gcd10p/Gcd14p complex is the essential two-subunit tRNA(1-methyladenosine) methyltransferase of *Saccharomyces cerevisiae*. *Proc. Natl. Acad. Sci. USA* **2000**, *97*, 5173–5178. [CrossRef] [PubMed]

276. Ohira, T.; Suzuki, T. Retrograde nuclear import of tRNA precursors is required for modified base biogenesis in yeast. *Proc. Natl. Acad. Sci. USA* **2011**, *108*, 10502–10507. [CrossRef] [PubMed]

277. Ohira, T.; Suzuki, T. Precursors of tRNAs are stabilized by methylguanosine cap structures. *Nat. Chem. Biol.* **2016**, *12*, 648–655. [CrossRef] [PubMed]

278. Chatterjee, K.; Nostramo, R.T.; Wan, Y.; Hopper, A.K. tRNA dynamics between the nucleus, cytoplasm and mitochondrial surface: Location, location, location. *Biochim. Biophys. Acta* **2018**, *1861*, 373–386. [CrossRef] [PubMed]

279. Kadaba, S.; Krueger, A.; Trice, T.; Krecic, A.M.; Hinnebusch, A.G.; Anderson, J. Nuclear surveillance and degradation of hypomodified initiator tRNA^{Met} in *S. cerevisiae*. *Genes Dev.* **2004**, *18*, 1227–1240. [CrossRef] [PubMed]

280. Alexandrov, A.; Chernyakov, I.; Gu, W.; Hiley, S.L.; Hughes, T.R.; Grayhack, E.J.; Phizicky, E.M. Rapid tRNA decay can result from lack of nonessential modifications. *Mol. Cell* **2006**, *21*, 87–96. [CrossRef] [PubMed]

281. Phizicky, E.M.; Hopper, A.K. tRNA biology charges to the front. *Genes Dev.* **2010**, *24*, 1832–1860. [CrossRef] [PubMed]

282. Durand, J.M.; Okada, N.; Tobe, T.; Watarai, M.; Fukuda, I.; Suzuki, T.; Nakata, N.; Komatsu, K.; Yoshikawa, M.; Sasakawa, C. vacC, a virulence-associated chromosomal locus of Shigella flexneri, is homologous to tgt, a gene encoding tRNA-guanine transglycosylase (Tgt) of Escherichia coli K-12. *J. Bacteriol.* **1994**, *176*, 4627–4634. [CrossRef] [PubMed]

283. Saga, A.E.; Vasil, A.I.; Vasil, M.L. Molecular characterization of mutants affected in the osmoprotectant-dependent induction of phspholipase C in *Pseudomonas aeruginosa* PAO1. *Mol. Microbiol.* **1997**, *23*, 43–56. [CrossRef]

284. Takano, Y.; Takayanagi, N.; Hori, H.; Ikeuchi, Y.; Suzuki, T.; Kimura, A.; Okuno, T. A gene involved in modifying transfer RNA is required for fungal pathogenicity and stress tolerance of *Colletotrichum lagenarium*. *Mol. Microbiol.* **2006**, *60*, 81–92. [CrossRef] [PubMed]

285. Sleiman, D.; Goldschmidt, V.; Barraud, P.; Marquet, R.; Paillart, J.C.; Tisné, C. Initiation of HIV-1 reverse transcription and functional role of nucleocapsid-mediated tRNA/viral genome interactions. *Virus Res.* **2012**, *169*, 324–339. [CrossRef] [PubMed]

286. Saadatmand, J.; Kleiman, L. Aspects of HIV-1 assembly that promote primer tRNA (Lys3) annealing to viral RNA. *Virus Res.* **2012**, *169*, 340–348. [CrossRef] [PubMed]

287. Gehrig, S.; Eberle, M.-E.; Botschen, F.; Rimbach, K.; Eberle, F.; Eigenbrod, T.; Kaiser, S.; Holmes, W.M.; Erdmann, V.A.; Sprinzl, M.; et al. Identification of modifications in microbial, native tRNA that suppress immunostimulatory activity. *J. Exp. Med.* **2012**, *209*, 225–233. [CrossRef] [PubMed]

288. Jöckel, S.; Nees, G.; Sommer, R.; Zhao, Y.; Cherkasov, D.; Hori, H.; Ehm, G.; Schnare, M.; Nain, M.; Kaufmann, A.; Bauer, S. The 2′-*O*-methylation status of a single guanosine controls transfer RNA-mediated Toll-like receptor 7 activation or inhibition. *J. Exp. Med.* **2012**, *209*, 235–241. [CrossRef] [PubMed]

289. Schmitt, F.C.F.; Freund, I.; Weigand, M.A.; Helm, M.; Dalpke, A.H.; Eigenbrod, T. Identification of an optimized 2′-*O*-methylated trinucleotide RNA motif inhibiting Toll-like receptors 7 and 8. *RNA* **2017**, *23*, 1344–1351. [CrossRef] [PubMed]

290. Keller, P.; Freund, I.; Marchand, V.; Bec, G.; Huang, R.; Motorin, Y.; Eigenbrod, T.; Dalpke, A.; Helm, M. Double methylation of tRNA-U54 to 2′-*O*-methylthymidine (Tm) synergistically decreases immune response by Toll-like receptor 7. *Nucleic Acids Res.* **2018**. [CrossRef] [PubMed]

291. Gu, X.R.; Nicoghosian, K.; Cedergren, R.J.; Wong, J.T. Sequences of halobacterial tRNAs and the paucity of U in the first position of their anticodons. *Nucleic Acids Res.* **1983**, *11*, 5433–5442. [CrossRef] [PubMed]

292. Grosjean, H.; Gaspin, C.; Marck, C.; Decatur, W.A.; de Crecy-Lagard, V. RNomics and Modomics in the halophilic archaea *Haloferax volcanii*: Identification of RNA modification genes. *BMC Genom.* **2008**, *9*, 470. [CrossRef] [PubMed]

293. Phillips, G.; de Crécy-Lagard, V. Biosynthesis and function of tRNA modifications in Archaea. *Curr. Opin. Microbiol.* **2011**, *14*, 335–341. [CrossRef] [PubMed]

294. Grosjean, H.; Gupta, R.; Maxwell, S. Modified nucleotides in arhcaeal RNAs. In *Archaea: New Models for Prokaryotic Biology*; Blum, P., Ed.; Caister Academic Press: Portland, OR, USA, 2008; pp. 171–196.

295. Rintala-Dempsey, A.C.; Kothe, U. Eukaryotic stand-alone pseudouridine synthases–RNA modifying enzymes and emerging regulators of gene expression? *RNA Biol.* **2017**, *14*, 1185–1196. [CrossRef] [PubMed]

296. Hori, H. Methylated nucleosides in tRNA and tRNA methyltransferases. *Front. Genet.* **2014**, *5*, 144. [CrossRef] [PubMed]

297. Grosjean, H.; Oshima, T. How nucleic acids cope with high temperature. In *Physiology and Biochemistry of Extremophiles*; Gerday, C., Glansdorff, N., Eds.; ASM Press: Washington, DC, USA, 2007; pp. 39–56.

298. Kumazawa, Y.; Yokogawa, T.; Tsurui, H.; Miura, K.; Watanabe, K. Effect of the higher-order structure of tRNAs on the stability of hybrids with oligodeoxyribonucleotides: Separation of tRNA by an efficient solution hybridization. *Nucleic Acids Res.* **1992**, *20*, 2223–2232. [CrossRef] [PubMed]

299. Tsurui, H.; Kumazawa, Y.; Sanokawa, R.; Watanabe, Y.; Kuroda, T.; Wada, A.; Watanabe, K.; Shirai, T. Batchwise purification of specific tRNAs by a solid-phase DNA probe. *Anal. Biochem.* **1994**, *221*, 166–172. [CrossRef] [PubMed]

300. Suzuki, T.; Ueda, T.; Watanabe, K. A new method for identifying the amino acid attached to a particular RNA in the cell. *FEBS Lett.* **1996**, *381*, 195–198. [CrossRef]

301. Yokogawa, T.; Kitamura, Y.; Nakamura, D.; Ohno, S.; Nishikawa, K. Optimization of the hybridization-based method for purification of thermostable tRNAs in the presence of tetraalkylammonium salts. *Nucleic Acids Res.* **2010**, *38*, e89. [CrossRef] [PubMed]

302. Kazayama, A.; Yamagami, R.; Yokogawa, T.; Hori, H. Improved solid-phase DNA probe method for tRNA purification: Large-scale preparation and alteration of DNA fixation. *J. Biochem.* **2015**, *157*, 411–418. [CrossRef] [PubMed]

303. Stanley, J.; Vassilenko, S. A different approach to RNA sequencing. *Nature* **1978**, *274*, 87–89. [CrossRef] [PubMed]

304. Kuchino, Y.; Kato, M.; Sugisaki, H.; Nishimura, S. Nucleotide sequence of starfish initiator tRNA. *Nucleic Acids Res.* **1979**, *6*, 3459–3469. [CrossRef] [PubMed]

305. Suzuki, T.; Ikeuchi, Y.; Noma, A.; Suzuki, T.; Sakaguchi, Y. Mass spectrometric identification and characterization of RNA-modifying enzymes. *Methods Enzymol.* **2007**, *425*, 211–229. [PubMed]

306. Suzuki, T.; Suzuki, T. A complete landscape of post-transcriptional modifications in mammalian mitochondrial tRNAs. *Nucleic Acids Res.* **2014**, *42*, 7346–7357. [CrossRef] [PubMed]

307. Mengel-Jorgensen, J.; Kirpekar, F. Detection of pseudouridine and other modifications in tRNA by cyanoethylation and MALDI mass spectrometry. *Nucleic Acids Res* **2002**, *30*, e135. [CrossRef] [PubMed]

308. Chen, C.; Huang, B.; Anderson, J.T.; Byström, A.S. Unexpected accumulation of ncm(5)U and ncm(5)S(2) (U) in a *trm9* mutant suggests an additional step in the synthesis of mcm(5)U and mcm(5)S(2)U. *PLoS ONE* **2011**, *6*, e20783.

309. Agris, P.F.; Koh, H.; Söll, D. The effect of growth temperatures on the in vivo ribose methylation of *Bacillus stearothermophilus* transfer RNA. *Arch Biochem Biophys.* **1973**, *154*, 277–282. [CrossRef]

310. Andachi, Y.; Yamao, F.; Muto, A.; Osawa, S. Codon recognition patterns as deduced from sequences of the complete set of transfer RNA species in *Mycoplasma capricolum*. Resemblance to mitochondria. *J. Mol. Biol.* **1989**, *209*, 37–54. [CrossRef]

311. Matsugi, J.; Jia, H.T.; Murao, K.; Ishikura, H. Nucleotide sequences of serine tRNAs from *Bacillus subtilis*. *Biochem. Biophys. Acta* **1992**, *1130*, 333–335. [CrossRef]

312. Hori, H. Transfer RNA methyltransferases with a SpoU-TrmD (SPOUT) fold and their modified nucleosides in tRNA. *Biomolecules* **2017**, *7*, 23. [CrossRef] [PubMed]

313. Reddy, D.M.; Crain, P.F.; Edmonds, C.G.; Gupta, R.; Hashizume, T.; Stetter, K.O.; Widdel, F.; McCloskey, J.A. Structure determination of two new amino acid-containing derivatives of adenosine from tRNA of thermophilic bacteria and archaea. *Nucleic Acids Res.* **1992**, *20*, 5607–5615. [CrossRef] [PubMed]

314. Cusack, S.; Yaremchuk, A.; Krikliviy, I.; Tukalo, M. tRNA^{Pro} anticodon recognition by *Thermus thermophilus* prolyl-tRNA synthetase. *Structure* **1998**, *6*, 101–108. [CrossRef]

315. Li, F.; Dong, J.; Hu, X.; Gong, W.; Li, J.; Shen, J.; Tian, H.; Wang, J. A covalent approach for site-specific RNA labeling in Mammalian cells. *Angew. Chem. Int. Ed. Engl.* **2015**, *54*, 4597–4602. [CrossRef] [PubMed]

316. Singh, S.K.; Gurha, P.; Tran, E.J.; Maxwell, E.S.; Gupta, R. Sequential 2′-*O*-methylation of archaeal pre-tRNA^{Trp} nucleotides is guided by the intron-encoded but trans-acting box C/D ribonucleoprotein of pre-tRNA. *J. Biol. Chem.* **2004**, *279*, 47661–47671. [CrossRef] [PubMed]

317. Haas, E.S.; Daniels, C.J.; Reeve, J.N. Genes encoding 5S rRNA and tRNAs in the extremely thermophilic archaebacterium *Methanothermus fervidus*. *Gene* **1989**, *77*, 253–263. [CrossRef]

318. Randau, L.; Münch, R.; Hohn, M.J.; Jahn, D.; Söll, D. *Nanoarchaeum equitans* creates functional tRNAs from separate genes for their 5′- and 3′-halves. *Nature* **2005**, *433*, 537–541. [CrossRef] [PubMed]

319. Randau, L.; Pearson, M.; Söll, D. The complete set of tRNA species in *Nanoarchaeum equitans*. *FEBS Lett.* **2005**, *579*, 2945–2947. [CrossRef] [PubMed]

320. Richter, H.; Mohr, S.; Randau, L. C/D box sRNA, CRISPR RNA and tRNA processing in an archaeon with a minimal fragmented genome. *Biochem. Soc. Trans.* **2013**, *41*, 411–415. [CrossRef] [PubMed]

321. Hardin, J.W.; Reyes, F.E.; Batey, R.T. Analysis of a critical interaction within the archaeal box C/D small ribonucleoprotein complex. *J. Biol. Chem.* **2009**, *284*, 15317–15324. [CrossRef] [PubMed]

322. McCloskey, J.A.; Crain, P.F.; Edmonds, C.G.; Gupta, R.; Hashizume, T.; Phillipson, D.W.; Stetter, K.O. Structure determination of a new fluorescent tricyclic nucleoside from archaebacterial tRNA. *Nucleic Acids Res.* **1987**, *15*, 683–693. [CrossRef] [PubMed]

323. Ushida, C.; Muramatsu, T.; Mizushima, H.; Ueda, T.; Watanabe, K.; Stetter, K.O.; Crain, P.F.; McCloskey, J.A.; Kuchino, Y. Structural feature of the initiator tRNA gene from *Pyrodictium occultum* and the thermal stability of its gene product, tRNA(imet). *Biochimie* **1996**, *79*, 847–855. [CrossRef]

324. McCloskey, J.A.; Liu, X.H.; Crain, P.F.; Bruenger, E.; Guymon, R.; Hashizume, T.; Stetter, K.O. Posttranscriptional modification of transfer RNA in the submarine hyperthermophile *Pyrolobus fumarii*. *Nucleic Acids Symp. Ser.* **2000**, *44*, 267–268. [CrossRef]

325. Zhou, S.; Sitaramaiah, D.; Noon, K.R.; Guymon, R.; Hashizume, T.; McCloskey, J.A. Structures of two new "minimalist" modified nucleosides from archaeal tRNA. *Bioorg. Chem.* **2004**, *32*, 82–91. [CrossRef] [PubMed]

326. Yang, Z.; Lin, J.; Ye, K. Box C/D guide RNAs recognize a maximum of 10 nt of substrates. *Proc. Natl. Acad. Sci. USA* **2016**, *27*, 10878–10883. [CrossRef] [PubMed]

327. Agari, Y.; Sato, S.; Wakamatsu, T.; Bessho, Y.; Ebihara, A.; Yokoyama, S.; Kuramitsu, S.; Shinkai, A. X-ray crystal structure of a hypothetical Sua5 protein from *Sulfolobus tokodaii* strain 7. *Proteins* **2008**, *70*, 1108–1111. [CrossRef] [PubMed]

328. Kuratani, M.; Kasai, T.; Akasaka, R.; Higashijima, K.; Terada, T.; Kigawa, T.; Shinkai, A.; Bessho, Y.; Yokoyama, S. Crystal structure of *Sulfolobus tokodaii* Sua5 complexed with L-threonine and AMPPNP. *Proteins* **2011**, *79*, 2065–2075. [CrossRef] [PubMed]

329. Parthier, C.; Goerlich, S.; Jaenecke, F.; Breithaupt, C.; Brauer, U.; Fandrich, U.; Clausnitzer, D.; Wehmeier, U.F.; Bottcher, C.; Scheel, D.; et al. The O-carbamoyltransferase TobZ catalyzes an ancient enzymatic reaction. *Angew. Chem. Int. Ed. Engl.* **2012**, *51*, 4046–4052. [CrossRef] [PubMed]

330. Klenk, H.P.; Schwass, V.; Zillig, W. Nucleotide sequence of the genes encoding proline tRNA(UGG) and threonine tRNA(GGU) and consensus promoter model of *Thermococcus celer*. *Biochim. Biophys. Acta* **1993**, *1172*, 236–238. [CrossRef]

331. Watanabe, K.; Oshima, T.; Saneyoshi, M.; Nishimura, S. Replacement of ribothymidine by 5-methyl-2-thiouridine in sequence GT psi C in tRNA of an extreme thermophile. *FEBS Lett.* **1974**, *43*, 59–63. [CrossRef]

332. Dalluge, J.J.; Hashizume, T.; Sopchik, A.E.; McCloskey, J.A.; Davis, D.R. Conformational flexibility in RNA: The role of dihydrouridine. *Nucleic Acids Res.* **1996**, *24*, 1073–1079. [CrossRef] [PubMed]

333. Sokołowski, M.; Klassen, R.; Bruch, A.; Schaffrath, R.; Glatt, S. Cooperativity between different tRNA modifications and their modification pathways. *Biochem. Biophys. Acta* **2018**, *1861*, 409–418. [CrossRef] [PubMed]

334. Liu, R.J.; Zhou, M.; Fang, Z.P.; Wang, M.; Zhou, X.L.; Wang, E.D. The tRNA recognition mechanism of the minimalist SPOUT methyltransferase, TrmL. *Nucleic Acids Res.* **2013**, *41*, 7828–7842. [CrossRef] [PubMed]

335. Morales, A.J.; Swairjo, M.A.; Schimmel, P. Structure-specific tRNA-binding protein from the extreme thermophile *Aquifex aeolicus*. *EMBO J.* **1999**, *18*, 3475–3483. [CrossRef] [PubMed]

336. Swairjo, M.A.; Morales, A.J.; Wang, C.C.; Ortiz, A.R.; Schimmel, P. Crystal structure of trbp111: A structure-specific tRNA-binding protein. *EMBO J.* **2000**, *19*, 6278–6298. [CrossRef] [PubMed]

337. Kushiro, T.; Schimmel, P. Trbp111 selectively binds a noncovalently assembled tRNA-like structure. *Proc. Natl. Acad. Sci. USA* **2002**, *99*, 16631–16635. [CrossRef] [PubMed]

338. Desai, K.K.; Cheng, C.L.; Bingman, C.A.; Phillips, G.N., Jr.; Raines, R.T. A tRNA splicing operon: Archease endows RtcB with dual GTP/ATP cofactor specificity and accelerates RNA ligation. *Nucleic Acids Res.* **2014**, *42*, 3931–3942. [CrossRef] [PubMed]

339. Popow, J.; Jurkin, J.; Schleiffer, A.; Martinez, J. Analysis of orthologous groups reveals archease and DDX1 as tRNA splicing factors. *Nature* **2014**, *511*, 104–107. [CrossRef] [PubMed]

340. Maraia, R.J.; Arimbasseri, A.G. Factors That Shape Eukaryotic tRNAomes: Processing, Modification and Anticodon-Codon Use. *Biomolecules* **2017**, *7*, 26. [CrossRef] [PubMed]

341. Blewett, N.H.; Maraia, R.J. La involvement in tRNA and other RNA processing events including differences among yeast and other eukaryotes. *Biochim. Biophys. Acta* **2018**, *1861*, 361–372. [CrossRef] [PubMed]

342. Gutgsell, N.; Englund, N.; Niu, L.; Kaya, Y.; Lane, B.G.; Ofengand, J. Deletion of the Escherichia coli pseudouridine synthase gene truB blocks formation of pseudouridine 55 in tRNA in vivo, does not affect exponential growth, but confers a strong selective disadvantage in competition with wild-type cells. *RNA* **2000**, *6*, 1870–1881. [CrossRef] [PubMed]

343. Keffer-Wilkes, L.C.; Veerareddygari, G.R.; Kothe, U. RNA modification enzyme TruB is a tRNA chaperone. *Proc. Natl. Acad. Sci. USA* **2016**, *113*, 14306–14311. [CrossRef] [PubMed]

344. Quigley, G.J.; Teeter, M.M.; Rich, A. Structural analysis of spermine and magnesium ion binding to yeast phenylalanine transfer RNA. *Proc. Natl. Acad. Sci. USA* **1978**, *75*, 64–68. [CrossRef] [PubMed]

345. Frydman, B.; de los Santos, C.; Frydman, R.B. A ^{13}C NMR study of [5,8-^{13}C$_2$]spermidine binding to tRNA and to *Escherichia coli* macromolecules. *J. Biol. Chem.* **1990**, *265*, 20874–20878. [PubMed]

346. Terui, Y.; Ohnuma, M.; Hiraga, K.; Kawashima, E.; Oshima, T. Stabilization of nucleic acids by unusual polyamines produced by an extreme thermophile, *Thermus thermophilus*. *Biochem. J.* **2005**, *388*, 427–433. [CrossRef] [PubMed]

347. Ouameur, A.A.; Bourassa, P.; Tajmir-Riahi, H.A. Probing tRNA interaction with biogenic polyamines. *RNA* **2010**, *16*, 1968–1979. [CrossRef] [PubMed]

348. Oshima, T. Unique polyamines produced by an extreme thermophiles, *Thermus thermophilus*. *Amino Acids* **2007**, *33*, 367–372. [CrossRef] [PubMed]

349. Oshima, T.; Moriya, T.; Terui, Y. Identification, chemical synthesis, and biological functions of unusual polyamines produced by extreme thermophiles. *Methods Mol. Biol.* **2011**, *720*, 81–111. [PubMed]

350. Hamana, K.; Niitsu, M.; Samejima, K.; Matsuzaki, S. Polyamine distributions in thermophilic eubacteria belonging to *Thermus* and *Acidothermus*. *J. Biochem.* **1991**, *109*, 444–449. [CrossRef] [PubMed]

351. Hamana, K.; Tanaka, T.; Hosoya, R.; Niitsu, M.; Itoh, T. Cellular polyamines of the acidophilic, thermophilic and thermoacidophilic archaebacteria, *Acidilobus, Ferroplasma, Pyrobaculum, Pyrococcus, Staphylothermus, Thermococcus, Thermodiscus* and *Vulcanisaeta*. *J. Gen. Appl. Microbiol.* **2003**, *49*, 287–293. [CrossRef] [PubMed]

352. Hayrapetyan, A.; Grosjean, H.; Helm, M. Effect of a quaternary pentamine on RNA stabilization and enzymatic methylation. *Biol. Chem.* **2009**, *390*, 851–861. [CrossRef] [PubMed]

353. Nakashima, M.; Yamagami, R.; Tomikawa, C.; Ochi, Y.; Moriya, T.; Asahara, H.; Fourmy, D.; Yoshizawa, S.; Oshima, T.; Hori, H. Long and branched polyamines are required for maintenance of the ribosome, tRNA^{His} and tRNA^{Tyr} in *Thermus thermophilus* cells at high temperatures. *Genes Cells* **2017**, *22*, 628–645. [CrossRef] [PubMed]

354. Yue, D.; Kintanar, A.; Horowitz, J. Nucleoside modifications stabilize Mg^{2+} binding in *Escherichia coli* tRNA(Val): An imino proton NMR investigation. *Biochemistry* **1994**, *33*, 8905–8911. [CrossRef] [PubMed]

355. Agris, P.F. The importance of being modified: Roles of modified nucleosides and Mg^{2+} in RNA structure and function. *Prog. Nucleic Acid Res. Mol. Biol.* **1996**, *53*, 79–129. [PubMed]

356. Nobles, K.N.; Yarian, C.S.; Liu, G.; Guenther, R.H.; Agris, P.F. Highly conserved modified nucleosides influence Mg^{2+}-dependent tRNA folding. *Nucleic Acids Res.* **2002**, *30*, 4751–4760. [CrossRef] [PubMed]

357. Hensel, R.; Konig, H. Thermoadaptation of metanogenic bacteria by intercellular ion concentration. *FEMS Microbiol. Lett.* **1988**, *49*, 75–79. [CrossRef]

358. Nawrot, B.; Sochacka, E.; Duchler, M. tRNA structural and functional changes induced by oxidative stress. *Cell Mol. Life Sci.* **2011**, *68*, 4023–4032. [CrossRef] [PubMed]

359. Preston, M.A.; D'Silva, S.; Kon, Y.; Phizicky, E.M. tRNA^{His} 5-methylcytidine levels increase in response to several growth arrest conditions in *Saccharomyces cerevisiae*. *RNA* **2013**, *19*, 243–256. [CrossRef] [PubMed]

360. Endres, L.; Dedon, P.C.; Begley, T.J. Codon-biased translation can be regulated by wobble-base tRNA modification systems during cellular stress responses. *RNA Biol.* **2015**, *12*, 603–614. [CrossRef] [PubMed]

361. Chan, C.T.; Deng, W.; Li, F.; DeMott, M.S.; Babu, I.R.; Begley, T.J.; Dedon, P.C. Highly Predictive Reprogramming of tRNA Modifications Is Linked to Selective Expression of Codon-Biased Genes. *Chem. Res. Toxicol.* **2015**, *28*, 978–988. [CrossRef] [PubMed]

362. Polikarpov, I.; Prade, R.; Caldana, C.; Paes Leme, A.F.; Mercadante, A.Z.; Riaño-Pachón, D.M.; Squina, F.M. Thermal adaptation strategies of the extremophile bacterium *Thermus filiformis* based on multi-omics analysis. *Extremophiles* **2017**, *21*, 775–788.

363. Kimura, S.; Suzuki, T. Iron-sulfur proteins responsible for RNA modifications. *Biochim. Biophys. Acta* **2015**, *1853*, 1272–1283. [CrossRef] [PubMed]

364. Ramamurthy, V.; Swann, S.L.; Spedaliere, C.J.; Mueller, E.G. Role of cysteine residues in pseudouridine synthases of different families. *Biochemistry* **1999**, *38*, 13106–13111. [CrossRef] [PubMed]

365. Urbonavicius, J.; Jäger, G.; Björk, G.R. Amino acid residues of the *Escherichia coli* tRNA(m^{5}U54)methyltransferase (TrmA) critical for stability, covalent binding of tRNA and enzymatic activity. *Nucleic Acids Res.* **2007**, *35*, 3297–3305. [CrossRef] [PubMed]

366. Alian, A.; Lee, T.T.; Griner, S.L.; Stroud, R.M.; Finer-Moore, J. Structure of a TrmA-RNA complex: A consensus RNA fold contributes to substrate selectivity and catalysis in m^{5}U methyltransferases. *Proc. Natl. Acad. Sci. USA* **2008**, *105*, 6876–6881. [CrossRef] [PubMed]

367. Byrne, R.T.; Jenkins, H.T.; Peters, D.T.; Whelan, F.; Stowell, J.; Aziz, N.; Kasatsky, P.; Rodnina, M.V.; Koonin, E.V.; Konevega, A.L.; et al. Major reorientation of tRNA substrates defines specificity of dihydrouridine synthases. *Proc. Natl. Acad. Sci. USA* **2015**, *112*, 6033–6037. [CrossRef] [PubMed]

368. Ryals, J.; Hsu, R.Y.; Lipsett, M.N.; Bremer, H. Isolation of single-site *Escherichia coli* mutants deficient in thiamine and 4-thiouridine syntheses: Identification of a *nuvC* mutant. *J. Bacteriol.* **1982**, *151*, 899–904. [PubMed]

369. Liu, Y.; Zhu, X.; Nakamura, A.; Orlando, R.; Söll, D.; Whitman, W.B. Biosynthesis of 4-thiouridine in tRNA in the methanogenic archaeon *Methanococcus maripaludis*. *J. Biol. Chem.* **2012**, *287*, 36683–36692. [CrossRef] [PubMed]

microorganisms

MDPI

Review

Thermophilic Proteins as Versatile Scaffolds for Protein Engineering

Anthony J. Finch *,† and Jin Ryoun Kim *

Department of Chemical and Biomolecular Engineering, New York University, 6 MetroTech Center, Brooklyn, NY 11201, USA
* Correspondence: anthonyjpfinch@gmail.com (A.J.F.); jin.kim@nyu.edu (J.R.K.); Tel.: +1-347-453-5165 (A.J.F.); +1-646-997-3719 (J.R.K.)
† Current address: Weill Cornell Medicine, 1300 York Avenue, New York, NY 10065, USA.

Received: 2 September 2018; Accepted: 23 September 2018; Published: 25 September 2018

Abstract: Literature from the past two decades has outlined the existence of a trade-off between protein stability and function. This trade-off creates a unique challenge for protein engineers who seek to introduce new functionality to proteins. These engineers must carefully balance the mutation-mediated creation and/or optimization of function with the destabilizing effect of those mutations. Subsequent research has shown that protein stability is positively correlated with "evolvability" or the ability to support mutations which bestow new functionality on the protein. Since the ultimate goal of protein engineering is to create and/or optimize a protein's function, highly stable proteins are preferred as potential scaffolds for protein engineering. This review focuses on the application potential for thermophilic proteins as scaffolds for protein engineering. The relatively high inherent thermostability of these proteins grants them a great deal of mutational robustness, making them promising scaffolds for various protein engineering applications. Comparative studies on the evolvability of thermophilic and mesophilic proteins have strongly supported the argument that thermophilic proteins are more evolvable than mesophilic proteins. These findings indicate that thermophilic proteins may represent the scaffold of choice for protein engineering in the future.

Keywords: thermophilic proteins; protein engineering; protein stability; evolvability

---

1. Introduction

The field of protein engineering focuses on rational modification or combinatorial design of proteins to enhance or otherwise alter protein functionality [1]. Each protein has a specific function which is dictated by its amino acid sequence and three-dimensional (or "folded") structure. Protein engineers intentionally modify these elements by introducing mutations which may alter the structure and functionality of the protein, improve existing functionality or introduce entirely new functionality [1]. This fact motivates the primary challenge facing protein engineers today: an apparent trade-off between protein function and stability [2]. Decades of research has shown that there is a negative correlation between protein stability and function; since one of the primary purposes of protein engineering is to create or optimize protein activity, protein engineers face the challenge of maintaining protein stability at the cost of desired functionality.

Over the course of the past decade and a half, a great deal of research has been done attempting to better understand this negative correlation between protein stability and functionality. The most prevalent explanation for this stems from the fact that the amino acid residues which promote functionality are often destabilizing, either inherently or as a function of their location in the folded protein [2]. It is known that functionality of proteins—enzymes in particular—largely depends upon the presence of reactive polar and/or hydrophilic residues in the active site [2]. Since the active site is often buried within the three-dimensional structure of the protein, it follows that functionality depends upon the inclusion of highly reactive residues in the interior of the protein [2,3]. In fact, protein functionality has been shown to depend directly on the flexibility of the active site rather than on its stability, indicating that the mobile active site attracts a substrate in order to reduce its free energy [3]. While flexibility and mobility at the active site enhances functionality, it also is potentially destabilizing to the protein tertiary structure as a whole [3].

To address the challenge associated with the trade-off between protein function and stability, protein engineers have spent the past decade and a half researching ways to manipulate highly stable proteins that are better able to tolerate the mutations necessary to confer new functionality. The concept of "mutational robustness" is one which has particularly interested researchers—a robust protein is better able to tolerate a given point mutation than is a less robust protein [4]. This "robustness" can be evaluated by introducing such a point mutation to the protein and evaluating the $\Delta\Delta G_{folding}$ between the native protein and the mutated protein [4]. $\Delta\Delta G_{folding}$ is defined as the change in $\Delta G_{folding}$ between the mutant protein and the parental, or wild-type, protein: $\Delta\Delta G_{folding} = \Delta G_{folding(mutant)} - \Delta G_{folding(wild\ type)}$. $\Delta\Delta G_{folding}$ values are direct measures of thermodynamic protein stability: once a mutation has been introduced, the higher (or more positive) the $\Delta\Delta G_{folding}$, the less stable the protein has been rendered as a result of that introduction. Since most function-inducing mutations increase the $\Delta G_{folding}$ of a protein, after the introduction of a functional mutation the more robust proteins will have a $\Delta\Delta G_{folding}$ lower than that of less robust proteins [4].

A positive correlation between protein robustness, thermodynamic stability, and "evolvability" has directed research towards utilizing thermophilic proteins as scaffolds for protein engineering [5]. Protein "evolvability" is defined by the ability of a protein to survive mutations which confer new functionality; it has been found that proteins with higher *thermostability* are much more evolvable than are less thermostable proteins. Protein engineers have explored various ways to capitalize upon the correlation between thermostability and evolvability by introducing mutations to thermophilic proteins [5]. The hope is that thermophilic proteins will be an effective and efficient scaffold for functional evolution.

The present article overviews the research that has been performed on thermophilic proteins as potential scaffolds for protein engineering. First, it reviews research which has defined the trade-off between protein activity and stability, and highlights the challenge this trade-off poses to the field of protein engineering. Next, it discusses research describing how protein stability and mutational robustness directly promote evolvability. It then proceeds to introduce thermophilic proteins to this discussion and outline the structural characteristics of these proteins that contribute to their increased stability and mutational robustness. Finally, it highlights recent research which has utilized highly thermostable proteins as effective scaffolds for engineered functional evolution and, when possible, compares the evolvability of these proteins with that of their mesophilic analogues.

2. The Trade-Off between Protein Activity and Stability

Over the last decade and a half, numerous studies have demonstrated that there is a distinct trade-off between protein activity and protein stability [2,3]. This trade-off is extremely relevant to the field of protein engineering because it implies that the process of mutating proteins to alter activity can be very destabilizing. In this section, we discuss a comparative analysis of the findings of three important studies which, taken together, demonstrate that mutations that enhance existing protein function may decrease protein stability. The discussion of a fourth study by Tokuriki et al. then demonstrates that those mutations which confer enhanced or new function ("functional" mutations) are more destabilizing than random "neutral" mutations.

Wang et al. studied the evolution of an enzyme, Temoniera-1 (TEM-1) β-lactamase, which is responsible for bacterial antibiotic resistance [6]. In this study, the authors selectively mutated the TEM-1 β-lactamase and observed both the resulting changes in enzymatic activity and in protein stability. Seven different TEM-1 β-lactamase mutants showing increased β-lactamase activity were studied; the authors found that these mutants exhibited a decrease in protein stability relative to the wild-type protein ranging from 0.3 kcal/mol to 4.2 kcal/mol [6]. X-ray crystallographic analysis of the mutant protein structures revealed that the mutations had enlarged the "active site cavity" of the protein—a structural alteration which apparently destabilized the entire three-dimensional structure. Wang et al. also found that the mutants which managed to survive this structural alteration featured secondary mutations which counterbalanced the destabilizing effect of the functional substitutions [6]. This data strongly indicates that a negative correlation between activity and stability may exist.

Bloom et al. computationally evaluated the destabilizing effects of functional mutations on protein structure. Specifically, the authors simulated protein evolution for improved ligand binding affinity by introducing mutations to 20 "lattice" proteins [5]. These lattice proteins were simplified protein models which could be used to simulate protein folding and evolution [5]. The authors found that among all mutated mesophilic proteins only 35% folded to their native three-dimensional structure and that the average $\Delta G_{folding}$ for these mutated proteins was higher than that observed for the wild-type protein [5]. The data indicate that proteins which mutate and evolve for new functionality are generally destabilized by the changes which confer the desired functionality. Many of these proteins lose the ability to fold to their native structure entirely, and of those proteins that do retain the ability to fold back to the native structure the average $\Delta G_{folding}$ is a less negative value than it had been for the original protein.

Liang et al. examined the tradeoff between protein activity and stability with two related metalloproteinases (MMP): MMP-3 and MMP-12. The authors evaluated the proteinase activity of these enzymes, and found that MMP-12 demonstrated a higher second order rate constant (k_{cat}/K_m) for the catalytic process than MMP-3 [7]. In contrast, MMP-3 was found to have a catalytic domain 2.8 kcal/mol more stable than that for MMP-12 [7]. The authors evaluated the structures of the two protease domains and noted that MMP-3 displayed a high number of proline residues located at "exposed turns" (or portions of the protein where the primary chain folds back on itself at the exterior of the tertiary structure); this feature accounted for 0.7 kcal/mol of the increased MMP-3 stability [7]. The remaining 2.1 kcal/mol of stability were attributed to the same structural characteristics evaluated by Wang et al.: the increased concentration of buried reactive residues in the vicinity of the active site destabilizing the tertiary network [7]. Liang et al. compared their findings to a previous study performed on MMP-13 (collagenase 3) and MMP-1 (collagenase 1) which similarly noted an increase in activity but a decrease in stability for MMP-13 with respect to MMP-1 [7]. Both of these studies confirm the correlation between increased activity and decreased stability.

While the findings of the previous three studies undoubtedly demonstrate that mutations that enhance function may reduce stability, none of these conclusively show that the "functional" mutations are more destabilizing than random, "neutral" mutations. Tokuriki et al. designed their study to address this [8]. The authors performed their analysis computationally using the folding simulation FoldX to analyze the $\Delta\Delta G_{folding}$ values of 22 mutated proteins with respect to the parental proteins. The proteins that were selected for this study demonstrated the ability to fold to their native structure when either neutral or functional mutations were introduced [8]. Interestingly, analysis of the proteins which underwent functional mutations and subsequently folded to their native state revealed that these proteins also uniformly evolved "other"—or secondary—mutations in addition to the functional mutations [8]. A closer examination of these "other" mutations revealed that they were stabilizing mutations which were largely located on the exterior surface of the protein [8]. It is important to note that the proteins which had undergone neutral mutations had far fewer such stabilizing mutations than did those that were subjected to functional mutations [8]. This discrepancy indicates that the functional mutations, most of which were substituted polar amino acids in the interior of the protein either near or within the active site, were inherently destabilizing to the protein. In order for proteins subjected to these mutations to retain their stability they required additional mutations to counterbalance the destabilizing influence of these substituted residues [8].

3. Stability Promotes Evolvability

It is well known that biological processes evolve to function optimally in the surrounding conditions [9]. The evolution of biological process depends greatly upon the evolution of proteins which regulate and control those processes. Thus, for the past two decades researchers have been examining how proteins evolve and how their evolution affects the biological processes which they catalyze and control. By definition, in order for a protein to evolve, it must acquire *new* functionality primarily through the process of mutation. However, as discussed above, this process of protein mutation is likely to be destabilizing. A protein that is stable enough to survive this process and support mutations that bestow new functionality is said to have higher "evolvability" [9]. The following three key papers demonstrate this positive correlation between protein stability and protein evolvability.

The first of these papers, a critical review written by Camps et al., consolidated previous research and identified three protein characteristics that strongly correlate with increased evolvability. The first of these characteristics is protein "promiscuity" which is defined as the ability of a protein "{to recognize} alternative substrates or catalysts or alternative chemical reactions" [9]. Camps et al. explained that a "promiscuous" protein will be more evolvable than other proteins because these proteins require fewer amino acid substitutions to develop new functionality. Clearly a protein that requires six amino acid substitutions to develop new activity will be more likely to evolve functionality than will a protein that requires 40 amino acid substitutions [9]. The second characteristic which Camps et al. correlated with increased evolvability is "modularity" which is defined as "the presence of functionally independent motifs" [9]. Camps et al. reported that a protein featuring many unique, independent subdomains might develop a functionality in one of the domains which is not associated with existing functions, such as enzyme activity [9]. The authors argued that an enzyme subdomain with no existing active site usually needs to undergo fewer destabilizing mutations to develop an active site for a new substrate than an existing active site must undergo to develop new substrate affinity [9].

The third and most important characteristic which the authors correlated with increased protein evolvability is mutational robustness, which is defined as the ability to tolerate mutations [9]. Camps et al. described that there are two subtypes of mutational robustness: "extrinsic" and "intrinsic" [9]. Extrinsic robustness is afforded a protein through "chaperone proteins" or interactions with other nearby proteins [9]. On the other hand, intrinsic robustness is derived from specific characteristics of the protein itself which allow it to more effectively withstand the potentially destabilizing influence of what Tokuriki et al. termed "new function" mutations [8,9]. Intrinsic robustness is thus a direct function of the inherent stability of the protein, so it follows that protein evolvability is directly correlated with protein stability.

This correlation is supported by data from a number of studies which have confirmed that protein mutational robustness and stability increases evolvability. Caetano-Anolles et al. employed an interesting experimental method to evaluate the influence of protein stability on evolvability. The authors inserted a "test protein" into the amino acid sequence of various β-lactamases, which were then expressed in *Escherichia coli*. [10]. It was found that the bacteria expressing more stable β-lactamases (i.e., lower $\Delta\Delta G_{folding}$ values on insertion of the test protein) became resistant to a wider range of penicillin derivatives [10]. Their result is supportive of the view that more stable proteins were better able to accommodate new functional mutations [10].

Another study confirming the existence of the positive correlation between protein stability and evolvability was performed by Philip et al. who experimentally manipulated the functionality of the photoactive yellow protein (PYP)—a 125 residue photoreceptor prototype of the period circadian protein-aryl hydrocarbon receptor nuclear translocator protein-single-minded protein, or PER-ARNT-SIM (PAS) signaling superfamily. PAS proteins are widely known to be extremely diverse—over 20,000 different individual PAS domains have been defined in thousands of signaling proteins in a diverse array of organisms ranging from bacteria to humans [11]. The authors explored the diversity of this class of proteins to determine which structural elements were responsible for its pronounced mutational robustness [11]. They systematically altered each of the 125 residues of this protein one by one, replacing all non-Ala residues with Ala, and all residues with Gly, and developed these mutants in 125 separate *E. coli* strains. Since PYP is a photoreceptor, the authors were able to effectively measure alterations in protein functionality based upon variations in four criteria: visible absorbance maximum, pKa, fluorescence quantum yield, and lifetime of the unstable "pB"—or blue-shifted absorbance—state [11]. These variations were assessed alongside two selected measures of protein stability: variation in ΔG_U values (a measure of thermodynamic stability against unfolding) and protein production level [11]. The authors found that many of the substitutions they introduced induced notable alteration in the four selected functional measures, confirming the strong mutational potential of the PYP protein [11]. They also found that alterations specifically at many of 23 specific residues known to be highly conserved throughout PAS domains—most of them far from the active site—either significantly lowered the ΔG_U of the PYP protein or decreased overall protein production [11]. From this, they concluded that the strong mutational robustness of the PAS domain derives from the fact that mutations at so many of its residues alters the functionality of the protein, and that a small, key group of highly conserved residues provides it with the stability necessary to support such mutations [11].

4. Thermophilic Proteins and Their High Thermostability

Due to the trade-off between protein stability and function, it is highly difficult to introduce mutations necessary for functional evolution without significant compromise of protein stability. Since proteins characterized by higher mutational robustness are more evolvable, as described above, those featuring high mutational robustness and stability could be an important structural framework for functional evolution. One group of proteins which could satisfy the stability requirements is the group of proteins stable at very high temperatures, namely thermophilic and hyperthermophilic proteins. Thermophilic proteins are defined as proteins derived from organisms with optimum growth temperature (OGT) from 45–80 °C, while hyperthermophilic proteins are derived from organisms with OGT above 80 °C (mostly from archaeal lineages) [12]. Mesophilic proteins, on the other hand, are associated with organisms with OGT between 15–45 °C [12]. A number of studies over the past decade have demonstrated that the enhanced stability of thermophilic and hyperthermophilic proteins is characterized by their ability to resist irreversible unfolding when subjected to denaturing conditions. For example, a thermophilic protein unfolded when subjected to heat and chemical denaturants but then refolded into a structure nearly identical to its native state after the denaturing agent was removed [13–15]. Many studies have been performed to tease out the factors contributing to enhanced stability of thermophilic proteins. Several comparative studies have recently pinpointed four significant structural differences between thermostable proteins and their mesophilic analogues which go a long way towards explaining the increased thermostability of the thermophilic proteins. These studies demonstrate that the etiology of this enhanced stability is multifactorial: thermophilic proteins feature an increased number of ion pairs, greater average surrounding hydrophobicity of buried side chains, more compact tertiary structure cores, and more hydrogen bonds bridging buried and exposed regions when compared with their mesophilic analogues [16–21].

The first major structural difference between thermophilic proteins and their mesophilic analogues is the number of ion pairs [16]. Szilagyi and Zavodszky compared the structures of mesophilic proteins with those of related "moderately thermophilic proteins" (corresponding to the proteins defined earlier in this article as "thermophilic") and "extremely thermophilic proteins" (corresponding to those defined earlier in this article as "hyperthermophilic"). They found a strong positive correlation between thermostability and the number of ion pairs included in the protein structure [16]. Szilagyi and Zavodszky also found that extremely thermophilic proteins are characterized by stronger ion pairs than their mesophilic counterparts (the strength of an ion pair was defined by the distance between ions: ion pairs separated by less than 4 Å were classified as strong). The authors noticed that many of the ion pairs in extremely thermophilic proteins were separated by distances less than this 4 Å cutoff [16]. By comparison, the separation between most of the ion pairs in moderately mesophilic proteins was on the order of 6 to 8 Å [16]. A follow-up study on the structural characteristics of thermophilic proteins by Gromiha et al. confirmed the findings of Szilagyi and Zavodszky, noting that 68% of thermophilic proteins featured significantly more ion pairs separated by less than 4 Å than did their mesophilic analogues [17].

The second major structural difference between thermophilic proteins and their mesophilic analogues is the average hydrophobicity of the amino acid side chains buried within the protein [17,18]. Gromiha et al. found that a "hydrophobic environment" was the greatest single defining structural element for all of the thermophilic proteins examined, noting that "80% of the thermophilic proteins examined were characterized by higher hydrophobicity than their mesophilic counterparts" [17]. In this study the "surrounding hydrophobicity" of an amino acid side chain was evaluated using a formula, which sums the "hydrophobic indices" of all of the residues within an 8 Å radius of that side chain. This sum was then normalized by the total number of residues in the protein to give the "average hydrophobicity." Gromiha et al. found that the average hydrophobicity of residues inside the protein is higher in thermophilic proteins (18.5 kcal/mol) than in corresponding mesophilic proteins (17.7 kcal/mol). In contrast, the average hydrophobicity of the exterior residues was similar between thermophilic and mesophilic proteins [17]. That same year, Takano et al. observed that esterase mutants isolated from thermophilic and hyperthermophilic archaea retain their stability at higher temperatures than do similar esterase enzymes derived from mesophilic bacteria [18]. Interestingly, some of the mutants derived from thermostable enzymes demonstrated up to 1.8 times greater relative activity compared to the corresponding wild type enzyme, while no mutants derived from mesophilic enzymes showed any increased activity [18]. The thermophilic enzymes were found to be characterized by buried residues with significantly greater average hydrophobicity than those of the mesophilic counterparts [18]. Unsurprisingly, given this finding, the mutants which were derived from thermophilic enzymes but found to be unstable were disproportionately those which underwent substitution of interior residues: 90% of the destabilized thermophilic mutant proteins in this study had interior residue substitution; only 30% of the thermophilic mutant proteins which retained stability had interior residue substitution [18]. Taken together, both of these studies clearly demonstrate the role of average hydrophobicity of interior residue side chains in the stabilization of thermophilic proteins. One potential explanation for these findings is that the hydrophobic residues packed into the core of the protein tightly adhere to one another and resist the destabilizing influence of the external environment.

The third major structural difference between thermophilic proteins and their mesophilic analogues is the presence of compact, stable structural cores [19–21]. Meruelo et al. found that, when compared to mesophilic analogues, thermophilic proteins consist of a higher quantity of small amino acids such as Gly, Ala, Ser, and Val and a smaller quantity of large and/or polar amino acids such as Cys, Asp, Glu, Gln, and Arg [19]. The authors note that the smaller number of bulky, reactive amino acid side chains allows for tighter packing as the protein folds into its tertiary structure, and greater resistance to unfolding and aggregation. This corresponds to the findings of Glyakina et al. where interior residues of thermophilic proteins were more tightly packed, when compared to their mesophilic counterparts [20]. Tompa et al. expanded on this work by demonstrating that even when residues with long, hydrophobic side chains are heavily represented in thermophilic proteins, these residues are packaged in the core of the tertiary structure in such a way where residue-residue contact is maximized, thereby tying the core into a tight, compact, and stable structure [21]. The importance of a compact tertiary core to protein stability is illustrated further by research which demonstrates that hyperthermophilic proteins have significantly more disulfide bonds than either mesophilic or thermophilic proteins [19,22,23]. This feature serves to maximize the tight binding arrangement of core protein residues and optimize tertiary stability. Taken together, the research cited above support the notion that tight residue packing promotes greater interactions between neighboring residues and thereby enhances thermostability.

The fourth major structural difference between thermophilic proteins and their mesophilic counterparts is the number of hydrogen bonds bridging a protein's buried and exposed regions. The study performed by Tompa et al. showed that the distribution of hydrogen bonds is different in thermophilic proteins compared to that seen in their mesophilic counterparts [21]. The researchers noted that 49% of the hydrogen bonds found in thermophilic proteins bridged residues that were buried within the core of the protein with residues located on the exterior of the protein—only 42% of hydrogen bonds in mesophilic proteins met those criteria [21]. Additionally, 49% of hydrogen bonds in thermophilic proteins were found to exist between the protein main chain (the amino acid backbone chain) and side chain residues; this compares with only 39% in mesophilic proteins [21]. Hydrogen bonds are a particularly strong and stable form of intermolecular force—the authors theorize that each of the described altered hydrogen bond distributions in thermophilic protein conveys enhanced stability in its own way. First, the higher percentage of hydrogen bonds between buried and exposed residues effectively tethers the more vulnerable and thus unstable exterior of the protein to its compact, protected core. Second, the higher percentage of hydrogen bonds between the backbone amino acid chains and the side chains effectively crosslinks all elements of the tertiary structure in a fashion less vulnerable to denaturation [21].

In summary, the increased thermostability of thermophilic proteins appears to be attributable to an increased number of strong ion pairs, an increased surrounding hydrophobicity of buried amino acid side chains, more compact cores, and a more consolidating distribution of hydrogen bonds with respect to similar mesophilic proteins (Figure 1). It might initially appear that an increased number of ion pairs is contradictory to an increased average hydrophobicity. However, the two characteristics are compatible; the non-ionic residues in thermophilic proteins have a significantly greater surrounding hydrophobicity than do those in their mesophilic counterparts [16].

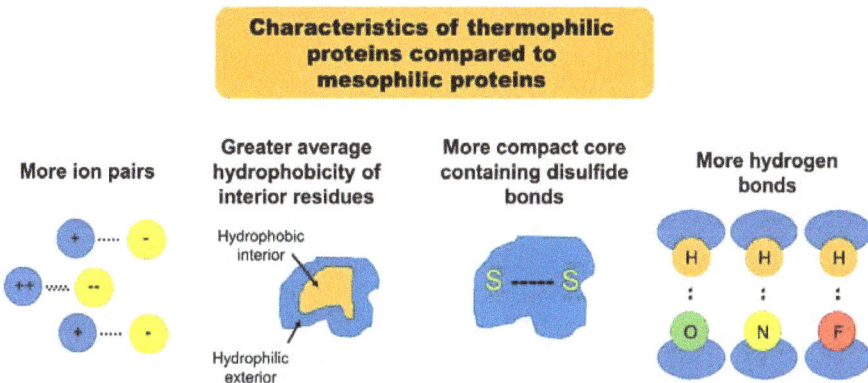

Figure 1. A schematic of characteristics of thermophilic proteins when compared to mesophilic proteins.

All of the above features contribute to the stabilization of thermophilic and hyperthermophilic proteins by increasing the rigidity of the corresponding tertiary structures. Rigidity and stability are correlated when it comes to protein structures; thermophilic and hyperthermophilic proteins can afford to be rigid, because they operate by definition in high temperature environments which develop sufficient interactions between substrates and active sites with limited flexibility [24]. These proteins are tightly packed, inflexible, and unfold much more slowly than do their less thermostable counterparts [25,26]. This discussion warrants mentioning psychrophilic proteins which stand on the other end of the rigidity/stability spectrum. These proteins have flexible tertiary structures which facilitate necessary interactions with substrates even at low temperatures [24,27]. Protein flexibility is a positive attribute under certain circumstances: it is vital for protein function in low temperature environments and it may permit singular proteins to interact with a wider range of substrates than can more rigid proteins. However, the proteins of most use to protein engineers are those able to withstand dramatic manipulation and retain functionality. Increased rigidity, and thus stability, make thermophilic and hyperthermophilic proteins more evolvable and subsequently invaluable in this capacity.

It is worth noting at this point that the magnitude of contribution from denaturation entropy changes to the enhanced stability of thermophilic and hyperthermophilic proteins is a topic of debate. This is an empirically observed phenomenon which specifies that the ΔS values associated with the denaturation of thermostable proteins are smaller than those associated with denaturation of mesophilic proteins [28]. The hypothesized rationale for this is that thermostable proteins have a higher baseline entropy in their native state compared to mesophilic proteins [28]. Wintrode et al. showed that thermophilic proteins have increased populations of low-frequency vibrational states largely due the high incidence of ion pairs and charge-charge networks on the surface of these proteins, and that this contributed to the baseline entropy level of the native state of thermophilic proteins [29]. Berezovsky et al. showed that thermophilic proteins are characterized by high numbers of lysine residues buried inside the protein core [30]. This is significant because buried lysine residues have a much higher density of variable rotamers than do other charged residues (such as arginine): this also contributes to the overall baseline entropy of the native state of thermophilic proteins [30]. These findings point to a potentially important entropic contribution to thermophilic protein stability. How much of overall thermophilic protein stability is attributable to enhanced structural rigidity (as discussed above) and how much is attributable to this entropic phenomenon is still under discussion. Regardless, the increased stability of thermophilic proteins provides some uniquely exciting opportunities in the protein engineering field. Most importantly, a protein characterized by increased thermostability may also feature a high level of mutational robustness and is thus better able to accommodate functional evolution than is a protein with lower thermostability.

5. Thermophilic Proteins as a Scaffold for Functional Evolution

Since thermophilic proteins generally exhibit a higher level of mutational robustness when compared with their mesophilic counterparts, it stands to reason that these proteins are natural candidates to become scaffolds for evolutionary engineering. A number of studies have therefore been performed over the past decade evaluating thermophilic proteins as scaffolds for protein engineering and exploring options for application. This literature has become more robust recently as studies have begun to directly compare the evolvability of thermophilic proteins with that of analogous mesophilic proteins.

Bloom et al. randomly mutated two separate variants of a cytochrome P450 BM3 heme domain peroxygenase and screened the resulting mutants for new functionality (in this instance, the hydroxylation of various antibiotic drugs) [5]. The two variants examined were the mesophilic 21B3 enzyme (the temperature at which the protein loses 50% activity (T_{50}) = 47 °C) and the thermophilic 5H6 enzyme (T_{50} = 62 °C) [5]. The authors found that the increased thermostability of the 5H6 enzyme allowed more mutants to fold (61% of the total) than did the 21B3 parent (34%) [5]. This data corroborates previous evidence that increased thermostability is correlated with increased ability to tolerate mutations. The authors also observed that the 5H6 mutants developed new functionality at a higher rate than did the 21B3 mutants. Citing this data, Bloom et al. argued that the increased mutational robustness of thermophilic proteins makes them more effective scaffolds for protein engineering than their mesophilic analogues [5]. A study performed by Bershtein et al. confirms the correlation asserted by Bloom et al. between protein thermostability and evolvability by mutating a thermophilic variant of TEM-1 β-lactamase to develop new functionality [31]. The authors used error-prone polymerase chain reaction (PCR) to induce "neutral drift" (the generation of active, mutant variants which retain the same function as the original protein) among the protein population [31]. The mutant variants which were characterized by higher T_{50} values were isolated from the rest and subsequently tested for new functionality. Interestingly, the isolated variants were found to have evolved new functionality against cephalosporin antibiotics: these thermophilic variants were roughly 800 times more efficient at degrading the cephalosporin cefotaxime than were the original parent TEM-1 β-lactamase [31]. Since the authors took great care to ensure a neutral drift among the TEM-1 β-lactamase variants, the parent proteins serve as mesophilic analogues to the thermophilic variants in this paradigm. Overall, their finding that the thermophilic variants exhibited new functionality which was lacking both in the parent protein and in mesophilic analogues demonstrates the benefit of high thermostability for facile functional evolution [31].

Another experimental example of thermophilic proteins being utilized as scaffolds for protein evolution was reported by Tokuriki and Tawfik [32]. In this study, the authors subjected GADPH and CAII to error-prone PCR in the presence and absence of GroEL/GroES heat-shock chaperonins isolated from *E. coli* bacteria [32]. Previous literature had established the selected chaperonins as vitally important in increasing the ability of proteins to survive at high temperatures [32]. The protein variants developed in the presence of the chaperonins were then compared with those developed in the absence of chaperonins. As anticipated, the protein's mutational robustness was enhanced upon addition of chaperonins; the variants evolved with chaperonins were able to accommodate the incorporation of destabilizing mutations with $\Delta\Delta G_{folding}$ of up to 3.5 kcal/mol, whereas those developed without chaperonins were only able to tolerate destabilizing mutations with a $\Delta\Delta G_{folding}$ of up to 1 kcal/mol [32]. In addition, the variants developed with chaperonins were identified as "adaptive functional" variants—that is, mutants which retained the ability to fold but demonstrated new function—at a higher frequency than those developed without chaperonins [32].

Subsequent research performed by Aledo et al. compared the thermostability of two mammalian protein analogues: cytochrome b and cyclooxygenase I (COX I) to examine a correlation between evolvability and thermostability. Whereas previous studies done to evaluate the relationship between thermostability and evolvability started with a known thermophilic protein and a known mesophilic protein and compared their evolvability, this study does the opposite. The authors started with a pair of proteins, cytochrome b and COX I, known respectively to be highly evolvable and poorly evolvable, and compared their thermostability [33]. Both proteins were virtually mutated and aligned on the ClustalX bioinformatics platform and the destabilizing effect of each mutation on the protein were evaluated [33]. Aledo et al. found that the average destabilizing effect of the mutations on the cytochrome b protein was ~1 kcal/mol, while that on the COX I protein ranged from 1.5–1.9 kcal/mol [33]. The average destabilizing effect of mutations on cytochrome b fell within a range of the degree by which mutations destabilize known thermophilic proteins. Likewise, the average destabilizing effect of mutations on COX I was quantitatively similar to the effect of mutations on known mesophilic proteins. Thus, the more evolvable protein exhibits similar mutational robustness to known thermophilic proteins whereas the less evolvable protein exhibits similar mutational robustness to known mesophilic proteins; these findings supports the assertion that thermophilic proteins are better scaffolds for protein engineering than are mesophilic proteins.

Takahashi et al. identified a thermophilic variant of D-amino acid oxidase (DAO) in the thermophilic bacterium *Rubrobacter xylanophilus* and explored its applicability as a stable substitute for the ubiquitous DAO variant found in eukaryotes [34]. DAO is a biotechnologically attractive enzyme utilized for a variety of applications in biomedical science; it is, however, extremely unstable. This fact circumscribes its applicability and limits its potential for introduced functionality. Importantly, the thermophilic variant of DAO remained active not only at an elevated temperature but also at a low pH unlike the eukaryotic, mesophilic DAO, and also retained enzyme activity in the presence of certain thiol-modifying reagents known to inhibit the mesophilic variant [34]. The authors assert that these findings render the thermophilic DAO a strong candidate for functional engineering which would expand the applicability of the DAO family beyond the current capabilities of the mesophilic DAO [34].

Thermophilic proteins have also proven themselves superior scaffolds for the branch of protein engineering that explores protein fragmentation and cooperative function. This is a particularly challenging area of protein engineering which involves the destruction of parent proteins in such a manner that the resulting fragments are capable of reassembling themselves into stable, functional tertiary structures similar to the parents [35]. It is unsurprising that hyperstable thermophilic and hyperthermophilic proteins have shown themselves to be superior scaffolds for such potentially destabilizing interruptions. Nguyen et al. first showed that split adenylate kinases from the hyperthermophilic *Thermotoga neapolitana* were able to form functional complements which efficiently supported the growth of *E. coli* in culture; similar fragments from the mesophilic *Bacillus subtilis* were not able to do the same [36]. The authors also found that the degree of enzyme fragment complementation and function was directly proportional to the midpoint for thermal denaturation (i.e., melting temperature (T_m)—a measure of protein thermostability) of the parent enzyme: a consequence of the fact that the enzyme fragments derived from the hyperthermophilic parents retained more of the original tertiary structure of their parents than did their mesophilic fragment counterparts [36]. A follow-up study performed by Segall-Shapiro et al. on the same enzymes characterized this dichotomy further by finding that 44% of hyperthermophilic fragments generated in their study were capable of forming composites that supported *E. coli* growth, whereas only 6% of mesophilic fragments were capable of doing the same [37]. The authors attributed this finding to the fact that truncation of the hyperthermophilic protein yielded both more functional fragments and more unique fragment variants overall (41% of total fragments) than did truncation of the mesophilic parent (30% of fragments) [37].

More recently, a series of studies performed by Kim and coworkers demonstrated that domain insertion of a guest protein may benefit from high thermostability of a host protein [38–41]. Insertional fusion has recently been highlighted as a novel means of creating multi-domain protein complexes, where functionalities are often integrated and coupled with each other. In this type of fusion, a guest protein domain is inserted into the middle of a host protein domain. Unfortunately, due to the disruption of a host protein's primary sequence upon domain insertion, the insertional fusion is energetically challenging and often leads to significant compromise of protein stability [39]. To overcome such energetic penalty, Kim and coworkers used a thermophilic maltodextrin-binding protein from *Pyrococcus furiosus* as a host protein, into which various guest enzyme domains, such as exoinulinase, TEM-1 β-lactamase and xylanase, were successfully inserted. The insertional fusions led to the creation of chimeric protein complexes, where thermostability of guest enzyme domains was improved by various mechanisms [38–42]. In contrast, similar insertional fusion into a mesophilic maltodextrin-binding protein from *E. coli* lowered thermostability and expression levels of a guest enzyme domain [39]. In addition, functional evolution for enhanced enzyme activity toward alkaline pH of *Bacillus circulans* xylanase was greatly assisted by its enhanced thermostability acquired by the insertional fusion [42]. The implication is that the thermophilic host protein could serve as a stabilizing scaffold allowing the introduction of many gain-of-function mutations to multiple guest proteins, which would otherwise be catastrophically destabilizing [42].

The research discussed above strongly supports the notion that high stability of thermophilic proteins makes them greatly suitable for protein engineering, when compared to their mesophilic analogues. The applications for these highly stable proteins are clearly manifold in the field of protein engineering as they provide a broad array of exciting opportunities for the protein engineers of the new century.

6. Conclusions

The field of protein engineering is one which endeavors to introduce new functionality to protein domains. Since there is an inherent trade-off between protein functionality and protein stability, protein engineers must constantly balance the destabilizing effect of the mutations they introduce to protein domains with the anticipated modifications they are attempting to make to the protein. Over the course of the past decade and a half, many studies have shown that one of the best ways to increase a protein's chance of accommodating a functional mutation is to have an increased stability or mutational robustness to begin with. Numerous studies first confirmed that proteins characterized by greater stability were better able to withstand mutations, and later demonstrated that they were better able to support the type of mutations required to enhance existing function or introduce new function. The positive correlation between protein stability and evolvability was thus established.

Thermophilic proteins are characterized by certain significant structural differences from mesophilic proteins and often are accompanied by "chaperone proteins" which help the protein maintain its three-dimensional structure at high temperatures. Over the past 20 years, a nascent but increasingly robust literature has been developed directly comparing the evolvability of thermophilic proteins to that of mesophilic proteins; this literature clearly demonstrates that thermophilic proteins are indeed more evolvable and more tolerant of mutations introduced to enhance or modify functionality. This renders the thermophilic proteins superior scaffolds for protein engineering.

Thermophilic proteins thus appear to be an efficient and effective solution to the challenge posed by the simultaneous manipulation of protein function and maintenance of protein stability. Functionality and stability do indeed trade off, stability does indeed promote evolvability, and the increased stability of thermophilic proteins does indeed make this kind of protein a uniquely effective scaffold for evolving protein function, as summarized in Figure 2. Thermophilic proteins will likely be the scaffold of choice for the protein engineers of the future.

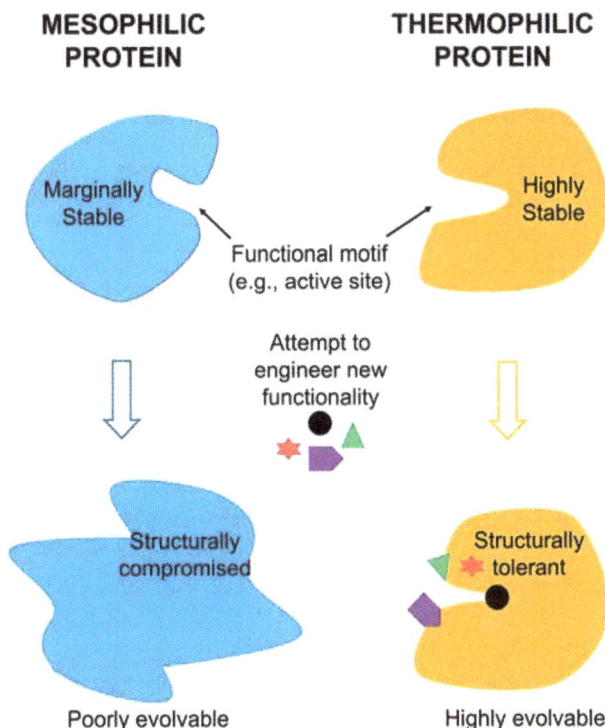

Figure 2. A schematic of facile functional evolution from thermophilic proteins assisted by their high thermostability. Insufficient stability of mesophilic proteins limits their evolvability.

Funding: This research was funded by the National Science Foundation, grant number CBET-1134247.

Conflicts of Interest: The authors declare no conflict of interest.

References

1. Rubingh, D.N. Protein engineering from a bioindustrial point of view. *Curr. Opin. Biotechnol.* **1997**, *8*, 417–422. [CrossRef]
2. Shoichet, B.K.; Baase, W.A.; Kuroki, R.; Matthews, B.W. A relationship between protein stability and protein function. *Proc. Natl. Acad. Sci. USA* **1995**, *92*, 452–456. [CrossRef] [PubMed]
3. Meiering, E.M.; Serran, L.; Fersht, A.R. Effect of active site residues in barnase on activity and stability. *J. Mol. Biol.* **1992**, *225*, 585–589. [CrossRef]
4. Taverna, D.M.; Goldstein, R.A. Why are proteins so robust to site mutations? *J. Mol. Biol.* **2002**, *315*, 479–484. [CrossRef] [PubMed]
5. Bloom, J.D.; Labthavikul, S.T.; Otey, C.R.; Arnold, F.H. Protein stability promotes evolvability. *Proc Natl Acad. Sci. USA* **2006**, *103*, 5869–5874. [CrossRef] [PubMed]
6. Wang, X.; Minasov, G.; Shoichet, B.K. Evolution of an Antibiotic Resistance Enzyme Constrained by Stability and Activity Trade-offs. *J. Mol. Biol.* **2002**, *320*, 85–95. [CrossRef]
7. Liang, X.; Arunima, A.; Zhao, Y.; Bhaksaran, R.; Shende, A.; Byrne, T.S.; Fleeks, J.; Palmier, M.O.; Van Doren, S.R. Apparent Tradeoff of Higher Activity in MMP-12 for Enhanced Stability and Flexibility in MMP-3. *Biophys. J.* **2010**, *99*, 273–283. [CrossRef] [PubMed]
8. Tokuriki, N.; Stricher, F.; Serrano, L.; Tawfik, D.S. How Protein Stability and New Functions Trade Off. *PLoS Comp. Biol.* **2008**, *4*, 1–7. [CrossRef] [PubMed]
9. Camps, M.; Herman, A.; Loh, E.; Loeb, L.A. Genetic Constraints on Protein Evolution. *Crit. Rev. Biochem. Mol. Biol.* **2007**, *42*, 313–326. [CrossRef] [PubMed]
10. Caetano-Anolles, G.; Mittenhall, J. Exploring the interplay of stability and function in protein evolution. *BioEssays* **2010**, *32*, 655–658. [CrossRef] [PubMed]
11. Philip, A.F.; Kumauchi, M.; Hoff, W.D. Robustness and evolvability in the functional anatomy of a PER-ARNT-SIM (PAS) domain. *Proc. Natl. Acad. Sci. USA* **2010**, *107*, 17986–17991. [CrossRef] [PubMed]
12. Taylor, T.J.; Vaisman, I.I. Discrimination of thermophilic and mesophilic proteins. *BMC Struct. Biol.* **2000**, *10* (Suppl. S1), S5.
13. Kim, H.; Kim, S.; Jung, Y.; Han, J.; Yun, J.H.; Chang, I.; Lee, W. Probing the folding-unfolding transition of a thermophilic protein. MTH1880. *PLoS ONE* **2016**, *11*, e0145853. [CrossRef] [PubMed]
14. Ota, N.; Kurahashi, R.; Sano, S.; Tankano, K. The direction of protein evolution is destined by the stability. *Biochimie* **2018**, *150*, 100–109. [CrossRef] [PubMed]
15. Yang, H.; Zhang, Y.; Li, X.; Bai, Y.; Xia, W.; Ma, R.; Juo, H.; Shi, P.; Yao, B. Impact of disulfide bonds on the folding and refolding capability of a novel thermostable GH45 cellulase. *Appl. Microbiol. Biotechnol.* **2018**. [CrossRef] [PubMed]
16. Szilagyi, A.; Zavodszky, P. Structural differences between mesophilic, moderately thermophilic, and extremely thermophilic protein subunits: Results of a comprehensive survey. *Structure* **2000**, *8*, 493–504. [CrossRef]
17. Gromiha, M.M.; Pathak, M.C.D.; Saraboji, K.; Ortlund, E.A.; Gaucher, E.A. Hydrophobic environment is a key factor for the stability of thermophilic proteins. *Proteins* **2013**, *81*, 715–721. [CrossRef] [PubMed]
18. Takano, K.; Aoi, A.; Koga, Y.; Kanaya, S. Evolvability of thermophilic proteins from archaea and bacteria. *Biochemistry* **2013**, *52*, 4774–4780. [CrossRef] [PubMed]
19. Meruelo, A.D.; Han, S.K.; Kim, S.; Bowie, J.U. Structural differences between thermophilic and mesophilic membrane proteins. *Protein Sci.* **2012**, *21*, 1746–1753. [CrossRef] [PubMed]
20. Glyakina, A.V.; Garbuzynskiy, S.O.; Lobanov, M.Y.; Galzitskaya, O.V. Different packing of external residues can explain differences in the thermostability of proteins from thermophilic and mesophilic organisms. *Bioinformatics* **2007**, *23*, 2231–2238. [CrossRef] [PubMed]
21. Tompa, D.R.; Gromiha, M.M.; Saraboji, K. Contribution of main chain and side chain atoms and their locations to the stability of thermophilic proteins. *J. Mol. Graph. Model.* **2016**, *64*, 85–93. [CrossRef] [PubMed]
22. Jorda, J.; Yeates, T.O. Widespread disulfide bonding in proteins from thermophilic archaea. *Archaea* **2011**, *2011*, 409156. [CrossRef] [PubMed]
23. Ladenstein, R.; Ren, B. Reconsideration of an early dogma, saying "there is no evidence for disulfide bonds in proteins from archaea". *Extremophiles* **2008**, *12*, 29–38. [CrossRef] [PubMed]

24. Georlette, D.; Damien, B.; Blaise, V.; Depiereux, E.; Uversky, V.N.; Gerday, C.; Feller, G. Structural and functional adaptations to extreme temperatures in psychrophilic, mesophilic, and thermophilic DNA ligases. *J. Biol. Chem.* **2003**, *278*, 37015–37023. [CrossRef] [PubMed]

25. Okada, J.; Okamoto, T.; Mukaiyama, A.; Tadokoro, T.; You, D.J.; Chon, H.; Koga, Y.; Takano, K.; Kanaya, S. Evolution and thermodynamics of the slow unfolding of hyperstable monomeric proteins. *BMC Evol. Biol.* **2010**, *10*, 207. [CrossRef] [PubMed]

26. Luke, K.A.; Higgins, C.L.; Wittung-Stafshede, P. Thermodyamic stability and folding of proteins from hyperthermophilic organisms. *FEBS J.* **2007**, *274*, 4023–4033. [CrossRef] [PubMed]

27. Collins, T.; Meuwis, M.A.; Gerday, C.; Feller, G. Activity, stability and flexibility in glycosidases adapted to extreme thermal environments. *J. Mol. Biol.* **2003**, *328*, 419–428. [CrossRef]

28. Pica, A.; Graziano, G. Shedding light on the extra thermal stability of thermophilic proteins. *Biopolymers* **2016**, *105*, 856–863. [CrossRef] [PubMed]

29. Wintrode, P.L.; Zhang, D.; Vaidehi, N.; Arnold, F.H.; Goddard, W.A., III. Protein dynamics in a family of laboratory evolved thermophilic enzymes. *J. Mol. Biol.* **2003**, *327*, 745–757. [CrossRef]

30. Berezovsky, I.N.; Chen, W.W.; Choi, P.J.; Shakhnovich, E.I. Entropic Stabilization of proteins and its proteomic consequences. *PLoS Comput. Biol.* **2005**, *1*, e47. [CrossRef] [PubMed]

31. Bershtein, S.; Goldin, K.; Tawfik, D.S. Intense Neutral Drifts Yield Robust and Evolvable Consensus Proteins. *J. Mol. Biol.* **2008**, *379*, 1029–1044. [CrossRef] [PubMed]

32. Tokuriki, N.; Tawfik, D.S. Chaperonin overexpression promotes genetic variation and enzyme evolution. *Nature* **2009**, *459*, 668–674. [CrossRef] [PubMed]

33. Aledo, J.C.; Valverde, H.; Ruiz-Camacho, M. Thermodynamic Stability Explains the Differential Evolutionary Dynamics of Cytochrome b and COX I in Mammals. *J. Mol. Evol.* **2012**, *74*, 69–80. [CrossRef] [PubMed]

34. Takahasi, S.; Furukawara, M.; Omae, K.; Tadokoro, N.; Saito, Y.; Abe, K.; Kera, Y. A highly stable D-amino oxidase of the thermophilic bacterium rubrobacter xylanophilus. *Appl. Environ. Microbiol.* **2014**, *80*, 7219–7229. [CrossRef] [PubMed]

35. Remy, I.; Michnick, S.W. Cloncal selection and in vivo quantitation of protein interactions with protein-fragment complementation assays. *Proc. Natl. Acad. Sci. USA* **1999**, *96*, 5394–5399. [CrossRef] [PubMed]

36. Nguyen, P.Q.; Liu, S.; Thompson, J.C.; Silberg, J.J. Thermostability promotes the cooperative function of split adenylate kinases. *Protein Eng. Des. Sel.* **2008**, *21*, 303–310. [CrossRef] [PubMed]

37. Segall-Shapiro, T.H.; Nguyen, P.Q.; Dos Santos, E.D.; Subedi, S.; Judd, J.; Suh, J.; Silberg, J. Mesophilic and hyperthermophilic adenylate kinases differ in their tolerance to random fragmentation. *J. Mol. Biol.* **2011**, *406*, 135–148. [CrossRef] [PubMed]

38. Kim, C.S.; Pierre, B.; Ostermeier, M.; Looger, L.L.; Kim, J.R. Enzyme stabilization by domain insertion into a thermophilic protein. *Protein Eng. Des. Sel.* **2009**, *22*, 615–623. [CrossRef] [PubMed]

39. Pierre, B.; Xiong, T.; Hayles, L.; Guntaka, V.R.; Kim, J.R. Stability of a guest protein depends on stability of a host protein in insertional fusion. *Biotechnol. Bioeng.* **2011**, *108*, 1011–1020. [CrossRef] [PubMed]

40. Pierre, B.; Labonte, J.W.; Xiong, T.; Aoraha, E.; Williams, A.; Shah, V.; Chau, E.; Helal, K.Y.; Gray, J.J.; Kim, J.R. Molecular determinants for protein stabilization by insertional fusion to a thermophilic host protein. *Chembiochem* **2015**, *16*, 2392–2402. [CrossRef] [PubMed]

41. Shah, V.; Pierre, B.; Kirtadze, T.; Shin, S.; Kim, J.R. Stabilization of bacillus circulans xylanase by combinatorial insertional fusion to a thermophilic host protein. *Protein Eng. Des. Sel.* **2017**, *30*, 281–290. [PubMed]

42. Shah, V.; Charlton, T.; Kim, J.R. Laboratory evolution of bacillus circulans xylanase inserted into pyrococcus furiosus maltodextrin-binding protein for increased xylanase activity and thermal stability toward alkaline pH. *Appl. Biochem. Biotechnol.* **2018**, *184*, 1232–1246. [CrossRef] [PubMed]

microorganisms

MDPI

Review

Cellulases from Thermophiles Found by Metagenomics

Juan-José Escuder-Rodríguez, María-Eugenia DeCastro, María-Esperanza Cerdán [ID],
Esther Rodríguez-Belmonte, Manuel Becerra [ID] and María-Isabel González-Siso *[ID]

Grupo EXPRELA, Centro de Investigacións Científicas Avanzadas (CICA), Departamento de Bioloxía,
Facultade de Ciencias, Universidade da Coruña, 15071 A Corunna, Spain; j.escuder@udc.es (J.-J.E.-R.);
m.decastro@udc.es (M.-E.D.); esper.cerdan@udc.es (M.-E.C.); esther.belmonte@udc.es (E.R.-B.);
manuel.becerra@udc.es (M.B.)
* Correspondence: migs@udc.es; Tel.: +34-981-167-000

Received: 21 June 2018; Accepted: 5 July 2018; Published: 10 July 2018

Abstract: Cellulases are a heterogeneous group of enzymes that synergistically catalyze the hydrolysis of cellulose, the major component of plant biomass. Such reaction has biotechnological applications in a broad spectrum of industries, where they can provide a more sustainable model of production. As a prerequisite for their implementation, these enzymes need to be able to operate in the conditions the industrial process requires. Thus, cellulases retrieved from extremophiles, and more specifically those of thermophiles, are likely to be more appropriate for industrial needs in which high temperatures are involved. Metagenomics, the study of genes and gene products from the whole community genomic DNA present in an environmental sample, is a powerful tool for bioprospecting in search of novel enzymes. In this review, we describe the cellulolytic systems, we summarize their biotechnological applications, and we discuss the strategies adopted in the field of metagenomics for the discovery of new cellulases, focusing on those of thermophilic microorganisms.

Keywords: cellulases; thermophiles; metagenomics; biotechnology

---

1. Introduction

Cellulose is a complex polymer that can be hydrolyzed into glucose by the synergetic action of a mixture of enzymes known as cellulases. Plants fix atmospheric CO_2 and incorporate about half of the carbon in structural polysaccharides and lignin (lignocellulose). This structural carbon can be used as an energy source by cellulolytic microorganisms [1]. The cellulolytic enzymes can form an enzyme complex known as the cellulosome, in which they are anchored to a common scaffold. This structure is mostly observed in anaerobes and exclusively in bacteria. They can also act as non-complexed extracellular free cellulase systems, more often associated to aerobes and present in fungi, bacteria, and archaea [1–4]. Additionally, other auxiliary enzymes like lytic polysaccharide monooxygenases have been reported to also contribute to the degradation of cellulose by cellulases by enhancing their activity [5–7]. An enhancer effect has also been proposed for hemicellulases such as xylanases, mannanases, galactosidases, and β-1,3-1,4-glycanases, which has activity on polysaccharides present in plant biomass by allowing cellulases to better reach the substrate [8].

Microorganisms adapted to live in harsh conditions (from a human standpoint) are known as extremophiles. Their enzymes, and especially the extracellular ones, have adopted mechanisms to maintain their function in such environments and are known as extremozymes. They are interesting from a biotechnological perspective, as many industrial applications involve conditions similar to those of extreme environments, and a more sustainable production model would require biocatalysts able to operate in such conditions [9–11].

Thermophiles are extremophiles that thrive at high temperatures ranging from moderate thermophiles (capable of growth at temperatures between 50 °C and 64 °C), extreme thermophiles (between 65 °C and 79 °C), and hyperthermophiles (over 80 °C) [12]. Extreme habitats where these microorganisms can be found include deep-sea hydrothermal vents, hot springs, volcanic fields, mud pots and deserts, and human-made environments like compost, among others. Many enzymes of industrial importance have been retrieved from thermophiles, including cellulases [11].

2. Modular Structure of Cellulases and their Classification

Most cellulases have a modular design, in which two or more discrete units have cooperative functions and are connected through linker sequences. Usually, this modular design includes the catalytic domain linked to a carbohydrate-binding module (CBM), but other non-catalytic domains can also be present, and multiple catalytic domains or CBMs can exist on the same enzyme. The CBM helps in the catalytic process by increasing the concentration of the enzyme near the polysaccharides they bind [5,13,14] and by disrupting the crystalline cellulose structure, increasing substrate accessibility [15]. As previously stated, some cellulases can form the enzyme complex known as the cellulosome, where they are anchored to a protein scaffold (composed of non-catalytic proteins known as scaffoldins). These cellulases contain dockerin domains that bind to the cohesin module of the scaffoldins, although these domains have also been described in proteins not related to the cellulosome [16]. In cellulosomes, the scaffolding proteins might also contain CBM modules [17].

The classic classification of cellulases is based on the mechanism of action of their catalytic domains and on their substrate specificity. This classification allows us to distinguish three major types of cellulases: β-1,4-endoglucanases (EC 3.2.1.4), exoglucanases [non-reducing end cellobiohydrolases (EC 3.2.1.91), reducing-end cellobiohydrolases (EC 3.2.1.176) and cellodextrinases (EC 3.2.1.74)], and β-glucosidases (EC 3.2.1.21) [3,7,18]. Endoglucanases act randomly cleaving internal glycosidic bonds of cellulose chains, releasing oligosaccharides of different length (like cellobiose and cellotriose). Cellobiohydrolases act processively on the reducing and non-reducing ends of cellulose, primarily releasing cellobiose but also other short oligosaccharides. Cellodextrinases act on soluble cellooligosaccharides, also releasing cellobiose. Lastly, β-glucosidases perform the hydrolysis of cellodextrins and cellobiose into glucose, enhancing both endoglucanase and exoglucanase activities by reducing the end product inhibition [3,6,7,9]. A schematic representation of cellulases acting on cellulose is depicted in Figure 1.

Due to the enormous variety of polysaccharides that exist in nature, and the fact that cellulases are not always easy to categorize as only endo- or exo-acting enzymes [19], an alternative classification based on amino acid sequence similarity was proposed [20]. Rather than substrate specificity, this classification addresses the structure-function relationships, substrate recognition and enzymatic reaction mechanisms, and evolutionary relationships between the enzymes. The publicly available Carbohydrate-Active Enzymes Database (CAZy, http://cazy.org) contains the classification of glycoside hydrolase (GH) families in which the cellulases are included. The database at the time of writing lists 149 different GH families [21]. Endoglucanases are mainly present in 12 GH families: GH5-9, GH12, GH44, GH45, GH48, GH51, GH74, and GH124; cellobiohydrolases acting on non-reducing ends can be found in families GH5, GH6, and GH9, whereas the reducing-end acting ones are mostly present in GH7, GH9, and GH48; cellodextrinases are distributed in families GH1, GH3, GH5, and GH9; and, lastly, β-glucosidases belong in families GH1-3, GH5, GH9, GH30, GH39, and GH116 [20].

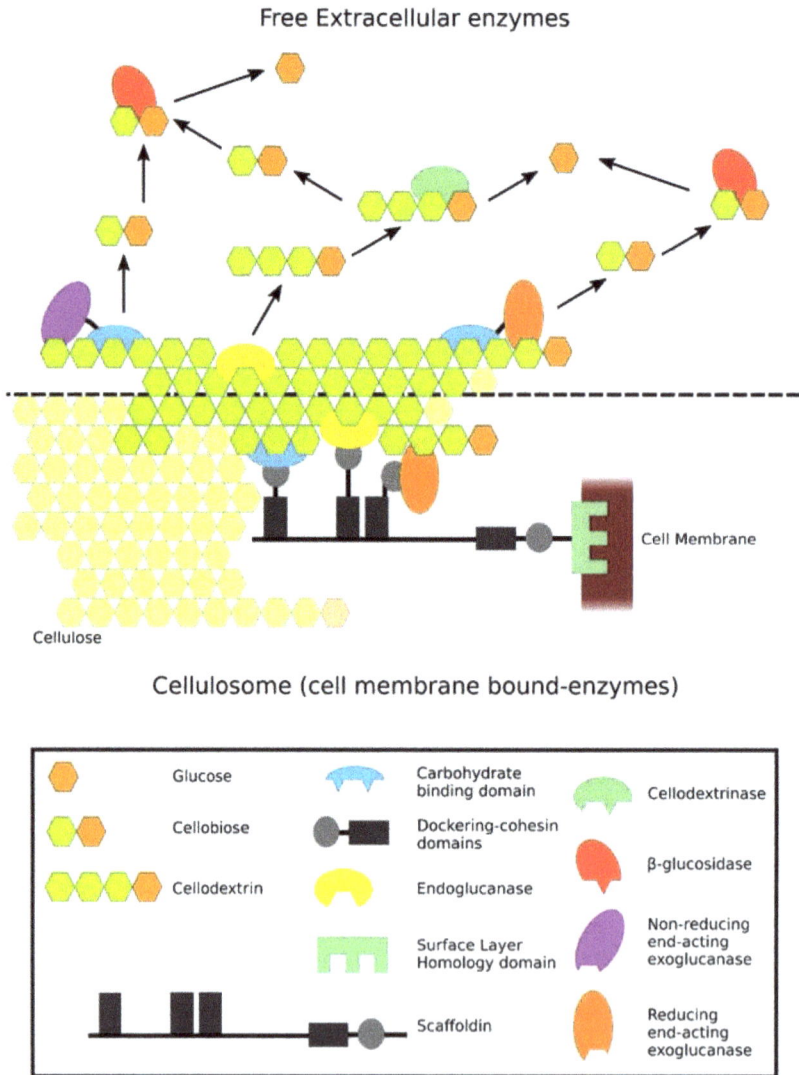

Figure 1. Overview of the two strategies (free or cell-bound cellulase systems) for degrading cellulose. In free extracellular systems, endoglucanases and exoglucanases act synergistically, with the endoglucanase cutting amorphous cellulose providing chain ends for exoglucanases to release cellobiose. Then, β-glucosidases complete the process of cellulose hydrolysis by releasing glucose. Also, cellodextrins released by endoglucanases can be further hydrolysed by cellodextrinases. The carbohydrate binding domain directs the enzymes to their specific substrates. In the cellulosome system, all cellulases are anchored to a common scaffold but are generally thought to follow the same synergic mode of action. The scaffolding is bound to the cell membrane through the surface layer homology domain, while a network of dockerin and cohesin domains amplifies the number of cellulases bound to the same scaffolding unit. Lastly, a carbohydrate binding domain is responsible for the targeting of the whole complex to the substrate.

Even if they share structural characteristics, members of the same GH family may differ widely in substrate specificity and their evolutionary history, and, due to their multidomain nature, some enzymes may contain sequences from different GH families [3,6,10]. As a further classification for GHs, some families are also grouped in clans in regard to their folding, as it is more conserved than their amino acid sequence [14]. Clans are designated by a letter, and some cellulases fall inside these groups: GH-A (with a $(\beta/\alpha)_8$ barrel) includes cellulases from families GH1, GH2, GH5, GH30, GH39, and GH51; GH-B (that fold in β-jelly roll) contains family GH7; GH-C (also folding with a β-jelly roll) includes family GH12; GH-M (folding with a $(\alpha/\alpha)_6$ barrel) comprises families GH8 and GH48; and GH-O [$(\alpha/\alpha)_6$ barrel folding] contains family GH116.

In regard to the catalytic mechanism, GHs (including cellulases) may perform the hydrolysis of the glycosidic bond by an inverting or retaining mechanism, whether the configuration of the substrate's anomeric carbon (C1) is changed or not after the cleavage. Retaining enzymes have a double nucleophilic displacement mechanism involving two carboxylate catalytic residues. Inverting enzymes act with a single nucleophilic displacement mechanism, also involving two carboxylate catalytic residues [1]. For cellulases, seven GH families have an inverting mechanism of catalysis (6, 8, 9, 45, 48, 74, and 124), whereas eleven act with a retaining mechanism (1–3, 5, 7, 12, 30, 39, 44, 51, and 116) [20,22].

3. Factors Influencing Thermostability of Thermophile Cellulases

As pointed out, a greater half-life of cellulases at high temperatures is a desirable trait for many industrial applications. In order to obtain more thermostable variants of cellulases, the molecular mechanisms behind thermostability have been studied. Some researchers argue that the study of smaller, single-domain enzymes would make it easier to pinpoint the mechanisms involved in a higher resistance to high temperature [23], while others have studied the effect of the number of domains and linker sequences and domain-removal on thermostability, though opposing stabilizing and destabilizing effects have been described in this regard [5].

Several stabilization factors have been proposed for the increased thermostability of thermozymes, such an increased number of ion pairs, a lower number of loops and cavities (thus making the protein more compact), a reduced ratio of protein surface area to protein volume, a higher number of proline residues in loops (limiting the conformational freedom of the protein), an increased amount of hydrophobic interactions, and a greater degree of oligomerization [24,25]. Despite that, a direct correlation between all these factors and protein thermostability cannot always be established; for example, for *Humicola insolens* exoglucanase Cel6A the addition of proline residues in the loop regions did not achieve greater stability and in some instances had the opposite effect [26]. It has been also proposed that proteins can undergo structure-based or sequence-based stabilization strategies through evolution. As thermophilic archaea emerged in already extreme environments, their enzymes would initially favour stable folding at high temperatures, whereas thermophilic bacteria would have to enhance the thermostability of their proteins by point mutations that increase the number of ion-pairs in order to colonize the new habitats. Despite this theory, it has been found that among archaea, the two different stabilization models can be adopted [24].

There are also reports on how hydrophobic and aromatic residues can play a major role in protein thermal stability, like in the endoglucanase from family GH12 from *Aspergillus niger* [27]. Other authors have described an increased percentage of the charged amino acid glutamic acid in thermophilic enzymes from family GH12 compared to mesophilic ones, which is thought to stabilize the protein's structure through salt bridges and hydrogen bonds [23]. Moreover, some key residues for protein stability have been already identified in this protein family [1]. When comparing mesophilic and thermophilic exoglucanases from family GH7, the potential disulphide bridge formation by the presence of cysteine residues could not be linked to an increased thermostability, whereas a higher number of charged residues and lower number of polar residues was observed in the more thermostable enzymes [28]. However, it was found that rational mutagenesis introducing disulphide bridges in an exoglucanase from this family did allow the mutant proteins to be more thermostable [29].

Lastly, eukaryotes' post-translational modifications (including glycosylation, phosphorylation, acetylation, and methylation) have been reported to account for protein thermostability [27], and heterologous expression of the enzyme in a yeast host can be a desirable production system for industrial applications.

The yeast *Pichia pastoris,* in particular, has been extensively employed due to this property, along with its relative ease for genetic manipulation and high level of protein expression [19,30–32], coupled with inexpensive production media and relatively simple protein processing protocols [33]. Nevertheless, most studies regarding the discovery and the characterization of new thermophilic cellulases have involved the model organism *Escherichia coli* [34–38], sometimes at the expense of thermostability [39].

4. Biotechnological Applications by Thermophile Cellulases

Thermozymes have general advantages over their mesophilic counterparts in regard to their application in various industries, as they are generally more stable towards extreme temperatures and pH, as well as in the presence of chemically destabilizing agents, and function at high temperatures with higher reaction rates [35] and higher mass-transfer rates that increase the substrates' solubility, as well as a lower risk of contamination [27]. Lastly, the process design gains flexibility (e.g., current process configurations with operations that needed pre-treatment of the substrates to lower the temperature can now be performed simultaneously without the requirement of a temperature modification between them), which in turn can reduce the cost of operation [27]. On the other hand, and as previously stated, preferred systems to produce these enzymes are not thermophilic, as thermophile production faces many technical challenges due to limited knowledge of their physiology and genetics, difficulty of growing and not being Generally Recognized As Safe [27] as defined by the US Food and Drug Administration under sections 201(s) and 409 of the Federal Food, Drug, and Cosmetic Act. In regard to the production process, extracellular enzymes are desirable, as they are easier to purify [27,33].

The range of industries in which degradation of cellulose by cellulases is required is considerably wide and includes biofuels (conversion of plant biomass in bioethanol), food and brewing, textiles (biostoning and biopolishing), laundry (in detergent formulations), pulp and paper (biopulping), and animal feeds [35]. Other uses include waste management, improvement of soils for agriculture [40], and extraction of compounds from plants such as olive oil, pigments, and bioactive molecules [4].

The full conversion of cellulose into glucose, which can later be converted into ethanol (named bioethanol to stress it being a biofuel, in contrast with the classic fossil fuels) has been previously stated to require the combined action of multiple cellulolytic enzymes (endo- and exoglucanases and β-glucosidases). This process has gained a lot of interest, as plant biomass poses a promising renewable substrate alternative to assess the increasing energy demands while limiting the use of fossil fuels [2,41]. In this regard, the use of non-food lignocellulosic waste from agriculture and forestry has replaced food crops as the substrate of choice, as the use of the latter would have the associated risk of raising basic foods prices and limiting their supply [42]. In general, biorefining (using biomass as a substrate to produce fuels, energy, or chemicals) benefits from thermostable enzymes, as heat treatment, is an important step for the pre-processing of the lignocellulosic material [43–45]. The use of thermostable cellulases for the treatment and pretreatment of the biomass reduces the energy cost of the process, improves the solubility of the substrate, reduces its viscosity, and reduces dependency on the use of environmentally harsh chemicals [39,45].

4.1. Endoglucanase-Specific Industrial Applications

Endoglucanases have been used in the textile industry for the process called biostoning. Biostoning achieves a wash-down look on denim cotton clothes, and represents an alternative to the chemical method using pumice stone. Biostoning has a number of advantages over the classical method, such as greater yields, less labor-intensive operations, more secure workplace, shorter time requirements, lower damage to the machinery, and a more environmentally friendly process [4].

Another textile industrial process in which endoglucanases are employed is the biopolishing of cotton products. This process removes the microfibrils from cottons' surfaces, enhancing the colour brightness and making them more resistant to pilling [40], as well as softening the product [46] and giving it a cleaner and smoother look [4]. Biopolishing is often performed after another enzymatic process called desizing (in which amylases remove starch from the fabrics). Desizing uses temperatures higher than 70 °C, so endoglucanases operating at such temperatures would be interesting for combining both processes and thus reducing the required time and energy costs [46]. Other textile processes in which endoglucanases are employed to remove cellulosic impurities, replacing chemical treatments, include bio-carbonization of polyester-cotton blends, wool scouring, and de-fibrillation of Lyocell [4].

In the brewing industry, the production of malt generates high molecular weight β-glucans. The presence of these molecules increases viscosity, lowering the efficiency and yield of the process due to the increased difficulty for pumping and also making filtration difficult [33]. As such, the addition of endoglucanases would alleviate those problems, allowing for the hydrolysis of β-glucans [33]. Also, endoglucanases may be used to increase the extraction of fermentable compounds both in brewing and fermentation industries [47].

In the laundry industry, the use of endoglucanases in detergent formulations is known to improve the colour brightness and soften cotton fabrics [4], similarly to the biopolishing in the textile industry.

In the animal feed industry, they enhance β-glucan digestibility and nutrient bioavailability [47], and have been shown to increase weight gain and milk production of ruminants [4].

Endoglucanases have been extensively used in the pulp and paper industry for the treatment of pulp wastes [4,47], deinking and removal of pollutants from paper without altering its brightness and strength [4], and in the pulping process (bio-pulping), reducing the energy cost of the process and improving the beatability of the pulp [4].

4.2. Exoglucanase-Specific Industrial Applications

As in nature, efficient degradation of cellulose from biomass in industrial applications requires the synergic action of a mixture of cellulases [26,48]. Synergism has been described between endoglucanases and exoglucanases, between reducing-end-acting and non-reducing-end-acting exoglucanases, between processive endoglucanases and endo- or exoglucanases, and between β-glucosidases and the other cellulases [48]. As such, the previously described industrial applications benefit from the addition of exoglucanases to enzyme mixtures already containing other cellulase classes.

4.3. β-glucosidase-Specific Industrial Applications

In addition to their application in the last step of cellulose hydrolysis to release glucose, β-glucosidases have several additional biotechnological applications.

In the food industry, they can be used to release aromatic compounds from fruit and fermentation products [49], like the release of terpenoids and phenylpropanoids in wine to enhance its aroma [50,51]. Other uses include juice clarification [32] and hydrolysis of bitter compounds in its extraction [52], and, in general, improvement of quality of beverages and foods [44] including colour, aroma, flavour, texture, and nutritional value [4].

In the pharmaceutical industry, they are used to deglycosylate ginsenosides, active compounds with many pharmaceutical uses, as the natural glycosylated ginsenosides from ginseng root are less active and less absorbable [50,52,53]. Similarly, they are used to convert the bioactive isoflavonoid-glucosides from soybean and other leguminous plants into aglycones with higher bioavailability and pharmaceutical activity [44,50,54]. Moreover, β-glucosidases can perform reverse hydrolysis or transglycosylation catalytic pathways for the formation of new glycosidic bonds, a property that makes them interesting for the production of functional compounds, and nutraceutical and pharmaceutical products [44]. For example gentibiose, a product of transglycosylation by β-glucanases, can be used as a prebiotic food additive [50]. These kinds of enzymatic transformations constitute important alternatives to chemical synthesis involving the use of organic solvents [55]. In this regard, the valorization of spent coffee grounds to produce isoflavone glycosides has also been proposed [54].

5. Metagenomics for the Search of Novel Cellulases

The metabolism of thermophiles holds great potential for several industrial applications, but due to the difficulty of growing extremophiles in the laboratory, culture-independent techniques constitute instrumental methods to have access to it. The use of metagenomics, the study of whole communities' genomes, has proven to be a useful tool for the discovery of novel cellulases, both in the functional and the sequence-based approaches [10,11]. Several studies had found cellulases in a wide variety of natural thermophilic environments, such as hydrothermal vents [56,57], continental geothermal pools and hotsprings [58,59], and man-made environments like vermicompost [60], compost [37,61,62], and biogas digesters [63]. Nevertheless, high-temperature acting enzymes have also been found by metagenomics on moderate-temperature samples like soils [40,64,65] and aquatic environments [66], and in microorganisms associated with animals like microbial communities in rabbit cecum [67], ruminants rumen [36,68,69], earthworm casts [70], and thermite guts [71,72].

The main limiting factor for the discovery of new thermophile cellulases by functional metagenomics is the host organism used for the metagenomic libraries, typically the mesophilic bacterium *E. coli*, which may have a limited or biased expression of gene products from thermophiles [3]. One of the proposed solutions for this problem is the use of an alternative thermophilic host for the metagenomic libraries that would increase the hit detection rate for cellulases [11]. It should also be noted that bacteria hosts are not able to express fungal enzymes, as the promoter and intron regions are not recognized [3]. Lastly, the discovery of novel cellobiohydrolases through metagenomics is limited due to the lack of specific substrates other than AVICEL that can discriminate between true cellobiohydrolases and other celullases, as AVICEL has the requirement of a synergy between an endoglucanase and an exoglucanase for detection of activity [3]. The other metagenomic approach, an analysis of the whole metagenome sequencing data, can overcome the problems that arise in the expression-based approach. Regardless, the discovery of gene products with novel characteristics is hindered due to the need of high amino acid homology with already known enzymes, and before assigning putative proteins a function, activities should be verified [11].

6. Thermophile Cellulases Characterized

Tables 1–5 list, respectively, endoglucanases, exoglucanases acting on non-reducing ends, exoglucanases acting on reducing ends, cellodextrinases and β-glucanases that can be considered thermophilic (optimum temperature at 50 °C or higher), and other key parameters for their industrial application, namely, pH optimum and temperature stability, their classification according to the CAZY database, and their source organism.

Table 1. Characterized endoglucanases (EC 3.2.1.4) from thermophiles. NM: not measured.

Enzyme	GH Family Domains	Optimum Temperature	Optimum pH	Temperature Stability [1]	Source	Reference
EGPh	5	>97 °C	5.4–6.0	80%; 97 °C; 3 h	Archaea (*Pyrococcus horikoshii*)	[46]
EG1	5	83 °C	5.0	20%; 90 °C; 2 h	Bacteria (*Acidothermus cellolyticus*)	[73]

Table 1. *Cont.*

Enzyme	GH Family Domains	Optimum Temperature	Optimum pH	Temperature Stability [1]	Source	Reference
EglII	5	50 °C	6.0	NM	Bacteria (*Bacillus amyloliquefaciens*)	[74]
EG	5	65 °C	6.0	72%; 55 °C; 42 h 50%; 65 °C; 12 min	Bacteria (*Bacillus licheniformis*)	[75]
CelA	5	60 °C	8.0	30%; 70 °C; 1 h	Bacteria (*Bacillus subtilis*)	[76]
TmCel5A	5	80 °C	6.0	50%; 80 °C; 18 h	Bacteria (*Thermotoga maritima*)	[77]
EglA	5	57 °C	4.0	NM	Fungi (*Aspergillus nidulans*)	[78]
EglB	5	52 °C	4.0	NM	Fungi (*Aspergillus nidulans*)	[78]
EBI-244	5	109 °C	5.5	50%; 100 °C; 4.5 h 50%; 105 °C; 0.57 h 108 °C; 50%; 0.17 h	Uncultured Archaea (Continental geothermal pool enrichment)	[58]
CelE1	5	50 °C	7.0	NM	Uncultured organism (Sugarcane field soil metagenome)	[64]
CelA10	5	55 °C	7.5	NM	Uncultured organism (Aquatic community and soil sample)	[66]
CelA24	5	55 °C	7.0	NM	Uncultured organism (Aquatic community and soil sample)	[66]
cMGL504	5	50 °C	5.5	NM	Uncultured organism (Vermicompost sample)	[60]
Cel5G	5	50 °C	4.8	>90%; 50 °C; 30 min	Uncultured organism (Soil metagenome)	[65]
En1	5	55 °C	5.5	87%; 45 °C; 16 h 67%; 50 °C; 6 h 42%; 55 °C; 30 min	Uncultured organism (Biogas digester metagenome)	[63]
RC1	5	55 °C	6.0–6.5	>90%; 50 °C; 30 min	Uncultured organism (Rabbit cecum metagenome)	[67]
RC3	5	50 °C	6.0–7.0	NM	Uncultured organism (Rabbit cecum metagenome)	[67]
RC5	5	50 °C	6.5–7.0	NM	Uncultured organism (Rabbit cecum metagenome)	[67]
CelL	6	50 °C	5.0	50%; 50 °C; 12 min	Bacteria (*Cellulosimicrobium funkei*)	[22]
Cel6A	6	58 °C	6.5	>80%; 56 °C; 18 h	Bacteria (*Thermobifida fusca*)	[79]
ThCel6A	6	55 °C	8.5	58%; 90 °C; 1 h	Bacteria (*Thermobifida halotolerans*)	[80]
Cel6A	6	50–55 °C	5.5	NM	Bacteria (*Xylanimicrobium pachnodae*)	[81]
HiCel6C	6	70 °C	6.5	>90%; 60 °C; 1 h	Fungi (*Humicola insolens*)	[82]
Cel6A	6	50 °C	4.8	>90%; 45 °C; 24 h 92%; 50 °C; 5 h	Fungi (*Orpinomyces* sp.)	[83]
C1	6	50 °C	6.0	100%; 60 °C; 30 min	Uncultured organism (Compost metagenome)	[61]
pre-LC-CelB	6	NM	NM	NM	Uncultured organism (Compost metagenome)	[62]
pre-LC-CelJ	6	NM	NM	NM	Uncultured organism (Compost metagenome)	[62]
EGI	7	55–60 °C	5.0	>80%; 60 °C; 10 min	Fungi (*Humicola grisea* var. *thermoidea*)	[84]
Cel7B	7	60 °C	4.0	>90%; 60 °C; 1 h	Fungi (*Penicillium decumbens*)	[85]
Cel7A	7	60 °C	5.0	100%; 60 °C; 1 h 16.1%; 70 °C; 1 h	Fungi (*Neosartorya fischeri*)	[33]
MtEG7	7	60 °C	5.0	50%; 70 °C; 9.96 h 50%; 80 °C; 6.5 h	Fungi (*Myceliophthora thermophila*)	[31]

Table 1. *Cont.*

Enzyme	GH Family Domains	Optimum Temperature	Optimum pH	Temperature Stability [1]	Source	Reference
EGL1	7	62 °C	4.8	NM	Fungi (*Trichoderma longibrachiatum*)	[51]
MaCel7A	7	65–70 °C	6.0	NM	Fungi (*Melanocarpus albomyces*)	[86]
CelC	8	50 °C	6.5	NM	Bacteria (*Salmonella typhimurium*)	[87]
Cel8Y	8	80 °C	7.0	50%; 90 °C; 4 h 50%; 100 °C; 2 h	Bacteria (*Aquifex geolicus*)	[88]
Egl-257	8	55 °C	8.5	100%; 50 °C; 15 min >50%; 60 °C; 15 min	Bacteria (*Bacillus circulans*)	[89]
CenC	9	70 °C	6.0	100%; 60 °C; 2 h 60%; 70 °C; 1 h	Bacteria (*Clostridium thermocellum*)	[90]
CelA	9 (endoglucanase) and 48 (cellobiohydrolase)	95 °C (endoglucanase) and 85 °C (cellobiohydrolase)	5.0–6.0	50%; 95 °C; 40 min (endoglucanase) 100%; 85 °C; 4 h (cellobiohydrolase)	Bacteria (*Caldicellulosiruptor bescii*)	[91]
Cel9A	9	65 °C	6.5	NM	Bacteria (*Lachnoclostridium phytofermentans*)	[92]
CelA20	9	55 °C	5.0	NM	Uncultured organism (Aquatic community and soil metagenome)	[66]
AcCel12B	12	75 °C	4.5	50%; 60 °C; 90 h 50%; 65 °C; 55 h 50%; 70 °C; 2 h	Bacteria (*Acidothermus cellulolyticus*)	[35]
CelA	12	95 °C	6.0	NM	Bacteria (*Thermotoga neapolitana*)	[8]
CelB	12	106 °C	6.0–6.6	50%; 106 °C; 130 min 50%; 110 °C; 26 min 73%; 100 °C; 4 h	Bacteria (*Thermotoga neapolitana*)	[8]
TmCel12A	12	90 °C	7.0	>40%; 85 °C; 48 h 50%; 90 °C; 3 h	Bacteria (*Thermotoga maritima*)	[93]
TmCel12B	12	85 °C	6.0	50%; 90 °C; 9 h	Bacteria (*Thermotoga maritima*)	[93]
CelA	12	>100 °C	6.0–7.0	45%; 90 °C; 8 h	Bacteria (*Rhodothermus marinus*)	[23]
EglA	12	100 °C	6.0	50%; 95 °C; 40 h	Archaea (*Pyrococcus furiosus*)	[94]
SSO1949	12	80 °C	1.8	50%; 80 °C; 8 h	Archeaea (*Sulfolobus solfataricus*)	[95]
SSO1354	12	90 °C	4.0	50%; 90 °C; 180 min	Archaea (*Sulfolobus solfataricus*)	[39]
EglS	12	65 °C	6.0	>40%; 60 °C; 30 min	Bacteria (*Streptomyces rochei*)	[96]
Cel12A	12	50 °C	5.0	NM	Fungi (*Trichoderma reseei*)	[97]
EG	12	70 °C	3.5	50%; 70 °C; 3 h 50%; 80 °C; 1 h	Fungi (*Aspergillus niger*)	[27]
Pre-LC-CelA	12	90 °C	5.0–9.0	100%; 90 °C; 30 min	Uncultured organism (Compost metagenome)	[62]
Pre-LC-CelD	12	NM	NM	NM	Uncultured organism (Compost metagenome)	[62]
Pre-LC-CelE	12	NM	NM	NM	Uncultured organism (Compost metagenome)	[62]
Cel12E	12	92 °C	5.5	>80%; 80 °C; 4.5 h	Uncharacterized archeon (deep sea vents metagenome enrichment)	[57]
GH44EG	44	55 °C	5.0	NM	Bacteria (*Clostridium acetobutylicum*)	[98]
CelA	44	60 °C	5.0–8.5	50%; 60 °C; 70 min	Bacteria (*Paenibacillus lautus*)	[99]
CelJ	44	70 °C	6.5	>90%; 80 °C; 10 min	Bacteria (*Ruminiclostridium thermocellum*)	[100]

Table 1. *Cont.*

Enzyme	GH Family Domains	Optimum Temperature	Optimum pH	Temperature Stability [1]	Source	Reference
pre-LC-CelH	44	NM	NM	NM	Uncultured organism (Compost metagenome)	[62]
Cel45A	45	60 °C	5.0	NM	Fungi (*Trichoderma reseei*)	[97]
PpCel45A	45	65 °C	4.8	70%; 65 °C; 48 h 60%; 80 °C; 4 h	Fungi (*Picchia pastoris*)	[5]
STCE1	45	60 °C	6.0	NM	Fungi (*Staphylotrichum coccosporum*)	[101]
BCC18080	45	70 °C	6.0	>70%; 70 °C; 2 h >50%; 70 °C; 4 h	Fungi (*Syncephalastrum racemosum*)	[102]
BCE1	45	55 °C	4.5	NM	Fungi (*Beltraniella portoricensis*)	[103]
MaCel45A	45	70 °C	7.0	NM	Fungi (*Melanocarpus albomyces*)	[86]
CelB	51	80 °C	4.0	60%; 80 °C; 1 h	Bacteria (*Alicyclobacillus acidocaldarius*)	[104]
CelA4	51	65 °C	2.6	>85%; 60 °C; 1 h	Bacteria (*Alicyclobacillus* sp. A4)	[47]
CelVA	51	80 °C	3.6–4.5	70%; 70 °C; 2 h	Bacteria (*Alicyclobacillus vulcanalis*)	[45]
pre-LC-CelC	51	NM	NM	NM	Uncultured organism (Compost metagenome)	[62]
TmCel74	74	90 °C	6.0	50%; 90 °C; 5 h	Bacteria (*Thermotoga maritima*)	[15]
CtCel124	124	NM	NM	NM	Bacteria (*Ruminiclostridium thermocellum*)	[105]

[1] Temperature stability is given as a percentage of activity (residual activity) after treatment at the specified temperature and time compared to the untreated enzyme.

Table 2. Characterized exoglucanases (1,4-β-cellobiosidase) acting on non-reducing ends from thermophiles (EC 3.2.1.91). NM: not measured.

Enzyme	GH Family Domains	Optimum Temperature	Optimum pH	Temperature Stability [1]	Source	Reference
CBHII	6	60 °C	4.0	30%; 100 °C; 10 min	Bacteria (*Streptomyces* sp. M23)	[106]
Cel6B	6	NM	7.0–8.0	100%; 55 °C; 16 h	Bacteria (*Thermobifida fusca*)	[107]
CBHII	6	57 °C	5.5	NM	Fungi (*Aspergillus nidulans*)	[78]
Cel6A	6	50 °C	4.0	50%; 70 °C; 30 min	Fungi (*Chaetomium thermophilum*)	[108]
CBHII (Cel6A)	6	60 °C	5.0–5.5	>90%; 50 °C; 5 h	Fungi (*Chrysosporium lucknowense*)	[109]
HiCel6A	6	60–65 °C	NM	50%; 75 °C; <25 min	Fungi (*Humicola insolens*)	[26]
Ex-4	6	50 °C	5.0	80%; 60 °C; 60 min	Fungi (*Irpex Lacteus*)	[110]
PoCel6A	6	50 °C	5.0	90%; 50 °C; 2 h 80%; 60 °C; 4 h	Fungi (*Penicillium oxalicum*)	[111]
PaCel6A	6	55 °C	5.0–9.0	100%; 35 °C; 24 h >20%; 45 °C; 24 h	Fungi (*Podospora anserina*)	[19]
CBHII	6	70 °C	5.0	NM	Fungi (*Trichoderma viride*)	[112]
G10-6	6	55 °C	9.5	NM	Uncultured organism (Eathworm casts metagenome)	[70]
Cbh9A	9	60 °C	6.5	NM	Bacteria (*Ruminiclostridium thermocellum*)	[113]
Cel9K	9	65 °C	6.0	97%; 60 °C; 200 h	Bacteria (*Ruminiclostridium thermocellum*)	[114]

[1] Temperature stability is given as a percentage of activity (residual activity) after treatment at the specified temperature and time compared to the untreated enzyme.

Table 3. Characterized exoglucanases (1,4-β-cellobiosidase) acting on reducing ends from thermophiles (EC 3.2.1.176). NM: not measured.

Enzyme	GH Family Domains	Optimum Temperature	Optimum pH	Temperature Stability [1]	Source	Reference
CelO	5	65 °C	6.6	NM	Bacteria (*Ruminiclostridium thermocellum*)	[115]
AtCel7A	7	60 °C	5.0	NM	Fungi (*Acremonium thermophilum*)	[28]
CBHI	7	60 °C	3.0	NM	Fungi (*Aspergillus aculeatus*)	[116]
CBHI	7	55 °C	NM	NM	Fungi (*Aspergillus fumigatus*)	[117]
CtCel7A	7	65 °C	4.0	NM	Fungi (*Chaetomium thermophilum*)	[28]
CBH3	7	65 °C	5.0	50%; 70 °C; 1 h 20%; 80 °C; 20 min	Fungi (*Chaetomium thermophilum*)	[118]
DpuCel7A	7	55 °C	5.0	NM	Metazoa (*Dictyostelium purpureum*)	[119]
CBHI	7	60 °C	5.0	>90%; 55 °C; 10 min	Fungi (*Humicola grisea* var. *thermoidea*)	[84]
EXO1	7	65 °C	5.0	>80%; 65 °C; 10 min	Fungi (*Humicola grisea* var. *thermoidea*)	[120]
MaCel7B	7	55 °C	NM	NM	Fungi (*Melanocarpus albomyces*)	[121]
TeCel7A	7	65 °C	4.0–5.0	50%; 70 °C; 30 min	Fungi (*Talaromyces emersonii*)	[29]
Cel7A	7	55 °C	3.7–5.2	50%; 50 °C; 2.5 h	Fungi (*Penicillium funiculosum*)	[122]
TaCel7A	7	65 °C	5.0	NM	Fungi (*Thermoascus aurantiacus*)	[28]
ThCBHI	7	50 °C	5.0	NM	Fungi (*Trichoderma harzianum*)	[123]
CBHI	7	60 °C	5.8	NM	Fungi (*Trichoderma viride*)	[112]
CelA	9 (endoglucanase) and 48 (cellobiohydrolase)	95 °C (endoglucanase) and 85 °C (cellobiohydrolase)	5.0–6.0	50%; 95 °C; 40 min (endoglucanase) 100%; 85 °C; 4 h (cellobiohydrolase)	Bacteria (*Caldicellulosiruptor bescii*)	[91]
CelY	48	70 °C	5.0–6.0	NM	Bacteria (*Clostridium stercorarium*)	[124]
CpCel48	48	55 °C	5.0–6.0	>70%; 50 °C; 30 min>20%; 55 °C; 30 min	Bacteria (*Lachnoclostridium phytofermentans*)	[125]
CelS	48	70 °C	5.5	NM	Bacteria (*Ruminiclostridium thermocellum*)	[126]

[1] Temperature stability is given as a percentage of activity (residual activity) after treatment at the specified temperature and time compared to the untreated enzyme.

Table 4. Characterized exoglucanases (cellodextrinases) acting on reducing ends from thermophiles (3.2.1.74).

Enzyme	GH family Domains	Optimum Temperature	Optimum pH	Temperature Stability [1]	Source	Reference
CcGH1	1	60 °C	6.5	61%; 50 °C; 30 min	Bacteria (*Clostridium Cellulolyticum*)	[127]
GghA	1	95 °C	6.5	85%; 90 °C; 9 h 88%; 95 °C; 1 h	Bacteria (*Thermotoga neapolitana*)	[128]

[1] Temperature stability is given as a percentage of activity (residual activity) after treatment at the specified temperature and time compared to the untreated enzyme.

Table 5. Characterized β-glucanases from thermophiles (3.2.1.21). NM: not measured.

Enzyme	GH family Domains	Optimum Temperature	Optimum pH	Temperature Stability [1]	Source	Reference
CelB	1	102–105 °C	5.0	50%; 100 °C; 85 h 50%; 110 °C; 13 h	Arquea (*Pyrococcus furiosus*)	[129]
Tpa-glu	1	75 °C	7.5	50%; 90 °C; 6 h	Arquea (*Thermococcus pacificus*)	[130]
BGPh	1	>100 °C	6.0	50%; 90 °C; 15 h	Arquea (*Pyrococcus horikoshii*)	[131]
LacS	1 (β-glucosidase and β-galactosidase)	90 °C	6.0	90%; 75 °C; 80 h	Arquea (*Sulfolobus solfataricus*)	[132]
O08324	1	78 °C	5.0–6.8	50%; 78 °C; 860 min	Arquea (*Thermococcus* sp.)	[133]
Bgl1	1	90 °C	6.5	67%; 90 °C; 1.5 h 78%; 50 °C; 24 h 68%; 60 °C; 24 h	Uncultured Arquea (hot spring metagenome)	[59]
GlyB	1 (multiple substrates)	85 °C	5.5	8%; 80 °C; 10 min >70%; 65 °C; 3 h	Bacteria (*Alicyclobacillus acidocaldarius*)	[134]
Bglp	1	60 °C	7.0	50%; 60 °C; 10 h	Bacteria (*Anoxybacillus flavithermus*)	[135]
BglA	1	55 °C	6.0–9.0	80%; 50 °C; 15 min 1%; 60 °C; 15 min	Bacteria (*Bacillus circulans* subsp. *Alkalophilus*)	[136]
BhbglA	1	50 °C	7.0	50%; 50 °C; 30 min	Bacteria (*Bacillus halodurans*)	[137]
BglA	1	85 °C	6.25	50%; 70 °C; 2280 min	Bacteria (*Caldicellulosiruptor saccharolyticus*)	[138]
BglA	1	50 °C	6.0	NM	Bacteria (*Clostridium cellulovorans*)	[139]
DtGH	1	90 °C	7.0	50%; 70 °C; 533 h 50%; 80 °C; 44 h 50%; 90 °C; 5 h	Bacteria (*Dictyoglomus thermophilum*)	[140]
DturβGlu	1	80 °C	5.4	70%; 70 °C; 2 h	Bacteria (*Dictyoglomus turgidum*)	[44]
FiBgl1A	1	90 °C	6.0–7.0	50%; 90 °C 25 min 50%; 100 °C; 15 min	Bacteria (*Fervidobacterium islandicum*)	[141]
BglA	1	60 °C	6.5	91%; 60 °C; 3 h 34%; 60 °C; 43 h	Bacteria (*Ruminiclostridium thermocellum*)	[142]
SdBgl1B	1	50 °C	6.0–7.5	NM	Bacteria (*Saccharophagus degradans*)	[143]
Bgl1	1	50 °C	5.1–5.7	60%; 40 °C; 4 h	Bacteria (*Sphingomonas paucimobilis*)	[144]
SGR_2426	1	69 °C	6.9	50%; 69 °C; 1.5 h	Bacteria (*Streptomyces griseus*)	[55]
Bgl3	1	50 °C	6.5	NM	Bacteria (*Streptomyces* sp. strain QM-B814)	[145]
CglT	1	75 °C	5.5	100%; 60 °C; 24 h	Bacteria (*Thermoanaerobacter brockii*)	[146]
TeBglA	1	80 °C	7.0	10%; 65 °C; 5 h	Bacteria (*Thermoanaerobacter ethanolicus*)	[147]
TmBglA	1	90 °C	6.2	>80%; 65 °C; 5 h	Bacteria (*Thermotoga maritima*)	[147]
Bgl	1	70 °C	6.4	50%; 68 °C; 1 h >80%; 60 °C; 2 h	Bacteria (*Thermoanaerobacterium thermosaccharolyticum*)	[148]
BglC	1	50 °C	7.0	NM	Bacteria (*Thermobifida fusca*)	[149]
BglB	1	60 °C	6.2	70%; 60 °C; 48 h	Bacteria (*Thermobispora bispora*)	[150]
BglA	1	80–90 °C	7.0–8.0	100%; 70 °C; 6 h	Bacteria (*Thermotoga petrophila*)	[151]
TcaBglA	1	90 °C	5.5–6.5	>40%; 80 °C; 30 min >20%; 80 °C; 30 min	Bacteria (*Thermus caldophilus*)	[152]
TnGly	1	90 °C	5.6	50%; 90 °C; 2.5 h	Bacteria (*Thermus nonproteolyticus*)	[153]

Table 5. *Cont.*

Enzyme	GH family Domains	Optimum Temperature	Optimum pH	Temperature Stability [1]	Source	Reference
BglA	1	70 °C	5.0–6.0	50%; 70 °C; 38 h 50%; 80 °C; <0.4 h 50%; 90 °C; <0.3 h	Bacteria (*Thermus* sp. IB-21)	[154]
BglB	1	80 °C	5.0–6.0	50%; 70 °C; 38 h 50%; 80 °C; 2.7 h 50%; 90 °C; 24 min	Bacteria (*Thermus* sp. IB-21)	[154]
Bgly	1	90 °C	5.4	100%; 80 °C; 2 h 50%; 90 °C; 1.5 h 50%; 95 °C; 20 min	Bacteria (*Thermus thermophilus*)	[155]
BglA	1	55 °C	6.5	82%; 50 °C; 60 min 20%; 55 °C; 60 min	Uncultured organism (soil metagenome)	[52]
AS-Esc10	1	60 °C	8.0	100%; 50 °C; 1 h	Uncultured organism (agricultural soil metagenome)	[40]
Bgl-gs1	1	90 °C	6.0	50%; 90 °C; 5 min 50%; 85 °C; 15 min 50%; 80 °C; 45 min	Uncultured organism (termite gut metagenome)	[71]
Bgl	1	60 °C	5.0	50%; 60 °C; 540 min	Fungi (*Fusarium oxysporum*)	[156]
Bgl4	1	55 °C	6.0	80%; 50 °C; 10 min	Fungi (*Humicola grisea* var. *thermoidea* IFO9854)	[157]
Bgl1	1	55 °C	5.5–7.5	100%; 50 °C; 8 h 50%; 55 °C; 8 h	Fungi (*Orpinomyces* sp. PC-2)	[158]
Bgl1G5	1	50 °C	6.0	50%; 50 °C; 6 h	Fungi (*Phialophora* sp. G5)	[159]
TaGH2	2	95 °C	6.5	100%; 90 °C; 3 h 50%; 70 °C; 22 h	Bacteria (*Thermus antranikianii*)	[160]
TbGH2	2	90 °C	6.5	17%; 80 °C; 3 h 50%; 70 °C; 12 h	Bacteria (*Thermus brockianus*)	[160]
TbBgl	3	90 °C	3.5	50%; 95 °C; 60 min	Arquea (*Thermofilum pendens*)	[161]
BlBG3	3	50 °C	6.0	NM	Bacteria (*Bifidobacterium longum*)	[162]
Cba2	3	70 °C	4.8	NM	Bacteria (*Cellulomonas biazotea*)	[163]
CfBgl3A	3	55 °C	7.5	NM	Bacteria (*Cellulomonas fimi*)	[164]
Bgl3Z	3	65 °C	5.5	50%; 60 °C; 5 h	Bacteria (*Clostridium stercorarium*)	[165]
Dtur_0219	3	85 °C	5.0	50%; 70 °C; 1575 min 50%; 75 °C; 854 min 50%; 80 °C; 524 min 50%; 85 °C; 334 min 50%; 90 °C; 20 min	Bacteria (*Dictyoglomus turgidum*)	[54]
Bgl	3	50 °C	4.2–5.0	NM	Bacteria (*Elizabethkingia meningoseptica*)	[166]
TmBglB	3	80 °C	4.2	>80%; 65 °C; 5 h	Bacteria (*Thermotoga maritima*)	[147]
Tpebgl3	3	90 °C	5.0	>90%; 70 °C; 3 h >50%; 90 °C; 3 h	Bacteria (*Thermotoga petrophila*)	[167]
Cel3A	3	50–60 °C	5.0	98%; 60 °C; 6 h >50%; 60 °C; 24 h >50%; 70 °C; 24 h	Fungi (*Amesia atrobrunnea*)	[168]
Cel3B	3	50–60 °C	5.0	88%; 60 °C; 6 h >50%; 60 °C; 24 h >50%; 70 °C; 24 h	Fungi (*Amesia atrobrunnea*)	[168]
Bgl3	3	60 °C	6.0	>50%; 70 °C; 1 h	Fungi (*Aspergillus fumigatus*)	[169]
BglB	3	52 °C	5.5	NM	Fungi (*Aspergillus nidulans*)	[78]
BglC	3	52 °C	6.0	NM	Fungi (*Aspergillus nidulans*)	[78]
Bgl	3	50 °C	5.0	100%; 50 °C; 30 min 60%; 60 °C; 30 min	Fungi (*Aspergillus oryzae*)	[49]
Bgl	3	60 °C	5.0	67.7%; 60 °C; 1 h 50%; 65 °C; 55 min 29.7%; 70 °C; 10 min	Fungi (*Chaetomium thermophilum*)	[170]

Table 5. *Cont.*

Enzyme	GH family Domains	Optimum Temperature	Optimum pH	Temperature Stability [1]	Source	Reference
Bxl5	3	75 °C	4.6	50%; 65 °C; 5 h 50%; 70 °C; 20 min 50%; 75 °C; 5 min	Fungi (*Chrysosporium lucknowense*)	[171]
MoCel3A	3	50 °C	5.0–5.5	NM	Fungi (*Magnaporthe oryzae*)	[41]
MoCel3B	3	50 °C	5.0–5.5	NM	Fungi (*Magnaporthe oryzae*)	[41]
Bgl2	3	60 °C	5.4	>50%; 40 °C; 2 h >45%; 50 °C; 2 h 25%; 55 °C; 1 h	Fungi (*Neurospora crassa*)	[172]
Bgl1	3	50 °C	3.5–5.0	100%; 45 °C; 30 min	Fungi (*Mucor circinelloides*)	[173]
Bgl2	3	55 °C	3.5–5.5	100%; 55 °C; 30 min	Fungi (*Mucor circinelloides*)	[173]
NfBGL1	3	80 °C	5.0	>80%; 70 °C; 2 h	Fungi (*Neosartorya fischeri*)	[174]
PtBglu3	3	65 °C	6.0	>85%; 60 °C; 30 min	Fungi (*Paecilomyces thermophila*)	[32]
Bgl1	3	70 °C	4.8	50%; 65 °C; 24 h	Fungi (*Penicillium brasilianum*)	[175]
pBGL1	3	65–70 °C	4.5–5.50	96.3%; 50 °C; 12 h 50%; 70 °C; 4 h	Fungi (*Penicillium decumbens*)	[176]
Bgl1	3	70 °C	5.0–6.0	60%; 70 °C; 1.5 h	Fungi (*Periconia* sp.)	[43]
RmBglu3B	3	50 °C	5.0	50%; 50 °C; 30 min	Fungi (*Rhizomucor miehei*)	[50]
Bgl1	3	50 °C	5.0	>70%; 50 °C; 30 min <10%; 60 °C; 30 min	Fungi (*Saccharomycopsis fibuligera*)	[177]
Bgl2	3	50 °C	5.0	>70%; 50 °C; 30 min <10%; 60 °C; 30 min	Fungi (*Saccharomycopsis fibuligera*)	[177]
β-glucosidase	3	75 °C	4.5	50%; 60 °C; 136 h 50%; 65 °C; 55 h 50%; 70 °C; 10 h 50%; 75 °C; 1 h	Fungi (*Talaromyces aculeatus*)	[53]
Cel3a	3	71.5 °C	4.02	50%; 65 °C; 62 min 50%; 75 °C; 18 min	Fungi (*Talaromyces emersonii*)	[178]
Bgl3A	3	75 °C	4.5	>65%; 60 °C; 1 h	Fungi (*Talaromyces leycettanus*)	[179]
Bgl1	3	70 °C	5.0	>70%; 60 °C; 1 h	Fungi (*Thermoascus auranticus*)	[180]
Bgl3a	3	70 °C	5.0	50%; 60 °C; 143 min	Fungi (*Myceliophthora thermophila*)	[181]
RG3	3	50–55 °C	5.5–6.0	NM	Uncultured organism (Rabbit cecum metagenome)	[67]
RG14	3	50–55 °C	5.5–7.0	NM	Uncultured organism (Rabbit cecum metagenome)	[67]
BGL7	3	50 °C	6.5	NM	Uncultured organism (Termite gut metagenome)	[72]
LAB25g2	3	55 °C	4.5	82%; 50 °C; 5 d	Uncultured organism (Cow rumen metagenome)	[68]
SRF2g14	3	55 °C	5.0	50%; 50 °C; 18.06 h	Uncultured organism (Cow rumen metagenome)	[68]
SRF2g18	3	50 °C	4.0	50%; 50 °C; 37.5 h	Uncultured organism (Cow rumen metagenome)	[68]
RuBGX1	3	50 °C	6.0	62%; 50 °C; 10 min	Uncultured organism (Yak rumen metagenome)	[36]
JMB19063	3	50–55 °C	6.5	NM	Uncultured organism (Compost metagenome)	[37]
GlyA1	3	55 °C	6.5	NM	Uncultured organism (Cow rumen metagenome)	[69]

Table 5. *Cont.*

Enzyme	GH family Domains	Optimum Temperature	Optimum pH	Temperature Stability [1]	Source	Reference
Bgx1	30	50 °C	4.0–6.0	NM	Oomycota (*Phytophthora infestans*)	[182]
SSO3039	116	>70 °C	4.0	>70%; 65 °C; 48 h >50%; 85 °C; 8 h	Arquea (*Sulfolobus solfataricus*)	[183]
TxGH116	116	85 °C	6.0	NM	Bacteria (*Thermoanaerobacterium xylanolyticum*)	[184]

[1] Temperature stability is given as a percentage of activity (residual activity) after treatment at the specified temperature and time compared to the untreated enzyme.

7. Conclusions

Cellulases retrieved from high-temperature environments are considered a valuable industrial resource for their vast biotechnological potential [35]. The use of culture-independent techniques such as metagenomics has allowed us to discover enzymes from unknown microorganisms thriving in extreme habitats [11]. Since the last decade, metagenomics has led to the discovery of almost half (46%) of the characterized thermophilic endoglucanases (Table 1) described in that period and a fraction (17% of each total) of the thermophilic cellobiosidases acting on the non-reducing end of cellulose (Table 2) and thermophilic β-glucosidases (Table 5). Nevertheless, metagenomics have yet to yield thermophilic cellobiosidases acting on the reducing end of cellulose (Table 3) or thermophilic cellodextrinases (Table 4). The lack of enzymes found by this strategy is likely a consequence of the mechanism of action of those enzymes, as the lack of substrates specific to those activities greatly limits its positive hit ratio. While thermophilic β-glucosidases discovered in the last 5 years still account for a similar proportion of the total (15%), no more thermophilic cellobiosidases acting on non-reducing ends have been characterized by this method. On the other hand, the proportion of thermophilic endoglucanases that have been characterized and identified by metagenomics have grown to account for more than half of the total (55%) in the last 5 years. In total, almost one fifth (18%) of all the thermophilic cellulases identified and characterized so far have been found by metagenomics. Functional metagenomic bottlenecks, like the lack of substrates for specific cellulases and problems associated with heterologous expression [3], and validation of sequence-based metagenomics annotation of cellulases [11], still need to be addressed to further increase the number of cellulases identified using these strategies. Biomining for novel thermophilic cellulases through metagenomic means is thus an ongoing challenge, with great potential as a source of commercially and environmentally important byocatalysts in all sorts of biotechnological applications.

Author Contributions: Writing—original draft preparation: J.-J.E.-R.; writing—review & editing, J.-J.E.-R., M.-E.D., M.-E.C., E.R.-B., M.B., M.-I.G.-S.; project administration: M.-I.G.-S.; funding acquisition: M.-E.C., M.-I.G.-S.

Funding: General support for EXPRELA (Universidade da Coruña, Spain) was funded by the Xunta de Galicia (Consolidación Grupos Referencia Competitiva Contract no. ED431C2016-012), co-financed by FEDER (EEC).

Acknowledgments: HOTDROPS (FP7/2007-2013, CN 324439).

Conflicts of Interest: The authors declare no conflict of interest. The funders had no role in the design of the study; in the collection, analyses, or interpretation of data; in the writing of the manuscript; or in the decision to publish the results.

References

1. Sandgren, M.; Ståhlberg, J.; Mitchinson, C. Structural and biochemical studies of GH family 12 cellulases: Improved thermal stability, and ligand complexes. *Prog. Biophys. Mol. Biol.* **2005**, *89*, 246–291. [CrossRef] [PubMed]

2. Blumer-Schuette, S.E.; Kataeva, I.; Westpheling, J.; Adams, M.W.; Kelly, R.M. Extremely thermophilic microorganisms for biomass conversion: Status and prospects. *Curr. Opin. Biotechnol.* **2008**, *19*, 210–217. [CrossRef] [PubMed]
3. Duan, C.-J.; Feng, J.-X. Mining metagenomes for novel cellulase genes. *Biotechnol. Lett.* **2010**, *32*, 1765–1775. [CrossRef] [PubMed]
4. Sharma, A.; Tewari, R.; Rana, S.S.; Soni, R.; Soni, S.K. Cellulases: Classification, Methods of Determination and Industrial Applications. *Appl. Biochem. Biotechnol.* **2016**, *179*, 1346–1380. [CrossRef] [PubMed]
5. Couturier, M.; Feliu, J.; Haon, M.; Navarro, D.; Lesage-Meessen, L.; Coutinho, P.M.; Berrin, J.-G. A thermostable GH45 endoglucanase from yeast: Impact of its atypical multimodularity on activity. *Microb. Cell Fact.* **2011**, *10*, 103. [CrossRef] [PubMed]
6. López-Mondéjar, R.; Zühlke, D.; Becher, D.; Riedel, K.; Baldrian, P. Cellulose and hemicellulose decomposition by forest soil bacteria proceeds by the action of structurally variable enzymatic systems. *Sci. Rep.* **2016**, *6*, 25279. [CrossRef] [PubMed]
7. Kaur, B.; Chadha, B.S. Approaches for Bioprospecting Cellulases. In *Extremophilic Enzymatic Processing of Lignocellulosic Feedstocks to Bioenergy*; Sani, R.K., Krishnaraj, R.N., Eds.; Springer International Publishing: Cham, Switzerland, 2017; pp. 53–71, ISBN 978-3-319-54683-4.
8. Bok, J.D.; Yernool, D.A.; Eveleigh, D.E. Purification, characterization, and molecular analysis of thermostable cellulases CelA and CelB from *Thermotoga neapolitana*. *Appl. Environ. Microbiol.* **1998**, *64*, 4774–4781. [PubMed]
9. Elleuche, S.; Schäfers, C.; Blank, S.; Schröder, C.; Antranikian, G. Exploration of extremophiles for high temperature biotechnological processes. *Curr. Opin. Microbiol.* **2015**, *25*, 113–119. [CrossRef] [PubMed]
10. Tiwari, R.; Nain, L.; Labrou, N.E.; Shukla, P. Bioprospecting of functional cellulases from metagenome for second generation biofuel production: A review. *Crit. Rev. Microbiol.* **2018**, *44*, 244–257. [CrossRef] [PubMed]
11. DeCastro, M.-E.; Rodríguez-Belmonte, E.; González-Siso, M.-I. Metagenomics of Thermophiles with a Focus on Discovery of Novel Thermozymes. *Front. Microbiol.* **2016**, *7*, 1521. [CrossRef] [PubMed]
12. Wagner, I.D.; Wiegel, J. Diversity of thermophilic anaerobes. *Ann. N. Y. Acad. Sci.* **2008**, *1125*, 1–43. [CrossRef] [PubMed]
13. Liu, Y.; Zhang, J.; Liu, Q.; Zhang, C.; Ma, Q. Molecular cloning of novel cellulase genes cel9A and cel12A from *Bacillus licheniformis* GXN151 and synergism of their encoded polypeptides. *Curr. Microbiol.* **2004**, *49*, 234–238. [CrossRef] [PubMed]
14. Crennell, S.J.; Hreggvidsson, G.O.; Nordberg Karlsson, E. The structure of *Rhodothermus marinus* Cel12A, a highly thermostable family 12 endoglucanase, at 1.8 A resolution. *J. Mol. Biol.* **2002**, *320*, 883–897. [CrossRef]
15. Chhabra, S.R.; Kelly, R.M. Biochemical characterization of *Thermotoga maritima* endoglucanase Cel74 with and without a carbohydrate binding module (CBM). *FEBS Lett.* **2002**, *531*, 375–380. [CrossRef]
16. Peer, A.; Smith, S.P.; Bayer, E.A.; Lamed, R.; Borovok, I. Noncellulosomal cohesin- and dockerin-like modules in the three domains of life. *FEMS Microbiol. Lett.* **2009**, *291*, 1–16. [CrossRef] [PubMed]
17. Bayer, E.A.; Belaich, J.-P.; Shoham, Y.; Lamed, R. The cellulosomes: Multienzyme machines for degradation of plant cell wall polysaccharides. *Annu. Rev. Microbiol.* **2004**, *58*, 521–554. [CrossRef] [PubMed]
18. Sathya, T.A.; Khan, M. Diversity of glycosyl hydrolase enzymes from metagenome and their application in food industry. *J. Food Sci.* **2014**, *79*, R2149–R2156. [CrossRef] [PubMed]
19. Poidevin, L.; Feliu, J.; Doan, A.; Berrin, J.-G.; Bey, M.; Coutinho, P.M.; Henrissat, B.; Record, E.; Heiss-Blanquet, S. Insights into exo- and endoglucanase activities of family 6 glycoside hydrolases from *Podospora anserina*. *Appl. Environ. Microbiol.* **2013**, *79*, 4220–4229. [CrossRef] [PubMed]
20. Henrissat, B. A classification of glycosyl hydrolases based on amino acid sequence similarities. *Biochem. J.* **1991**, *280*, 309–316. [CrossRef] [PubMed]
21. Lombard, V.; Golaconda Ramulu, H.; Drula, E.; Coutinho, P.M.; Henrissat, B. The carbohydrate-active enzymes database (CAZy) in 2013. *Nucleic Acids Res.* **2014**, *42*, D490–D495. [CrossRef] [PubMed]
22. Kim, D.Y.; Lee, M.J.; Cho, H.-Y.; Lee, J.S.; Lee, M.-H.; Chung, C.W.; Shin, D.-H.; Rhee, Y.H.; Son, K.-H.; Park, H.-Y. Genetic and functional characterization of an extracellular modular GH6 endo-β-1,4-glucanase from an earthworm symbiont, *Cellulosimicrobium funkei* HY-13. *Antonie Van Leeuwenhoek* **2016**, *109*, 1–12. [CrossRef] [PubMed]

23. Halldórsdóttir, S.; Thórólfsdóttir, E.T.; Spilliaert, R.; Johansson, M.; Thorbjarnardóttir, S.H.; Palsdottir, A.; Hreggvidsson, G.O.; Kristjánsson, J.K.; Holst, O.; Eggertsson, G. Cloning, sequencing and overexpression of a *Rhodothermus marinus* gene encoding a thermostable cellulase of glycosyl hydrolase family 12. *Appl. Microbiol. Biotechnol.* **1998**, *49*, 277–284. [CrossRef] [PubMed]

24. Ausili, A.; Cobucci-Ponzano, B.; Di Lauro, B.; D'Avino, R.; Perugino, G.; Bertoli, E.; Scirè, A.; Rossi, M.; Tanfani, F.; Moracci, M. A comparative infrared spectroscopic study of glycoside hydrolases from extremophilic archaea revealed different molecular mechanisms of adaptation to high temperatures. *Proteins* **2007**, *67*, 991–1001. [CrossRef] [PubMed]

25. Aguilar, C.F.; Sanderson, I.; Moracci, M.; Ciaramella, M.; Nucci, R.; Rossi, M.; Pearl, L.H. Crystal structure of the β-glycosidase from the hyperthermophilic archeon *Sulfolobus solfataricus*: Resilience as a key factor in thermostability. *J. Mol. Biol.* **1997**, *271*, 789–802. [CrossRef] [PubMed]

26. Wu, I.; Arnold, F.H. Engineered thermostable fungal Cel6A and Cel7A cellobiohydrolases hydrolyze cellulose efficiently at elevated temperatures. *Biotechnol. Bioeng.* **2013**, *110*, 1874–1883. [CrossRef] [PubMed]

27. Rawat, R.; Kumar, S.; Chadha, B.S.; Kumar, D.; Oberoi, H.S. An acidothermophilic functionally active novel GH12 family endoglucanase from *Aspergillus niger* HO: Purification, characterization and molecular interaction studies. *Antonie Van Leeuwenhoek* **2015**, *107*, 103–117. [CrossRef] [PubMed]

28. Voutilainen, S.P.; Puranen, T.; Siika-Aho, M.; Lappalainen, A.; Alapuranen, M.; Kallio, J.; Hooman, S.; Viikari, L.; Vehmaanperä, J.; Koivula, A. Cloning, expression, and characterization of novel thermostable family 7 cellobiohydrolases. *Biotechnol. Bioeng.* **2008**, *101*, 515–528. [CrossRef] [PubMed]

29. Voutilainen, S.P.; Murray, P.G.; Tuohy, M.G.; Koivula, A. Expression of *Talaromyces emersonii* cellobiohydrolase Cel7A in *Saccharomyces cerevisiae* and rational mutagenesis to improve its thermostability and activity. *Protein Eng. Des. Sel.* **2010**, *23*, 69–79. [CrossRef] [PubMed]

30. Li, Y.-L.; Li, H.; Li, A.-N.; Li, D.-C. Cloning of a gene encoding thermostable cellobiohydrolase from the thermophilic fungus *Chaetomium thermophilum* and its expression in *Pichia pastoris*. *J. Appl. Microbiol.* **2009**, *106*, 1867–1875. [CrossRef] [PubMed]

31. Karnaouri, A.C.; Topakas, E.; Christakopoulos, P. Cloning, expression, and characterization of a thermostable GH7 endoglucanase from *Myceliophthora thermophila* capable of high-consistency enzymatic liquefaction. *Appl. Microbiol. Biotechnol.* **2014**, *98*, 231–242. [CrossRef] [PubMed]

32. Yan, Q.; Hua, C.; Yang, S.; Li, Y.; Jiang, Z. High level expression of extracellular secretion of a β-glucosidase gene (PtBglu3) from *Paecilomyces thermophila* in *Pichia pastoris*. *Protein Expr. Purif.* **2012**, *84*, 64–72. [CrossRef] [PubMed]

33. Liu, Y.; Dun, B.; Shi, P.; Ma, R.; Luo, H.; Bai, Y.; Xie, X.; Yao, B. A Novel GH7 Endo-β-1,4-Glucanase from *Neosartorya fischeri* P1 with Good Thermostability, Broad Substrate Specificity and Potential Application in the Brewing Industry. *PLoS ONE* **2015**, *10*, e0137485. [CrossRef] [PubMed]

34. Dougherty, M.J.; D'haeseleer, P.; Hazen, T.C.; Simmons, B.A.; Adams, P.D.; Hadi, M.Z. Glycoside hydrolases from a targeted compost metagenome, activity-screening and functional characterization. *BMC Biotechnol.* **2012**, *12*, 38. [CrossRef] [PubMed]

35. Wang, J.; Gao, G.; Li, Y.; Yang, L.; Liang, Y.; Jin, H.; Han, W.; Feng, Y.; Zhang, Z. Cloning, Expression, and Characterization of a Thermophilic Endoglucanase, AcCel12B from *Acidothermus cellulolyticus* 11B. *Int. J. Mol. Sci.* **2015**, *16*, 25080–25095. [CrossRef] [PubMed]

36. Zhou, J.; Bao, L.; Chang, L.; Liu, Z.; You, C.; Lu, H. Beta-xylosidase activity of a GH3 glucosidase/xylosidase from yak rumen metagenome promotes the enzymatic degradation of hemicellulosic xylans. *Lett. Appl. Microbiol.* **2012**, *54*, 79–87. [CrossRef] [PubMed]

37. McAndrew, R.P.; Park, J.I.; Heins, R.A.; Reindl, W.; Friedland, G.D.; D'haeseleer, P.; Northen, T.; Sale, K.L.; Simmons, B.A.; Adams, P.D. From soil to structure, a novel dimeric β-glucosidase belonging to glycoside hydrolase family 3 isolated from compost using metagenomic analysis. *J. Biol. Chem.* **2013**, *288*, 14985–14992. [CrossRef] [PubMed]

38. Lee, C.-M.; Lee, Y.-S.; Seo, S.-H.; Yoon, S.-H.; Kim, S.-J.; Hahn, B.-S.; Sim, J.-S.; Koo, B.-S. Screening and Characterization of a Novel Cellulase Gene from the Gut Microflora of *Hermetia illucens* Using Metagenomic Library. *J. Microbiol. Biotechnol.* **2014**, *24*, 1196–1206. [CrossRef] [PubMed]

39. Girfoglio, M.; Rossi, M.; Cannio, R. Cellulose degradation by *Sulfolobus solfataricus* requires a cell-anchored endo-β-1-4-glucanase. *J. Bacteriol.* **2012**, *194*, 5091–5100. [CrossRef] [PubMed]

40. Biver, S.; Stroobants, A.; Portetelle, D.; Vandenbol, M. Two promising alkaline β-glucosidases isolated by functional metagenomics from agricultural soil, including one showing high tolerance towards harsh detergents, oxidants and glucose. *J. Ind. Microbiol. Biotechnol.* **2014**, *41*, 479–488. [CrossRef] [PubMed]

41. Takahashi, M.; Konishi, T.; Takeda, T. Biochemical characterization of *Magnaporthe oryzae* β-glucosidases for efficient β-glucan hydrolysis. *Appl. Microbiol. Biotechnol.* **2011**, *91*, 1073–1082. [CrossRef] [PubMed]

42. Bhalla, A.; Bansal, N.; Kumar, S.; Bischoff, K.M.; Sani, R.K. Improved lignocellulose conversion to biofuels with thermophilic bacteria and thermostable enzymes. *Bioresour. Technol.* **2013**, *128*, 751–759. [CrossRef] [PubMed]

43. Harnpicharnchai, P.; Champreda, V.; Sornlake, W.; Eurwilaichitr, L. A thermotolerant beta-glucosidase isolated from an endophytic fungi, *Periconia* sp., with a possible use for biomass conversion to sugars. *Protein Expr. Purif.* **2009**, *67*, 61–69. [CrossRef] [PubMed]

44. Fusco, F.A.; Fiorentino, G.; Pedone, E.; Contursi, P.; Bartolucci, S.; Limauro, D. Biochemical characterization of a novel thermostable β-glucosidase from *Dictyoglomus turgidum*. *Int. J. Biol. Macromol.* **2018**, *113*, 783–791. [CrossRef] [PubMed]

45. Boyce, A.; Walsh, G. Characterisation of a novel thermostable endoglucanase from *Alicyclobacillus vulcanalis* of potential application in bioethanol production. *Appl. Microbiol. Biotechnol.* **2015**, *99*, 7515–7525. [CrossRef] [PubMed]

46. Ando, S.; Ishida, H.; Kosugi, Y.; Ishikawa, K. Hyperthermostable Endoglucanase from *Pyrococcus horikoshii*. *Appl. Environ. Microbiol.* **2002**, *68*, 430–433. [CrossRef] [PubMed]

47. Bai, Y.; Wang, J.; Zhang, Z.; Shi, P.; Luo, H.; Huang, H.; Feng, Y.; Yao, B. Extremely acidic beta-1,4-glucanase, CelA4, from thermoacidophilic *Alicyclobacillus* sp. A4 with high protease resistance and potential as a pig feed additive. *J. Agric. Food Chem.* **2010**, *58*, 1970–1975. [CrossRef] [PubMed]

48. Vuong, T.V.; Wilson, D.B. Processivity, synergism, and substrate specificity of *Thermobifida fusca* Cel6B. *Appl. Environ. Microbiol.* **2009**, *75*, 6655–6661. [CrossRef] [PubMed]

49. Tang, Z.; Liu, S.; Jing, H.; Sun, R.; Liu, M.; Chen, H.; Wu, Q.; Han, X. Cloning and expression of *A. oryzae* β-glucosidase in *Pichia pastoris*. *Mol. Biol. Rep.* **2014**, *41*, 7567–7573. [CrossRef] [PubMed]

50. Guo, Y.; Yan, Q.; Yang, Y.; Yang, S.; Liu, Y.; Jiang, Z. Expression and characterization of a novel β-glucosidase, with transglycosylation and exo-β-1,3-glucanase activities, from *Rhizomucor miehei*. *Food Chem.* **2015**, *175*, 431–438. [CrossRef] [PubMed]

51. Villanueva, A.; Ramón, D.; Vallés, S.; Lluch, M.A.; MacCabe, A.P. Heterologous Expression in *Aspergillus nidulans* of a *Trichoderma longibrachiatum* Endoglucanase of Enological Relevance. *J. Agric. Food Chem.* **2000**, *48*, 951–957. [CrossRef] [PubMed]

52. Kim, S.-J.; Lee, C.-M.; Kim, M.-Y.; Yeo, Y.-S.; Yoon, S.-H.; Kang, H.-C.; Koo, B.-S. Screening and characterization of an enzyme with beta-glucosidase activity from environmental DNA. *J. Microbiol. Biotechnol.* **2007**, *17*, 905–912. [PubMed]

53. Lee, G.-W.; Yoo, M.-H.; Shin, K.-C.; Kim, K.-R.; Kim, Y.-S.; Lee, K.-W.; Oh, D.-K. β-glucosidase from *Penicillium aculeatum* hydrolyzes exo-, 3-*O*-, and 6-*O*-β-glucosides but not 20-*O*-β-glucoside and other glycosides of ginsenosides. *Appl. Microbiol. Biotechnol.* **2013**, *97*, 6315–6324. [CrossRef] [PubMed]

54. Kim, Y.-S.; Yeom, S.-J.; Oh, D.-K. Characterization of a GH3 family β-glucosidase from *Dictyoglomus turgidum* and its application to the hydrolysis of isoflavone glycosides in spent coffee grounds. *J. Agric. Food Chem.* **2011**, *59*, 11812–11818. [CrossRef] [PubMed]

55. Kumar, P.; Ryan, B.; Henehan, G.T.M. β-Glucosidase from *Streptomyces griseus*: Nanoparticle immobilisation and application to alkyl glucoside synthesis. *Protein Expr. Purif.* **2017**, *132*, 164–170. [CrossRef] [PubMed]

56. Placido, A.; Hai, T.; Ferrer, M.; Chernikova, T.N.; Distaso, M.; Armstrong, D.; Yakunin, A.F.; Toshchakov, S.V.; Yakimov, M.M.; Kublanov, I.V.; et al. Diversity of hydrolases from hydrothermal vent sediments of the Levante Bay, Vulcano Island (Aeolian archipelago) identified by activity-based metagenomics and biochemical characterization of new esterases and an arabinopyranosidase. *Appl. Microbiol. Biotechnol.* **2015**, *99*, 10031–10046. [CrossRef] [PubMed]

57. Leis, B.; Heinze, S.; Angelov, A.; Pham, V.T.T.; Thürmer, A.; Jebbar, M.; Golyshin, P.N.; Streit, W.R.; Daniel, R.; Liebl, W. Functional Screening of Hydrolytic Activities Reveals an Extremely Thermostable Cellulase from a Deep-Sea Archaeon. *Front. Bioeng. Biotechnol.* **2015**, *3*, 95. [CrossRef] [PubMed]

58. Graham, J.E.; Clark, M.E.; Nadler, D.C.; Huffer, S.; Chokhawala, H.A.; Rowland, S.E.; Blanch, H.W.; Clark, D.S.; Robb, F.T. Identification and characterization of a multidomain hyperthermophilic cellulase from an archaeal enrichment. *Nat. Commun.* **2011**, *2*, 375. [CrossRef] [PubMed]
59. Schröder, C.; Elleuche, S.; Blank, S.; Antranikian, G. Characterization of a heat-active archaeal β-glucosidase from a hydrothermal spring metagenome. *Enzyme Microb. Technol.* **2014**, *57*, 48–54. [CrossRef] [PubMed]
60. Yasir, M.; Khan, H.; Azam, S.S.; Telke, A.; Kim, S.W.; Chung, Y.R. Cloning and functional characterization of endo-β-1,4-glucanase gene from metagenomic library of vermicompost. *J. Microbiol.* **2013**, *51*, 329–335. [CrossRef] [PubMed]
61. Kwon, E.J.; Jeong, Y.S.; Kim, Y.H.; Kim, S.K.; Na, H.B.; Kim, J.; Yun, H.D.; Kim, H. Construction of a Metagenomic Library from Compost and Screening of Cellulase- and Xylanase-positive Clones. *J. Korean Soc. Appl. Biol. Chem.* **2010**, *53*, 702–708. [CrossRef]
62. Okano, H.; Ozaki, M.; Kanaya, E.; Kim, J.-J.; Angkawidjaja, C.; Koga, Y.; Kanaya, S. Structure and stability of metagenome-derived glycoside hydrolase family 12 cellulase (LC-CelA) a homolog of Cel12A from *Rhodothermus marinus*. *FEBS Open Bio* **2014**, *4*, 936–946. [CrossRef] [PubMed]
63. Yan, X.; Geng, A.; Zhang, J.; Wei, Y.; Zhang, L.; Qian, C.; Wang, Q.; Wang, S.; Zhou, Z. Discovery of (hemi-)cellulase genes in a metagenomic library from a biogas digester using 454 pyrosequencing. *Appl. Microbiol. Biotechnol.* **2013**, *97*, 8173–8182. [CrossRef] [PubMed]
64. Alvarez, T.M.; Paiva, J.H.; Ruiz, D.M.; Cairo, J.P.L.F.; Pereira, I.O.; Paixão, D.A.A.; de Almeida, R.F.; Tonoli, C.C.C.; Ruller, R.; Santos, C.R.; et al. Structure and function of a novel cellulase 5 from sugarcane soil metagenome. *PLoS ONE* **2013**, *8*, e83635. [CrossRef] [PubMed]
65. Liu, J.; Liu, W.-D.; Zhao, X.-L.; Shen, W.-J.; Cao, H.; Cui, Z.-L. Cloning and functional characterization of a novel endo-β-1,4-glucanase gene from a soil-derived metagenomic library. *Appl. Microbiol. Biotechnol.* **2011**, *89*, 1083–1092. [CrossRef] [PubMed]
66. Pottkämper, J.; Barthen, P.; Ilmberger, N.; Schwaneberg, U.; Schenk, A.; Schulte, M.; Ignatiev, N.; Streit, W.R. Applying metagenomics for the identification of bacterial cellulases that are stable in ionic liquids. *Green Chem.* **2009**, *11*, 957. [CrossRef]
67. Feng, Y.; Duan, C.-J.; Pang, H.; Mo, X.-C.; Wu, C.-F.; Yu, Y.; Hu, Y.-L.; Wei, J.; Tang, J.-L.; Feng, J.-X. Cloning and identification of novel cellulase genes from uncultured microorganisms in rabbit cecum and characterization of the expressed cellulases. *Appl. Microbiol. Biotechnol.* **2007**, *75*, 319–328. [CrossRef] [PubMed]
68. Del Pozo, M.V.; Fernández-Arrojo, L.; Gil-Martínez, J.; Montesinos, A.; Chernikova, T.N.; Nechitaylo, T.Y.; Waliszek, A.; Tortajada, M.; Rojas, A.; Huws, S.A.; et al. Microbial β-glucosidases from cow rumen metagenome enhance the saccharification of lignocellulose in combination with commercial cellulase cocktail. *Biotechnol. Biofuels* **2012**, *5*, 73. [CrossRef] [PubMed]
69. Ramírez-Escudero, M.; Del Pozo, M.V.; Marín-Navarro, J.; González, B.; Golyshin, P.N.; Polaina, J.; Ferrer, M.; Sanz-Aparicio, J. Structural and Functional Characterization of a Ruminal β-Glycosidase Defines a Novel Subfamily of Glycoside Hydrolase Family 3 with Permuted Domain Topology. *J. Biol. Chem.* **2016**, *291*, 24200–24214. [CrossRef] [PubMed]
70. Beloqui, A.; Nechitaylo, T.Y.; López-Cortés, N.; Ghazi, A.; Guazzaroni, M.-E.; Polaina, J.; Strittmatter, A.W.; Reva, O.; Waliczek, A.; Yakimov, M.M.; et al. Diversity of glycosyl hydrolases from cellulose-depleting communities enriched from casts of two earthworm species. *Appl. Environ. Microbiol.* **2010**, *76*, 5934–5946. [CrossRef] [PubMed]
71. Wang, Q.; Qian, C.; Zhang, X.-Z.; Liu, N.; Liu, N.; Yan, X.; Zhou, Z. Characterization of a novel thermostable β-glucosidase from a metagenomic library of termite gut. *Enzyme Microb. Technol.* **2012**, *51*, 319–324. [CrossRef] [PubMed]
72. Zhang, M.; Liu, N.; Qian, C.; Wang, Q.; Wang, Q.; Long, Y.; Huang, Y.; Zhou, Z.; Yan, X. Phylogenetic and functional analysis of gut microbiota of a fungus-growing higher termite: Bacteroidetes from higher termites are a rich source of β-glucosidase genes. *Microb. Ecol.* **2014**, *68*, 416–425. [CrossRef] [PubMed]
73. Himmel, M.E.; Adney, W.S.; Tucker, M.P.; Grohmann, K. Thermostable Purified Endoglucanas from *Acidothermus cellulolyticus* ATCC 43068. U.S. Patent 5,275,944, 4 January 1994.
74. Nurachman, Z.; Kurniasih, S.D.; Puspitawati, F.; Hadi, S.; Radjasa, O.K.; Natalia, D. Cloning of the Endoglucanase Gene from a *Bacillus amyloliquefaciens* PSM 3.1 in *Escherichia coli* Revealed Catalytic Triad Residues Thr-His-Glu. *Am. J. Biochem. Biotechnol.* **2010**, *6*, 268–274. [CrossRef]

75. Bischoff, K.M.; Rooney, A.P.; Li, X.-L.; Liu, S.; Hughes, S.R. Purification and characterization of a family 5 endoglucanase from a moderately thermophilic strain of *Bacillus licheniformis*. *Biotechnol. Lett.* **2006**, *28*, 1761–1765. [CrossRef] [PubMed]

76. Jung, Y.-J.; Lee, Y.-S.; Park, I.-H.; Chandra, M.S.; Kim, K.-K.; Choi, Y.-L. Molecular cloning, purification and characterization of thermostable beta-1,3-1,4 glucanase from *Bacillus subtilis* A8-8. *Indian J. Biochem. Biophys.* **2010**, *47*, 203–210. [PubMed]

77. Chhabra, S.R.; Shockley, K.R.; Ward, D.E.; Kelly, R.M. Regulation of endo-acting glycosyl hydrolases in the hyperthermophilic bacterium *Thermotoga maritima* grown on glucan- and mannan-based polysaccharides. *Appl. Environ. Microbiol.* **2002**, *68*, 545–554. [CrossRef] [PubMed]

78. Bauer, S.; Vasu, P.; Persson, S.; Mort, A.J.; Somerville, C.R. Development and application of a suite of polysaccharide-degrading enzymes for analyzing plant cell walls. *Proc. Natl. Acad. Sci. USA* **2006**, *103*, 11417–11422. [CrossRef] [PubMed]

79. Calza, R.E.; Irwin, D.C.; Wilson, D.B. Purification and characterization of two β-1,4-endoglucanases from *Thermomonospora fusca*. *Biochemistry* **1985**, *24*, 7797–7804. [CrossRef]

80. Yin, Y.-R.; Zhang, F.; Hu, Q.-W.; Xian, W.-D.; Hozzein, W.N.; Zhou, E.-M.; Ming, H.; Nie, G.-X.; Li, W.-J. Heterologous expression and characterization of a novel halotolerant, thermostable, and alkali-stable GH6 endoglucanase from *Thermobifida halotolerans*. *Biotechnol. Lett.* **2015**, *37*, 857–862. [CrossRef] [PubMed]

81. Cazemier, A.E.; Verdoes, J.C.; Op den Camp, H.J.M.; Hackstein, J.H.P.; van Ooyen, A.J. A beta-1,4-endoglucanase-encoding gene from *Cellulomonas pachnodae*. *Appl. Microbiol. Biotechnol.* **1999**, *52*, 232–239. [CrossRef] [PubMed]

82. Xu, X.; Li, J.; Zhang, W.; Huang, H.; Shi, P.; Luo, H.; Liu, B.; Zhang, Y.; Zhang, Z.; Fan, Y.; et al. A Neutral Thermostable β-1,4-Glucanase from *Humicola insolens* Y1 with Potential for Applications in Various Industries. *PLoS ONE* **2015**, *10*, e0124925. [CrossRef] [PubMed]

83. Li, X.L.; Chen, H.; Ljungdahl, L.G. Two cellulases, CelA and CelC, from the polycentric anaerobic fungus *Orpinomyces* strain PC-2 contain N-terminal docking domains for a cellulase-hemicellulase complex. *Appl. Environ. Microbiol.* **1997**, *63*, 4721–4728. [PubMed]

84. Takashima, S.; Nakamura, A.; Hidaka, M.; Masaki, H.; Uozumi, T. Cloning, sequencing, and expression of the cellulase genes of *Humicola grisea* var. *thermoidea*. *J. Biotechnol.* **1996**, *50*, 137–147. [CrossRef]

85. Wei, X.-M.; Qin, Y.; Qu, Y. Molecular Cloning and Characterization of Two Major Endoglucanases from *Penicillium decumbens*. *J. Microbiol. Biotechnol.* **2010**, *20*, 265–270. [CrossRef] [PubMed]

86. Miettinen-Oinonen, A.; Londesborough, J.; Joutsjoki, V.; Lantto, R.; Vehmaanperä, J. Three cellulases from *Melanocarpus albomyces* for textile treatment at neutral pH. *Enzyme Microb. Technol.* **2004**, *34*, 332–341. [CrossRef]

87. Yoo, J.-S.; Jung, Y.-J.; Chung, S.-Y.; Lee, Y.-C.; Choi, Y.-L. Molecular cloning and characterization of CMCase gene (celC) from *Salmonella typhimurium* UR. *J. Microbiol.* **2004**, *42*, 205–210. [PubMed]

88. Kim, J.O.; Park, S.R.; Lim, W.J.; Ryu, S.K.; Kim, M.K.; An, C.L.; Cho, S.J.; Park, Y.W.; Kim, J.H.; Yun, H.D. Cloning and characterization of thermostable endoglucanase (Cel8Y) from the hyperthermophilic *Aquifex aeolicus* VF5. *Biochem. Biophys. Res. Commun.* **2000**, *279*, 420–426. [CrossRef] [PubMed]

89. Hakamada, Y.; Endo, K.; Takizawa, S.; Kobayashi, T.; Shirai, T.; Yamane, T.; Ito, S. Enzymatic properties, crystallization, and deduced amino acid sequence of an alkaline endoglucanase from *Bacillus circulans*. *Biochim. Biophys. Acta* **2002**, *1570*, 174–180. [CrossRef]

90. Ul Haq, I.; Akram, F.; Khan, M.A.; Hussain, Z.; Nawaz, A.; Iqbal, K.; Shah, A.J. CenC, a multidomain thermostable GH9 processive endoglucanase from *Clostridium thermocellum*: Cloning, characterization and saccharification studies. *World J. Microbiol. Biotechnol.* **2015**, *31*, 1699–1710. [CrossRef] [PubMed]

91. Zverlov, V.; Mahr, S.; Riedel, K.; Bronnenmeier, K. Properties and gene structure of a bifunctional cellulolytic enzyme (CelA) from the extreme thermophile "*Anaerocellum thermophilum*" with separate glycosyl hydrolase family 9 and 48 catalytic domains. *Microbiology* **1998**, *144*, 457–465. [CrossRef] [PubMed]

92. Zhang, X.-Z.; Sathitsuksanoh, N.; Zhang, Y.-H.P. Glycoside hydrolase family 9 processive endoglucanase from *Clostridium phytofermentans*: Heterologous expression, characterization, and synergy with family 48 cellobiohydrolase. *Bioresour. Technol.* **2010**, *101*, 5534–5538. [CrossRef] [PubMed]

93. Liebl, W.; Ruile, P.; Bronnenmeier, K.; Riedel, K.; Lottspeich, F.; Greif, I. Analysis of a *Thermotoga maritima* DNA fragment encoding two similar thermostable cellulases, CelA and CelB, and characterization of the recombinant enzymes. *Microbiology* **1996**, *142*, 2533–2542. [CrossRef] [PubMed]

94. Bauer, M.W.; Driskill, L.E.; Callen, W.; Snead, M.A.; Mathur, E.J.; Kelly, R.M. An Endoglucanase, EglA, from the Hyperthermophilic Archaeon *Pyrococcus Furiosus* Hydrolyzes β-1,4 Bonds in Mixed-Linkage (1→3),(1→4)-β-D-Glucans and Cellulose. *J. Bacteriol.* **1999**, *181*, 284–290. [PubMed]

95. Huang, Y.; Krauss, G.; Cottaz, S.; Driguez, H.; Lipps, G. A highly acid-stable and thermostable endo-beta-glucanase from the thermoacidophilic archaeon *Sulfolobus solfataricus*. *Biochem. J.* **2005**, *385*, 581–588. [CrossRef] [PubMed]

96. Irdani, T.; Perito, B.; Mastromei, G. Characterization of a *Streptomyces rochei* endoglucanase. *Ann. N. Y. Acad. Sci.* **1996**, *782*, 173–181. [CrossRef] [PubMed]

97. Karlsson, J.; Siika-aho, M.; Tenkanen, M.; Tjerneld, F. Enzymatic properties of the low molecular mass endoglucanases Cel12A (EG III) and Cel45A (EG V) of *Trichoderma reesei*. *J. Biotechnol.* **2002**, *99*, 63–78. [CrossRef]

98. Warner, C.D.; Hoy, J.A.; Shilling, T.C.; Linnen, M.J.; Ginder, N.D.; Ford, C.F.; Honzatko, R.B.; Reilly, P.J. Tertiary structure and characterization of a glycoside hydrolase family 44 endoglucanase from *Clostridium acetobutylicum*. *Appl. Environ. Microbiol.* **2010**, *76*, 338–346. [CrossRef] [PubMed]

99. Hansen, C.K.; Diderichsen, B.; Jørgensen, P.L. celA from *Bacillus lautus* PL236 encodes a novel cellulose-binding endo-beta-1,4-glucanase. *J. Bacteriol.* **1992**, *174*, 3522–3531. [CrossRef] [PubMed]

100. Ahsan, M.M.; Matsumoto, M.; Karita, S.; Kimura, T.; Sakka, K.; Ohmiya, K. Purification and Characterization of the Family J Catalytic Domain Derived from the *Clostridium thermocellum* Endoglucanase CelJ. *Biosci. Biotechnol. Biochem.* **1997**, *61*, 427–431. [CrossRef] [PubMed]

101. Koga, J.; Baba, Y.; Shimonaka, A.; Nishimura, T.; Hanamura, S.; Kono, T. Purification and characterization of a new family 45 endoglucanase, STCE1, from *Staphylotrichum coccosporum* and its overproduction in *Humicola insolens*. *Appl. Environ. Microbiol.* **2008**, *74*, 4210–4217. [CrossRef] [PubMed]

102. Wonganu, B.; Pootanakit, K.; Boonyapakron, K.; Champreda, V.; Tanapongpipat, S.; Eurwilaichitr, L. Cloning, expression and characterization of a thermotolerant endoglucanase from *Syncephalastrum racemosum* (BCC18080) in *Pichia pastoris*. *Protein Expr. Purif.* **2008**, *58*, 78–86. [CrossRef] [PubMed]

103. Baba, Y.; Shimonaka, A.; Koga, J.; Murashima, K.; Kubota, H.; Kono, T. Purification and characterization of a new endo-1,4-beta-D-glucanase from *Beltraniella portoricensis*. *Biosci. Biotechnol. Biochem.* **2005**, *69*, 1198–1201. [CrossRef] [PubMed]

104. Eckert, K.; Schneider, E. A thermoacidophilic endoglucanase (CelB) from *Alicyclobacillus acidocaldarius* displays high sequence similarity to arabinofuranosidases belonging to family 51 of glycoside hydrolases. *Eur. J. Biochem.* **2003**, *270*, 3593–3602. [CrossRef] [PubMed]

105. Brás, J.L.A.; Cartmell, A.; Carvalho, A.L.M.; Verzé, G.; Bayer, E.A.; Vazana, Y.; Correia, M.A.S.; Prates, J.A.M.; Ratnaparkhe, S.; Boraston, A.B.; et al. Structural insights into a unique cellulase fold and mechanism of cellulose hydrolysis. *Proc. Natl. Acad. Sci. USA* **2011**, *108*, 5237–5242. [CrossRef] [PubMed]

106. Park, C.-S.; Kawaguchi, T.; Sumitani, J.-I.; Takada, G.; Izumori, K.; Arai, M. Cloning and sequencing of an exoglucanase gene from *Streptomyces* sp. M 23, and its expression in *Streptomyces lividans* TK-24. *J. Biosci. Bioeng.* **2005**, *99*, 434–436. [CrossRef] [PubMed]

107. Zhang, S.; Lao, G.; Wilson, D.B. Characterization of a *Thermomonospora fusca* exocellulase. *Biochemistry* **1995**, *34*, 3386–3395. [CrossRef] [PubMed]

108. Liu, S.-A.; Li, D.-C.; Er, S.-J.; Zhang, Y. Cloning and expressing of cellulase gene (cbh2) from thermophilic fungi *Chaetomium thermophilum* CT2. *Sheng Wu Gong Cheng Xue Bao* **2005**, *21*, 892–899. [PubMed]

109. Bukhtojarov, F.E.; Ustinov, B.B.; Salanovich, T.N.; Antonov, A.I.; Gusakov, A.V.; Okunev, O.N.; Sinitsyn, A.P. Cellulase complex of the fungus *Chrysosporium lucknowense*: Isolation and characterization of endoglucanases and cellobiohydrolases. *Biochemistry* **2004**, *69*, 542–551. [CrossRef] [PubMed]

110. Toda, H.; Nagahata, N.; Amano, Y.; Nozaki, K.; Kanda, T.; Okazaki, M.; Shimosaka, M. Gene cloning of cellobiohydrolase II from the white rot fungus *Irpex lacteus* MC-2 and its expression in *Pichia pastoris*. *Biosci. Biotechnol. Biochem.* **2008**, *72*, 3142–3147. [CrossRef] [PubMed]

111. Gao, L.; Wang, F.; Gao, F.; Wang, L.; Zhao, J.; Qu, Y. Purification and characterization of a novel cellobiohydrolase (PdCel6A) from *Penicillium decumbens* JU-A10 for bioethanol production. *Bioresour. Technol.* **2011**, *102*, 8339–8342. [CrossRef] [PubMed]

112. Song, J.; Liu, B.; Liu, Z.; Yang, Q. Cloning of two cellobiohydrolase genes from *Trichoderma viride* and heterogenous expression in yeast *Saccharomyces cerevisiae*. *Mol. Biol. Rep.* **2010**, *37*, 2135–2140. [CrossRef] [PubMed]

113. Singh, R.N.; Akimenko, V.K. Isolation of a Cellobiohydrolase of *Clostridium thermocellum* Capable of Degrading Natural Crystalline Substrates. *Biochem. Biophys. Res. Commun.* **1993**, *192*, 1123–1130. [CrossRef] [PubMed]

114. Kataeva, I.; Li, X.-L.; Chen, H.; Choi, S.-K.; Ljungdahl, L.G. Cloning and sequence analysis of a new cellulase gene encoding CelK, a major cellulosome component of *Clostridium thermocellum*: Evidence for gene duplication and recombination. *J. Bacteriol.* **1999**, *181*, 5288–5295. [PubMed]

115. Zverlov, V.V.; Velikodvorskaya, G.A.; Schwarz, W.H. A newly described cellulosomal cellobiohydrolase, CelO, from *Clostridium thermocellum*: Investigation of the exo-mode of hydrolysis, and binding capacity to crystalline cellulose. *Microbiology* **2002**, *148*, 247–255. [CrossRef] [PubMed]

116. Takada, G.; Kawaguchi, T.; Sumitani, J.; Arai, M. Expression of *Aspergillus aculeatus* No. F-50 cellobiohydrolase I (cbhI) and beta-glucosidase 1 (bgl1) genes by *Saccharomyces cerevisiae*. *Biosci. Biotechnol. Biochem.* **1998**, *62*, 1615–1618. [CrossRef] [PubMed]

117. Moroz, O.V.; Maranta, M.; Shaghasi, T.; Harris, P.V.; Wilson, K.S.; Davies, G.J. The three-dimensional structure of the cellobiohydrolase Cel7A from *Aspergillus fumigatus* at 1.5 Å resolution. *Acta Crystallogr. Sect. F Struct. Biol. Commun.* **2015**, *71*, 114–120. [CrossRef] [PubMed]

118. Li, Y.; Li, D.; Teng, F. Purification and characterization of a cellobiohydrolase from the thermophilic fungus *Chaetomium thermophilus* CT2. *Wei Sheng Wu Xue Bao* **2006**, *46*, 143–146. [PubMed]

119. Hobdey, S.E.; Knott, B.C.; Haddad Momeni, M.; Taylor, L.E.; Borisova, A.S.; Podkaminer, K.K.; VanderWall, T.A.; Himmel, M.E.; Decker, S.R.; Beckham, G.T.; et al. Biochemical and Structural Characterizations of Two *Dictyostelium* Cellobiohydrolases from the Amoebozoa Kingdom Reveal a High Level of Conservation between Distant Phylogenetic Trees of Life. *Appl. Environ. Microbiol.* **2016**, *82*, 3395–3409. [CrossRef] [PubMed]

120. Takashima, S.; Iikura, H.; Nakamura, A.; Hidaka, M.; Masaki, H.; Uozumi, T. Isolation of the gene and characterization of the enzymatic properties of a major exoglucanase of *Humicola grisea* without a cellulose-binding domain. *J. Biochem.* **1998**, *124*, 717–725. [CrossRef] [PubMed]

121. Voutilainen, S.P.; Boer, H.; Linder, M.B.; Puranen, T.; Rouvinen, J.; Vehmaanperä, J.; Koivula, A. Heterologous expression of *Melanocarpus albomyces* cellobiohydrolase Cel7B, and random mutagenesis to improve its thermostability. *Enzyme Microb. Technol.* **2007**, *41*, 234–243. [CrossRef]

122. Texier, H.; Dumon, C.; Neugnot-Roux, V.; Maestracci, M.; O'Donohue, M.J. Redefining XynA from *Penicillium funiculosum* IMI 378536 as a GH7 cellobiohydrolase. *J. Ind. Microbiol. Biotechnol.* **2012**, *39*, 1569–1576. [CrossRef] [PubMed]

123. Colussi, F.; Serpa, V.; Delabona, P.D.S.; Manzine, L.R.; Voltatodio, M.L.; Alves, R.; Mello, B.L.; Pereira, N.; Farinas, C.S.; Golubev, A.M.; et al. Purification, and biochemical and biophysical characterization of cellobiohydrolase I from *Trichoderma harzianum* IOC 3844. *J. Microbiol. Biotechnol.* **2011**, *21*, 808–817. [CrossRef] [PubMed]

124. Bronnenmeier, K.; Rücknagel, K.P.; Staudenbauer, W.L. Purification and properties of a novel type of exo-1,4-beta-glucanase (avicelase II) from the cellulolytic thermophile *Clostridium stercorarium*. *Eur. J. Biochem.* **1991**, *200*, 379–385. [CrossRef] [PubMed]

125. Zhang, X.-Z.; Zhang, Z.; Zhu, Z.; Sathitsuksanoh, N.; Yang, Y.; Zhang, Y.-H.P. The noncellulosomal family 48 cellobiohydrolase from *Clostridium phytofermentans* ISDg: Heterologous expression, characterization, and processivity. *Appl. Microbiol. Biotechnol.* **2010**, *86*, 525–533. [CrossRef] [PubMed]

126. Kruus, K.; Wang, W.K.; Ching, J.; Wu, J.H.D. Exoglucanase activities of the recombinant *Clostridium thermocellum* CelS, a major cellulosome component. *J. Bacteriol.* **1995**, *177*, 1641–1644. [CrossRef] [PubMed]

127. Liu, W.; Bevan, D.R.; Zhang, Y.-H.P. The family 1 glycoside hydrolase from *Clostridium cellulolyticum* H10 is a cellodextrin glucohydrolase. *Appl. Biochem. Biotechnol.* **2010**, *161*, 264–273. [CrossRef] [PubMed]

128. Yernool, D.A.; McCarthy, J.K.; Eveleigh, D.E.; Bok, J.D. Cloning and characterization of the glucooligosaccharide catabolic pathway β-glucan glucohydrolase and cellobiose phosphorylase in the marine hyperthermophile *Thermotoga neapolitana*. *J. Bacteriol.* **2000**, *182*, 5172–5179. [CrossRef] [PubMed]

129. Kengen, S.W.M.; Luesink, E.J.; Stams, A.J.M.; Zehnder, A.J.B. Purification and characterization of an extremely thermostable β-glucosidase from the hyperthermophilic archaeon *Pyrococcus furiosus*. *Eur. J. Biochem.* **1993**, *213*, 305–312. [CrossRef] [PubMed]

130. Kim, Y.J.; Lee, J.E.; Lee, H.S.; Kwon, K.K.; Kang, S.G.; Lee, J. Novel substrate specificity of a thermostable β-glucosidase from the hyperthermophilic archaeon, *Thermococcus pacificus* P-4. *Korean J. Microbiol.* **2015**, *51*, 68–74. [CrossRef]

131. Matsui, I.; Sakai, Y.; Matsui, E.; Kikuchi, H.; Kawarabayasi, Y.; Honda, K. Novel substrate specificity of a membrane-bound beta-glycosidase from the hyperthermophilic archaeon *Pyrococcus horikoshii*. *FEBS Lett.* **2000**, *467*, 195–200. [CrossRef]

132. Wu, Y.; Yuan, S.; Chen, S.; Wu, D.; Chen, J.; Wu, J. Enhancing the production of galacto-oligosaccharides by mutagenesis of *Sulfolobus solfataricus* β-galactosidase. *Food Chem.* **2013**, *138*, 1588–1595. [CrossRef] [PubMed]

133. Sinha, S.K.; Datta, S. β-Glucosidase from the hyperthermophilic archaeon *Thermococcus* sp. is a salt-tolerant enzyme that is stabilized by its reaction product glucose. *Appl. Microbiol. Biotechnol.* **2016**, *100*, 8399–8409. [CrossRef] [PubMed]

134. Di Lauro, B.; Rossi, M.; Moracci, M. Characterization of a beta-glycosidase from the thermoacidophilic bacterium *Alicyclobacillus acidocaldarius*. *Extremophiles* **2006**, *10*, 301–310. [CrossRef] [PubMed]

135. Liu, Y.; Li, R.; Wang, J.; Zhang, X.; Jia, R.; Gao, Y.; Peng, H. Increased enzymatic hydrolysis of sugarcane bagasse by a novel glucose- and xylose-stimulated β-glucosidase from *Anoxybacillus flavithermus* subsp. *yunnanensis* E13T. *BMC Biochem.* **2017**, *18*, 4. [CrossRef] [PubMed]

136. Paavilainen, S.; Hellman, J.; Korpela, T. Purification, characterization, gene cloning, and sequencing of a new β-glucosidase from *Bacillus circulans* subsp. *alkalophilus*. *Appl. Environ. Microbiol.* **1993**, *59*, 927–932. [PubMed]

137. Xu, H.; Xiong, A.-S.; Zhao, W.; Tian, Y.-S.; Peng, R.-H.; Chen, J.-M.; Yao, Q.-H. Characterization of a glucose-, xylose-, sucrose-, and D-galactose-stimulated β-glucosidase from the alkalophilic bacterium *Bacillus halodurans* C-125. *Curr. Microbiol.* **2011**, *62*, 833–839. [CrossRef] [PubMed]

138. Love, D.R.; Fisher, R.; Bergquist, P.L. Sequence structure and expression of a cloned β-glucosidase gene from an extreme thermophile. *MGG Mol. Gen. Genet.* **1988**, *213*, 84–92. [CrossRef] [PubMed]

139. Kosugi, A.; Arai, T.; Doi, R.H. Degradation of cellulosome-produced cello-oligosaccharides by an extracellular non-cellulosomal beta-glucan glucohydrolase, BglA, from *Clostridium cellulovorans*. *Biochem. Biophys. Res. Commun.* **2006**, *349*, 20–23. [CrossRef] [PubMed]

140. Zou, Z.-Z.; Yu, H.-L.; Li, C.-X.; Zhou, X.-W.; Hayashi, C.; Sun, J.; Liu, B.-H.; Imanaka, T.; Xu, J.-H. A new thermostable β-glucosidase mined from *Dictyoglomus thermophilum*: Properties and performance in octyl glucoside synthesis at high temperatures. *Bioresour. Technol.* **2012**, *118*, 425–430. [CrossRef] [PubMed]

141. Jabbour, D.; Klippel, B.; Antranikian, G. A novel thermostable and glucose-tolerant β-glucosidase from *Fervidobacterium islandicum*. *Appl. Microbiol. Biotechnol.* **2012**, *93*, 1947–1956. [CrossRef] [PubMed]

142. Gefen, G.; Anbar, M.; Morag, E.; Lamed, R.; Bayer, E.A. Enhanced cellulose degradation by targeted integration of a cohesin-fused β-glucosidase into the *Clostridium thermocellum* cellulosome. *Proc. Natl. Acad. Sci. USA* **2012**, *109*, 10298–10303. [CrossRef] [PubMed]

143. Brognaro, H.; Almeida, V.M.; de Araujo, E.A.; Piyadov, V.; Santos, M.A.M.; Marana, S.R.; Polikarpov, I. Biochemical Characterization and Low-Resolution SAXS Molecular Envelope of GH1 β-Glycosidase from *Saccharophagus degradans*. *Mol. Biotechnol.* **2016**, *58*, 777–788. [CrossRef] [PubMed]

144. Marques, A.R.; Coutinho, P.M.; Videira, P.; Fialho, A.M.; Sá-Correia, I. *Sphingomonas paucimobilis* β-glucosidase Bgl1: A member of a new bacterial subfamily in glycoside hydrolase family 1. *Biochem. J.* **2003**, *370*, 793–804. [CrossRef] [PubMed]

145. Perez-Pons, J.A.; Cayetano, A.; Rebordosa, X.; Lloberas, J.; Guasch, A.; Querol, E. A beta-glucosidase gene (bgl3) from *Streptomyces* sp. strain QM-B814. Molecular cloning, nucleotide sequence, purification and characterization of the encoded enzyme, a new member of family 1 glycosyl hydrolases. *Eur. J. Biochem.* **1994**, *223*, 557–565. [CrossRef] [PubMed]

146. Breves, R.; Bronnenmeier, K.; Wild, N.; Lottspeich, F.; Staudenbauer, W.L.; Hofemeister, J. Genes encoding two different β-glucosidases of *Thermoanaerobacter brockii* are clustered in a common operon. *Appl. Environ. Microbiol.* **1997**, *63*, 3902–3910. [PubMed]

147. Song, X.; Xue, Y.; Wang, Q.; Wu, X. Comparison of three thermostable β-glucosidases for application in the hydrolysis of soybean isoflavone glycosides. *J. Agric. Food Chem.* **2011**, *59*, 1954–1961. [CrossRef] [PubMed]

148. Pei, J.; Pang, Q.; Zhao, L.; Fan, S.; Shi, H. *Thermoanaerobacterium thermosaccharolyticum* β-glucosidase: A glucose-tolerant enzyme with high specific activity for cellobiose. *Biotechnol. Biofuels* **2012**, *5*, 31. [CrossRef] [PubMed]

149. Spiridonov, N.A.; Wilson, D.B. Cloning and biochemical characterization of BglC, a beta-glucosidase from the cellulolytic actinomycete *Thermobifida fusca*. *Curr. Microbiol.* **2001**, *42*, 295–301. [CrossRef] [PubMed]

150. Wright, R.M.; Yablonsky, M.D.; Shalita, Z.P.; Goyal, A.K.; Eveleigh, D.E. Cloning, characterization, and nucleotide sequence of a gene encoding *Microbispora bispora* BglB, a thermostable beta-glucosidase expressed in Escherichia coli. *Appl. Environ. Microbiol.* **1992**, *58*, 3455–3465. [PubMed]

151. Haq, I.U.; Khan, M.A.; Muneer, B.; Hussain, Z.; Afzal, S.; Majeed, S.; Rashid, N.; Javed, M.M.; Ahmad, I. Cloning, characterization and molecular docking of a highly thermostable β-1,4-glucosidase from *Thermotoga petrophila*. *Biotechnol. Lett.* **2012**, *34*, 1703–1709. [CrossRef] [PubMed]

152. Oh, E.-J.; Lee, Y.-J.; Chol, J.J.; Seo, M.S.; Lee, M.S.; Kim, G.A.; Kwon, S.-T. Mutational analysis of *Thermus caldophilus* GK24 beta-glycosidase: Role of His119 in substrate binding and enzyme activity. *J. Microbiol. Biotechnol.* **2008**, *18*, 287–294. [PubMed]

153. Xiangyuan, H.; Shuzheng, Z.; Shoujun, Y. Cloning and expression of thermostable beta-glycosidase gene from *Thermus nonproteolyticus* HG102 and characterization of recombinant enzyme. *Appl. Biochem. Biotechnol.* **2001**, *94*, 243–255. [CrossRef]

154. Kang, S.K.; Cho, K.K.; Ahn, J.K.; Bok, J.D.; Kang, S.H.; Woo, J.H.; Lee, H.G.; You, S.K.; Choi, Y.J. Three forms of thermostable lactose-hydrolase from *Thermus* sp. IB-21: Cloning, expression, and enzyme characterization. *J. Biotechnol.* **2005**, *116*, 337–346. [CrossRef] [PubMed]

155. Nam, E.S.; Kim, M.S.; Lee, H.B.; Ahn, J.K. β-Glycosidase of *Thermus thermophilus* KNOUC202: Gene and biochemical properties of the enzyme expressed in *Escherichia coli*. *Appl. Biochem. Microbiol.* **2010**, *46*, 515–524. [CrossRef]

156. Zhao, Z.; Ramachandran, P.; Kim, T.-S.; Chen, Z.; Jeya, M.; Lee, J.-K. Characterization of an acid-tolerant β-1,4-glucosidase from *Fusarium oxysporum* and its potential as an animal feed additive. *Appl. Microbiol. Biotechnol.* **2013**, *97*, 10003–10011. [CrossRef] [PubMed]

157. Takashima, S.; Nakamura, A.; Hidaka, M.; Masaki, H.; Uozumi, T. Molecular Cloning and Expression of the Novel Fungal -Glucosidase Genes from *Humicola grisea* and *Trichoderma reesei*. *J. Biochem.* **1999**, *125*, 728–736. [CrossRef] [PubMed]

158. Li, X.-L.; Ljungdahl, L.G.; Ximenes, E.A.; Chen, H.; Felix, C.R.; Cotta, M.A.; Dien, B.S. Properties of a recombinant beta-glucosidase from polycentric anaerobic fungus *Orpinomyces* PC-2 and its application for cellulose hydrolysis. *Appl. Biochem. Biotechnol.* **2004**, *113*, 233–250. [CrossRef]

159. Li, X.; Zhao, J.; Shi, P.; Yang, P.; Wang, Y.; Luo, H.; Yao, B. Molecular cloning and expression of a novel β-glucosidase gene from *Phialophora* sp. G5. *Appl. Biochem. Biotechnol.* **2013**, *169*, 941–949. [CrossRef] [PubMed]

160. Schröder, C.; Blank, S.; Antranikian, G. First Glycoside Hydrolase Family 2 Enzymes from *Thermus antranikianii* and *Thermus brockianus* with β-Glucosidase Activity. *Front. Bioeng. Biotechnol.* **2015**, *3*, 76. [CrossRef] [PubMed]

161. Li, D.; Li, X.; Dang, W.; Tran, P.L.; Park, S.-H.; Oh, B.-C.; Hong, W.-S.; Lee, J.-S.; Park, K.-H. Characterization and application of an acidophilic and thermostable β-glucosidase from *Thermofilum pendens*. *J. Biosci. Bioeng.* **2013**, *115*, 490–496. [CrossRef] [PubMed]

162. Yan, S.; Wei, P.; Chen, Q.; Chen, X.; Wang, S.; Li, J.; Gao, C. Functional and structural characterization of a β-glucosidase involved in saponin metabolism from intestinal bacteria. *Biochem. Biophys. Res. Commun.* **2018**, *496*, 1349–1356. [CrossRef] [PubMed]

163. Lau, A.T.Y.; Wong, W.K.R. Purification and characterization of a major secretory cellobiase, Cba2, from *Cellulomonas biazotea*. *Protein Expr. Purif.* **2001**, *23*, 159–166. [CrossRef] [PubMed]

164. Gao, J.; Wakarchuk, W. Characterization of five β-glycoside hydrolases from *Cellulomonas fimi* ATCC 484. *J. Bacteriol.* **2014**, *196*, 4103–4110. [CrossRef] [PubMed]

165. Bronnenmeier, K.; Staudenbauer, W.L. Purification and properties of an extracellular β-glucosidase from the cellulolytic thermophile *Clostridium stercorarium*. *Appl. Microbiol. Biotechnol.* **1988**, *28*, 380–386. [CrossRef]

166. Li, Y.-K.; Lee, J.-A. Cloning and expression of β-glucosidase from *Flavobacterium meningosepticum*: A new member of family B β-glucosidase. *Enzyme Microb. Technol.* **1999**, *24*, 144–150. [CrossRef]

167. Xie, J.; Zhao, D.; Zhao, L.; Pei, J.; Xiao, W.; Ding, G.; Wang, Z. Overexpression and characterization of a Ca^{2+} activated thermostable β-glucosidase with high ginsenoside Rb1 to ginsenoside 20(S)-Rg3 bioconversion productivity. *J. Ind. Microbiol. Biotechnol.* **2015**, *42*, 839–850. [CrossRef] [PubMed]

168. Colabardini, A.C.; Valkonen, M.; Huuskonen, A.; Siika-aho, M.; Koivula, A.; Goldman, G.H.; Saloheimo, M. Expression of Two Novel β-Glucosidases from *Chaetomium atrobrunneum* in *Trichoderma reesei* and Characterization of the Heterologous Protein Products. *Mol. Biotechnol.* **2016**, *58*, 821–831. [CrossRef] [PubMed]

169. Liu, D.; Zhang, R.; Yang, X.; Zhang, Z.; Song, S.; Miao, Y.; Shen, Q. Characterization of a thermostable β-glucosidase from *Aspergillus fumigatus* Z5, and its functional expression in *Pichia pastoris* X33. *Microb. Cell Fact.* **2012**, *11*, 25. [CrossRef] [PubMed]

170. Xu, R.; Teng, F.; Zhang, C.; Li, D. Cloning of a Gene Encoding β-Glucosidase from *Chaetomium thermophilum* CT2 and Its Expression in *Pichia pastoris*. *J. Mol. Microbiol. Biotechnol.* **2011**, *20*, 16–23. [CrossRef] [PubMed]

171. Dotsenko, G.S.; Sinitsyna, O.A.; Hinz, S.W.A.; Wery, J.; Sinitsyn, A.P. Characterization of a GH family 3 β-glycoside hydrolase from *Chrysosporium lucknowense* and its application to the hydrolysis of β-glucan and xylan. *Bioresour. Technol.* **2012**, *112*, 345–349. [CrossRef] [PubMed]

172. Pei, X.; Zhao, J.; Cai, P.; Sun, W.; Ren, J.; Wu, Q.; Zhang, S.; Tian, C. Heterologous expression of a GH3 β-glucosidase from Neurospora crassa in Pichia pastoris with high purity and its application in the hydrolysis of soybean isoflavone glycosides. *Protein Expr. Purif.* **2016**, *119*, 75–84. [CrossRef] [PubMed]

173. Kato, Y.; Nomura, T.; Ogita, S.; Takano, M.; Hoshino, K. Two new β-glucosidases from ethanol-fermenting fungus *Mucor circinelloides* NBRC 4572: Enzyme purification, functional characterization, and molecular cloning of the gene. *Appl. Microbiol. Biotechnol.* **2013**, *97*, 10045–10056. [CrossRef] [PubMed]

174. Yang, X.; Ma, R.; Shi, P.; Huang, H.; Bai, Y.; Wang, Y.; Yang, P.; Fan, Y.; Yao, B. Molecular characterization of a highly-active thermophilic β-glucosidase from *Neosartorya fischeri* P1 and its application in the hydrolysis of soybean isoflavone glycosides. *PLoS ONE* **2014**, *9*, e106785. [CrossRef] [PubMed]

175. Krogh, K.B.R.M.; Harris, P.V.; Olsen, C.L.; Johansen, K.S.; Hojer-Pedersen, J.; Borjesson, J.; Olsson, L. Characterization and kinetic analysis of a thermostable GH3 β-glucosidase from *Penicillium brasilianum*. *Appl. Microbiol. Biotechnol.* **2010**, *86*, 143–154. [CrossRef] [PubMed]

176. Chen, M.; Qin, Y.; Liu, Z.; Liu, K.; Wang, F.; Qu, Y. Isolation and characterization of a β-glucosidase from *Penicillium decumbens* and improving hydrolysis of corncob residue by using it as cellulase supplementation. *Enzyme Microb. Technol.* **2010**, *46*, 444–449. [CrossRef] [PubMed]

177. Machida, M.; Ohtsuki, I.; Fukui, S.; Yamashita, I. Nucleotide sequences of *Saccharomycopsis fibuligera* genes for extracellular β-glucosidases as expressed in *Saccharomyces cerevisiae*. *Appl. Environ. Microbiol.* **1988**, *54*, 3147–3155. [PubMed]

178. Murray, P.; Aro, N.; Collins, C.; Grassick, A.; Penttilä, M.; Saloheimo, M.; Tuohy, M. Expression in *Trichoderma reesei* and characterisation of a thermostable family 3 beta-glucosidase from the moderately thermophilic fungus *Talaromyces emersonii*. *Protein Expr. Purif.* **2004**, *38*, 248–257. [CrossRef] [PubMed]

179. Xia, W.; Xu, X.; Qian, L.; Shi, P.; Bai, Y.; Luo, H.; Ma, R.; Yao, B. Engineering a highly active thermophilic β-glucosidase to enhance its pH stability and saccharification performance. *Biotechnol. Biofuels* **2016**, *9*, 147. [CrossRef] [PubMed]

180. Hong, J.; Tamaki, H.; Kumagai, H. Cloning and functional expression of thermostable beta-glucosidase gene from *Thermoascus aurantiacus*. *Appl. Microbiol. Biotechnol.* **2007**, *73*, 1331–1339. [CrossRef] [PubMed]

181. Karnaouri, A.; Topakas, E.; Paschos, T.; Taouki, I.; Christakopoulos, P. Cloning, expression and characterization of an ethanol tolerant GH3 β-glucosidase from *Myceliophthora thermophila*. *PeerJ* **2013**, *1*, e46. [CrossRef] [PubMed]

182. Brunner, F.; Wirtz, W.; Rose, J.K.C.; Darvill, A.G.; Govers, F.; Scheel, D.; Nürnberger, T. A β-glucosidase/xylosidase from the phytopathogenic oomycete, *Phytophthora infestans*. *Phytochemistry* **2002**, *59*, 689–696. [CrossRef]

183. Ferrara, M.C.; Cobucci-Ponzano, B.; Carpentieri, A.; Henrissat, B.; Rossi, M.; Amoresano, A.; Moracci, M. The identification and molecular characterization of the first archaeal bifunctional exo-β-glucosidase/N-acetyl-β-glucosaminidase demonstrate that family GH116 is made of three functionally distinct subfamilies. *Biochim. Biophys. Acta* **2014**, *1840*, 367–377. [CrossRef] [PubMed]

184. Sansenya, S.; Mutoh, R.; Charoenwattanasatien, R.; Kurisu, G.; Ketudat Cairns, J.R. Expression and crystallization of a bacterial glycoside hydrolase family 116 β-glucosidase from *Thermoanaerobacterium xylanolyticum. Acta Crystallogr. Sect. F Struct. Biol. Commun.* **2015**, *71*, 41–44. [CrossRef] [PubMed]

MDPI

St. Alban-Anlage 66

4052 Basel

Switzerland

Tel. +41 61 683 77 34

Fax +41 61 302 89 18

www.mdpi.com

Microorganisms Editorial Office

E-mail: microorganisms@mdpi.com

www.mdpi.com/journal/microorganisms